"There is now no doubt regarding the epochal, world-shaping significance of the curatorial practices of late eighteenth and nineteenth century museums. With a matching boldness of vision, Fiona Cameron now calls on museums to play a world-saving role by 'curating for planet habitability.' Better still, in identifying the intellectual and institutional challenges this entails, she also shows how these might best be met. A timely manifesto for the contribution museums might make to addressing the crises produced by our relations to the more-than-human worlds that press upon us with increasing force."

Tony Bennett, *Emeritus Professor, Social and Cultural Theory, Institute for Culture and Society, Western Sydney University, Australia*

"This book offers a timely posthumanist provocation for students of, and practitioners in, museums. A fierce critique of humanist museum practices and theories, the book challenges us to take account of emerging practices in museums in the 21st century. Neither what museums are nor what they are becoming remain the same after its reading. With a deep commitment to more-than-human worlds, the book offers theoretical grounding for museum activism in the face of climate and planetary crisis. This book is a testament to Fiona Cameron's longstanding engagement with difficult topics in museums and provides researchers, practitioners, and students alike with new tools for analyses and action."

Brita Brenna, *Professor of Museology, University of Oslo, Norway*

"Museum Practices and the Posthumanities: Curating for Planetary Habitability is a deeply felt plea and argument for the need to get beyond our human centered approaches for dealing with ecological crisis. Museums, Cameron argues, are institutions that were central to the humanist project that produced the current ecological crisis. They are therefore also central to undoing that project. Doing so involves a radical rethinking of the central categories of thought that underpin modern society. In doing so, this book opens an alternative future by showing us what we need to overcome and how to go about it. Using the concept of viral contagion as both idea and reality, Cameron opens the possibility that we might be able to move beyond our humanist centered perspectives and productively deal with current threats to planetary wellbeing. This book is a magnificent tour de force in how museums might become part of a viral contagion that works to undo our current understandings of our place on this planet."

Andrea Witcomb, *Alfred Deakin Professor, Faculty of Arts and Education, Deakin University, Australia*

MUSEUM PRACTICES AND THE POSTHUMANITIES

This book critiques modern museologies and curatorial practices that have been complicit in emerging existential crises. It confidently presents novel, more-than-human curatorial visions, methods, frameworks, policies, and museologies radically refiguring the epistemological foundations of curatorial, museological thinking, and practice for a habitable planet.

Modern curatorial and museological practices are dominated by modern humanism in which capital growth, social, technological advancement, hubris, extraction, speciest logics, and colonial domination predominate, often without reflection. While history, science, and technology museums and their engagement with non-human worlds have always been ecological as an empirical reality, the human-centred frameworks and forms of human agency that institutions deploy tend to be non-cognizant of this reality. *Museum Practices and the Posthumanities: Curating for Planetary Habitability* reveals how these practices are ill-equipped to deal with the contemporary world of rapid digital transformations, post-Covid living, climate change, and its impacts among other societal changes, and it shows how museums might best meet these challenges by thinking with and in more-than-human worlds.

This book is aimed at museological scholars and museum professionals, and it will provide them with the inspiration to conduct research on and curate from a different ecological reference point to promote a world good enough for all things to thrive in radical co-existence.

Fiona R. Cameron is Associate Professor, Principal Research Fellow, Contemporary Museologies at the Institute for Culture and Society, Western Sydney University, Australia. Fiona is also Professor Dr. at the Rachel Carson Center, Ludwig Maximilian University of Munich, Germany and visiting Professor, Linköping University, Sweden.

Routledge Environmental Humanities

The *Routledge Environmental Humanities* series is an original and inspiring venture recognising that today's world agricultural and water crises, ocean pollution and resource depletion, global warming from greenhouse gases, urban sprawl, overpopulation, food insecurity, and environmental justice are all *crises of culture.*

The reality of understanding and finding adaptive solutions to our present and future environmental challenges has shifted the epicentre of environmental studies away from an exclusively scientific and technological framework to one that depends on the human-focused disciplines and ideas of the humanities and allied social sciences.

We thus welcome book proposals from all humanities and social sciences disciplines for an inclusive and interdisciplinary series. We favour manuscripts aimed at an international readership and written in a lively and accessible style. The readership comprises scholars and students from the humanities and social sciences and thoughtful readers concerned about the human dimensions of environmental change.

For more information about this series, please visit: www.routledge.com/Routledge-Environmental-Humanities/book-series/REH

MUSEUM PRACTICES AND THE POSTHUMANITIES

Curating for Planetary Habitability

Fiona R. Cameron

Routledge
Taylor & Francis Group
LONDON AND NEW YORK

earthscan
from Routledge

Designed cover image: Liminal landscape in the universe of oneness. An artwork by Audrey Rhoda, audreyrhoda.art.

First published 2024
by Routledge
4 Park Square, Milton Park, Abingdon, Oxon OX14 4RN

and by Routledge
605 Third Avenue, New York, NY 10158

Routledge is an imprint of the Taylor & Francis Group, an informa business

British Library Cataloguing-in-Publication Data
A catalogue record for this book is available from the British Library

Library of Congress Cataloging-in-Publication Data
Names: Cameron, Fiona.
Title: Museum practices and the posthumanities : curating for planetary habitability / Fiona R. Cameron.
Other titles: Curating for planetary habitability
Description: Abingdon, Oxon ; New York, NY : Routledge, 2024. | Series: Routledge environmental humanities | Includes bibliographical references and index.
Identifiers: LCCN 2023018289 (print) | LCCN 2023018290 (ebook) | ISBN 9780415792011 (hbk) | ISBN 9780367196844 (pbk) | ISBN 9781315212067 (ebk)
Subjects: LCSH: Museums—Social aspects. | Museums—Environmental aspects. | Climatic changes.
Classification: LCC AM7 .C35 2024 (print) | LCC AM7 (ebook) | DDC 069—dc23/eng/20230424
LC record available at https://lccn.loc.gov/2023018289
LC ebook record available at https://lccn.loc.gov/2023018290

ISBN: 978-0-415-79201-1 (hbk)
ISBN: 978-0-367-19684-4 (pbk)
ISBN: 978-1-315-21206-7 (ebk)

DOI: 10.4324/9781315212067

Typeset in ITC Galliard Pro
by codeMantra

Printed and bound in Great Britain by
CPI Group (UK) Ltd, Croydon, CR0 4YY

In celebration of the establishment of my Create with Love retreats for artists and writers in Galatone, Salento, Italy and to Salento with Love and Davide Mengoli who made it all happen.

CONTENTS

FIGURES

ACKNOWLEDGEMENTS

This book, *Museum Practices and the Posthumanities: Curating for Planetary Habitability*, and the insights, arguments and speculative possibilities for establishing more-than-human museum, curatorial practices, and museologies, has been a passion of mine since the mid-2000s. This work has sought to develop new roles and agencies for museums in contemporary societies, to refigure curatorial practice in crisis, and to expand sustaining future prospects for the planet and all its inhabitants – both human and non-human. This monograph is the outcome of many years of creative, conceptual, and empirical work and publication. It has been a thrilling but also a very demanding undertaking, in which many iterations of the book's content preceded this final production.

Emerging from my early work on museums, collections, controversial topics, complexity theory, and climate action, I became aware of how limiting humanist curatorial practice and the new museology is in the face of digital, societal transformation, and existential crises, and indeed how these dispositions and practices have contributed to impending socio-ecological collapse. I have actively considered what new work museum professional practice and scholarly museology might be able to do in partial recuperation, and in respect to the future crafting of attuned interventions in a world suffering from more than two centuries of capitalist and colonial devastation. So rather than just grabbing onto new keywords, the Anthropocene being one example, I sought to enrich museologies and present them as plural figurations encompassing many forms of ecological thinking and acting in attunement. That is, by extending understandings of the non-human world in creative ways, beyond objects of taxonomy, species identification, or as ecosystem and ecosystem services, I sought to develop new figurations of history in museum practices. Other acts along this pathway also included a radical re-working of nature-culture relations in museology and museum practice, the application of theories of humanisms, and an exploration of relationships and their entanglement with material situations. Through such changes, the collections, curatorial work, and curating processes that museums engage in can become enriched, vital, and communal. For this end, I also rethought discipline roles and agencies within communitarian constellations in order to understand the multifarious agencies of all manner of things beyond the sovereign human and to rework understandings of the nature of time, space, and historicity. I considered how the extension of the coordinates of museum practice and museology might change the work of curation.

Central to these processes of refiguration was a critique of the new museology and its humanist, social constructivist, and post-structural foci, and its obsession with substances as a defining logic for analysis, whereby discipline silos dictate the work of museum practice. I also considered the notion of physical entities such as buildings and other structures, as well as the neo-liberal capitalist logic and extractivist mindset that often drives museum programming.

Inspired by feminist theorists Donna Haraway and Rosi Braidotti and their work in rethinking humanisms and the potential for partial recuperation through new knowledge practices and experimental work, and Jane Bennett's concept of new materialisms and the vibrancy of all manner of things, I sought to gesture towards a new museological regime that can add to the existing repertoire of museologies and that has the capacity to work alongside others for example, Indigenous practices. This book is the second in a series. My first monograph of the series, *The Future of Digital Data, Heritage and Curation in a More-Than-Human World* (Routledge, Taylor and Francis, 2021) considered more-than-human digital heritage practices and museologies.

During the course of this monograph's development, I have greatly benefited from the rich collegial interdisciplinary and intellectual culture of the Institute for Culture and Society at Western Sydney University. I would especially like to thank esteemed scholars and colleagues Tony Bennett and Ned Rossiter, and also Conal McCarthy from the Museum Studies program at Victoria University, Wellington. All have provided generous support and advice throughout this book's development.

I have been fortunate to hold a number of fellowships at leading intellectual centres for the advancement of the environmental humanities and museum research. In 2010, I was awarded a Rachel Carson Fellowship and spent six months at the Rachel Carson Center, a joint research centre between Munich's Ludwig-Maximilians-Universität and the German Deutsches Museum of Science and Technology. There I was under the inspiring leadership of Professor Drs Christof Mauch and Helmuth Trischler. I was also fortunate to have a research residency at the Deutsches Museum Research Institute over two consecutive years and it was here I began to sketch out the monograph's conceptual framework. For this opportunity, I would like to thank Director Helmut Trischler for his support.

In 2018, I also held a visiting fellow position at KTH Environmental Humanities Lab, Stockholm, under the directorship of Dr Marco Armiero, and this gifted me the time to think and write. This visiting fellowship opportunity allowed me to further develop my skills and expertise in new materialisms, media ecology, and the environmental and posthumanities humanities, all of which informed the book's theoretical framework.

Also, in 2018, I was fortunate enough to receive a professorial appointment at Linköping University in my capacity as Co-Investigator on the project "In Orbit: Distributed Curatorial Agency when Collections go Online" (2017–2021) funded by the Bank of Sweden Jubilee fund. This was with colleagues: Bodil Axellson (Principal Investigator), Sheenaugh Pietrobruno and Katherine Hauptman, Director of the Swedish History Museum. This position and grant emerged out of long-term, ongoing collaborations with TEMAQ Research Institute and, more recently, SEEDBOX staff (2015–2022), a transdisciplinary program in Environmental Humanities funded by MISTRA and FORMAS. These connections, alongside the grant, advanced my research in the digital and environmental humanities and facilitated the completion of three monographs. The sole-authored monograph, ***The Future of Digital Data, Heritage and Curation in a More-Than-Human World*** (Routledge, Taylor and Francis, 2021), the multi-authored monograph, with authors Bodil Axellson, Sheenagh Pietrobruno, Katherine Haputman, and myself, ***Museum Collections and Emerging Curatorial Agencies Online: Vikings in the Digital Age*** (London: Palgrave Macmillian, 2022), and this publication.

This book emerged through work conducted on two Australian Research Council (ARC) grants of which I was Chief Investigator and Project leader. These were Linkage Projects "Reconceptualising Heritage Collections" (2005–2007, LP0561202), and "Hot Science, Global Citizens: The Agencies of Museums in Climate Change Interventions" (2008–2013, LP0882088). The Bank of Sweden Jubilee's funds for the project titled "In Orbit: Distributed Curatorial Agency when Museum Objects and Knowledge go Online" (2018–2021) mentioned above, also was invaluable for developing this publication. A very special thanks to both of these agencies for funding these digital heritage, climate change, museums, and environmental humanities research projects.

I would also like to thank the Powerhouse, Sydney in which I was fortunate to be awarded a Research Fellow position at the Museum in 2019. Also, I wish to express gratitude for the work of collaborators David Ellsworth and Ben Dibley, at Western Sydney University both of whom were investigators on the Australian Research Council funded Linkage Project "Curating Museum Collections for Climate Change Mitigation", with the Powerhouse, Swinburne University of Technology, Melbourne and the New Institute

(2022–2024, LP200100103) on which I am the Chief Investigator and Project Leader.

Much of the final writing was conducted while in Salento, in the region of Puglia, Italy. Special thanks to the staff of Salento with Love: Davide Mengoli, Anna Maria Mengoli, and Toti Rosso, who all kindly supported my work with printouts of the chapters. Special thanks to Davide who facilitated my new and exciting Salento journey.

A special mention and sincere thanks to esteemed colleagues and museum scholars: Professors Tony Bennett (Institute for Culture and Society, Western Sydney University), Britta Brenna (Museum Studies, University of Oslo), and Andrea Witcomb (Deakin University, Melbourne) who all kindly provided generous endorsements for the book. I would also like pay my respects and appreciation to the Powerhouse First Nations Directorate (FND), to the Ernabella Arts Centre and artist Anne Thompson for supporting my request to include an image of the painted sign, "Mamanya mukulya kura tjuta kulira wantima (Look after one another, love one another)" in Chapter 6.

I would also like to express my thanks to Melinda Jewell for her exceptional editorial work on the final draft and to Marie-Louise Taylor for her precision on the book's referencing.

My dear friend, and talented artist Audrey Rhoda supplied an image of her work for the book cover, titled "Liminal landscape in the universe of oneness" for which I am most grateful. And to photographer John McRae who supplied the high-resolution digital image of Audrey's work.

I would also like to express my appreciation to colleague Ned Rossiter who supplied several excellent images of data centres and infrastructures, gathered as part of his research for the Australian Research Council Discovery Project, "Data Centres and the Governance of Labour and Territory" (DP160103307). Special thanks to Museum Victoria for providing a high-resolution copy of Bill Putt's bucket and to vale Bill Putt who donated the object to the collection.

Thank you to Iain McCalman and Libby Robin, inaugural editors for the Environmental Humanities series at Routledge, and for their unwavering enthusiasm for the book project from its outset in 2015. It was a time when few saw the merit of a publication of this kind. Special thanks to the commissioning editors of the Environmental Humanities list, Grace Harrison and Rosie Anderson, and especially Matthew Shobbrook, Senior Editorial Assistant who steered the manuscript through the production process.

And finally, I would like to thank my beautiful friends, Audrey Rhoda, Corrine Shaw, Michael Saunders, Rob Marshall, Patsy Lewis, Louise and Rossario Grasso, Tracey and Stu Nome, Karen Soldatić, and Bowie and Sting, my two Russian Blue and Russian White cats who all offered support and love throughout this long book journey.

1

INTRODUCTION

Curating for planetary habitability

Introduction

Historians, Zoltan Boldizsár Simon and Marek Tamm, remind us that our future prospects in the western world, "whether that be existential catastrophes or desired futures related to technoscientific revolutions, anthropogenic changes in the Earth's systems, or socio-political transformations require urgent action."[1] Western modernity and its repertoire of human-centred knowledge practices, economies, and cultural infrastructures (including museums), alongside the scaling up of the modern project through the engineering of nature with geo-engineering, better climate scientific and digital technological advancements and vaccines, assume that societies have the ability to steer the planet towards desired futures. Global policy initiatives, such as Agenda 2030 and the Paris Agreement (PA), are a part of this modern project and, informed by humanist and neo-capitalist directives, seek to support and direct actions to achieve these preferred futures. Despite all these initiatives, the changes wrought by existential crises are escaping the ability of modern societies to control.[2]

Current ways of thinking and living in the western world, and the value judgements and actions that are borne out of modern humanism, capital growth, and extraction – in which museums are complicit – are contributing to the devastating consequences of global warming and related more-than-human disasters. Such circumstances require a rethinking of museum, curatorial practices, and the anthropocentric conception of strong agency founded on epistemological subject–object models and the mindsets that emerge from these relational figurations.[3] It is increasingly apparent that all things are

DOI: 10.4324/9781315212067-1

interconnected, and human agency is just one among many others in complex and often unruly processes.

If the climate crisis, or the Anthropocene concept, fails to adequately highlight our deep entanglements with the non-human world, on which the future of the planet depends, then contagion certainly has and within a short timeframe and in a dramatic way. The Covid crisis political theorist, Paul Neuman, argues has rendered humans vulnerable and impotent because of the interdependence between bodies, viral worlds, and social and economic systems.[4] Viral contagion, however, also opens the potential for both exploring and making visible the all-embracing interdependences between modern populations and non-human agencies and thus produce different worlds from that of modern society. Museums can take a lead role in this conceptual and material transformative processes by playing a part in debunking the illusion of anthropocentricism as a locus of control.

While modern societies and their engagement with the non-human world have always been ecological as an empirical reality, the human subject-based object frameworks that western modern societies have chosen for themselves have become habitual and institutionalised and non-cognizant of this reality. These frameworks have divested modern communities of their ecological sensibilities.[5] On the other hand, the newly visible powers and agencies of a more-than-human world are making such entanglements increasingly apparent and forcing modern society to confront the material consequences of advanced capitalism and two centuries of industrialisation.[6]

The emerging planetary dimensions of life and destruction are becoming broader and deeper. The types of modern humanist ecological thoughts and the acts that such optics promote manifest through an often, unacknowledged empirical reality of the radical interconnectedness of all things and processes. These thoughts and acts have led to the inscription of destructive human activity in geological substrates – in air, water, and living organisms, including ourselves. The consequences include climate change – an outcome of carbon combustion, digital waste, and extraction across continents as a result of rapid technological innovation – and the Covid-19 pandemic – a speciest, capitalist, exploitative mindset, and event. Modern institutions have been complicit in cementing the relations and values through which such occasions and events are mobilised and made material.

Both the PA of 2015 and Agenda 2030 maintain modern humanism and strong human agency as the most important lever for action, thereby promoting the belief that these policy frameworks can represent solutions that incur less harm to the environment. But capital modernity is harmful at its core, and globalising policies, such as the PA and Agenda 2030, enrol other societies and their living practices into humanist and capitalist formations.

Acts of curating, the organisation of collections, associated research, interpretative and documentation procedures, and exhibitory practices, whether driven by intentional action or the unexpected and profoundly radical asymmetrical interactions of more-than and other-than-human agential coordinates, are influential conceptually and materially. Typically, these material enactments are mobilised by privileging certain types of knowledge, detailing how desired futures are imagined and might be realised, and who is deemed to be involved. Values are ascribed to different forms of life, non-life, and human-others through the curation of exhibitions, programs, and policy frameworks, and through collecting, interpreting, and documenting objects.

Throughout this monograph, ***Museum Practices and the Posthumanities: Curating for Planetary Habitability*** and in each of the chapters herein, I argue that museum and curatorial practices have historically been dominated by discourses and practices of modern humanism in which the concerns of capital growth, progress, social and technological advancement, hubris, extraction, speciest logics and colonial domination predominate, and often without reflection.

In modern museum spaces, for example, the human and non-human worlds are separated, and the latter often then further divided into life and non-life galleries of the earth and the biological sciences. The division between animate and inanimate things has had a profound impact on the way life and agency are visibly rendered, valued and conceptualised. In Chapter 6, "Viral museologies: Curating human-species-viral worlds in sympoiesis," I contend that museums have historically perpetuated injustice. Just like different forms of discrimination, such as racism, sexism, or colonialism, museum representations of speciesism render certain non-human and sentient lives inferior and therefore suitable for exploitation and subjugation.

Museums of modernity and colonialism have long taken it for granted that their subjects of study – non-human others, other peoples, and earthly processes – are there to satisfy the needs of modern populations. This epistemology has and continues to remain largely unchallenged in mainstream institutions. Many humanist-centred practices continue to inform the frameworks and procedures that guide curatorial work. Cartesian figurations, such as subject–object, nature–culture, virtual–real, and self–other, are all modern concepts central to curatorial practice.[7] While ecological thinking of the modern human-centred kind, as represented by Cartesian thinking, is still relational, it positions the non-human world in unequal relations. These reified categories of non-human life and processes are no longer tenable and now demand respect and humility as I variously argue across several chapters. In Chapter 3 – "Collections and curating human–non-human climates" – I consider this notion in respect to the entanglement of humans and climate change and its consequences, including wildfire. In Chapter 4 – "Museums, climate policy frameworks and the problem of humanist-driven solutions" – and Chapter 7 – "Curating sustaining practices in more-than-human worlds" – I consider the implementation of humanist global policy frameworks. All these policy strategies disregard not only other knowledge practices but also the material agencies of others.

Furthermore, the figure of the human and humanism, as a reified series of categories and an all-encompassing set of values and philosophies underlaying museum practice and policy, activates a limited range of coordinates. These relations of comparison are self-interested and place modern populations, and in particular male privilege, needs, desires, aspirations, and capital accumulation, at the centre. Such conceptual truisms have directed thinking and acting in specific types of relationships such that strong human agency comes to act as a lever for framing post-carbon futures. This is discussed in Chapter 4, in respect to climate action and museum exhibition and program activities.

Representational, documentary, and pedagogical practices in museums are often acts of negation, that is the negation of agency of others, the narcissistic use of them, and the promotion of a humanistic vision of man as the supreme authority. Such humanist-centric practices have manifested in a number of material ways. In Chapter 3, I illustrate the example of how strong human agency in scientific research and technological innovation is presented in exhibitions and through collecting activities as a means of solving existential, social, and environmental problems. The domination of Nature is often achieved through collecting, ordering, and arranging specimens. For example, geological samples extracted from the Earth's strata are being used as resources by modern populations for their flourishing. Underpinned by advanced capitalism museum curation is also framed around what cultural theorists Nick Axel, Daniel Barber, Nikolaus Hirsch, and Anton Vidokle call fundamental value relations of modernity in which "nature" becomes "resource" and modern populations become accumulators of raw, unruly materials that contaminate such as plastics in the oceans and carbon in the atmosphere.[8] These relations of nature, resource, and accumulation are dramatically illustrated in Chapter 2, "Technospheric heritage: Curating more-than-digital heritages in and for planetary durations." In this chapter, I discuss Donald Trump's tweets as digital cultural heritage and explain how they are complicit in the vagaries of digital transformation and expansion and produce heritage forms and wastes from deep within the Earth's strata to high elevations in the atmosphere. That is, how they penetrate the very foundations of life itself. In Chapter 3, I discuss how museums accumulate and display material goods that were once consumer items or aspirational objects, and how some are made from coal, natural gas, and crude oil such as plastic and by the machines of industrial processes driven by fossil fuel combustion. For example, Bill Putt's green plastic bucket was used to save his house from the Black Saturday bushfires in Victoria in 2009. The Boulton and Watt rotative steam engine, built and first steamed in 1784, was first deployed for the industrial expansion of consumer culture and capital by "climate men" entrepreneur Matthew Boulton and engineer James Watt. The humanist frameworks, presented through collection, exhibitory, interpretive, and documentary practices and, as I illustrate in Chapter 7, in respect to Agenda 2030 and the Sustainable Development Goals (SDGs), cement the idea that societal well-being is dependent on the ideal form of social and economic structure inherent in the ownership of land, material accumulation, profit maximisation, individualism, waged labour, and unlimited economic growth. All these styles of thought work against coexistence and interconnectedness.

Museum practices also promulgate the idea that all things have an intrinsic identity. For example, museums articulate a reality based on the notion of the human species as being positioned first and foremost at the centre of worldly life and represent human diversity as a series of biologically and culturally discrete individuals organised into societies and cultures regulated by cultural conventions. The non-human world, including earthly processes, on the other hand, is routinely represented by the all-encompassing concept of Nature. Discipline imperatives co-opt the non-human world into historical narratives of events as being tied to human stories, and as scientific, natural, and historical specimens, or phenomena, to be understood and organised according to species categorisations such as hominds, vertebrates, and invertebrates, and so forth. Research is often directed to knowing, and therefore representing, the capacity of humans, routinely western and male to harness and culture the non-human world.

The curatorial framing of relations between human species, cultures, animals, plants, and technology as being in separate domains maintains the status of modern human populations as commanders and controllers rather being part of a shared world characterised by radical copresence, shared events, and entangled processes of different intensities, temporalities, and durations. Agency is typically restricted to thinking human subjects and the non-human world is relegated to a collection of objects on which a superior human subject–object lens of interrogation and interpretation is directed. Capitalist modernity, and its figuration of agency, is, according to humanities scholar Jeffrey Scott Marchand,[9] driven by intention and rational thought, is exercised by voice as an expression of subjectivity, and is empowered by a one-dimensional determination of action and impact. Modern heritage time, discussed in Chapters 2, 3, 6, and 7, predominates with all temporalities and durations of humans, the more-than-human and non-humans being folded into and presented as a progressive human-centred narrative.

The way the majority of museums approach current crises, whether that be climate change, as discussed in Chapter 4, or the Covid-19 pandemic, examined in Chapter 6, “Viral museologies: Curating human species-viral worlds in sympoiesis ” is also problematic because institutions primarily operate from a humanist orientation and represent non-human worlds and their enmeshments with human populations from a position of control or, in the case of policy implementation, through strong human agency. For example, the activation across the museum sector of Agenda 2030, and its SDGs, discussed in Chapter 7, envisions a humanist and neo-capitalist future in which human populations are distinct from earth others through which new worlds are produced and new forms of museological globalism are enacted. In this way, Agenda 2030 and the SDGs become a western and global museological solution.

Drawing on Rosi Braidotti's critique of humanism, I argue that since the 1970s museums have engaged multiple forms of critique put forward by critical, cultural, and social theories as well as social constructivism, to conceptualise power relations, diversity, and inclusion and guide the development of the new and critical museology, and curatorial practices. Processes of decolonisation upscaled in recent years and methods that seek to include marginalised communities through plural perspectives and social discourses seek to empower through an emphasis on voice and subjectivity.

If museums are to avoid compounding existing problems and creating new ones, they require new cognitive frames and practices of life that promote ethical forms of conjoint action and that acknowledge humans' deep entanglements, shared vulnerabilities, and futures with not only other humans, but with the more-than-human world on which we rely.[10] In western scholarly circles, there has been a renewed interest in a reconstitution of the human subject. This is in part a response to the climate crisis, but also more recently the pandemic and is an entry point for promoting new theories and practices of agential life. Inspired by First Nations epistemologies, established over millennia, and through insights offered by eighteenth-century philosopher Baruch Spinoza, the world is coming to be viewed as an entire interconnected cosmos.[11] These perspectives seek to re-institute relational and agentic connections between humans, animate life, and non-animate things so they can be understood as part of one dynamic system and non-linear processes.[12]

Despite advances in the inclusion of First Nations and marginalised communities in museums – in terms of voice, perspective, and even epistemologies, much of which is still yet to be fully realised – the non-human world still often remains relegated to the categorical definition of Nature but without its own agency. Furthermore, representations of the environment, as concept and figuration, operate as a background for human activities. Both "Nature" and "Environment" are forms of identification that have prevented the development of styles of ecological thought required for, what I call, planetary habitability.

Capitalism, and its accumulator consumerism enabled through the exploitation of resources penetrating deep into earthly processes and planetary geological substrates, has taken modern humans further away from ecological thinking of the embedded and embodied kind and therefore the potential for recuperative and attuned thoughts and acts. In curatorial practice, the environment and Nature, as I have argued previously, and examined further in Chapters 2 and 5, "Communitarian design and eco-curatorial methods," must be replaced by a broader range of coordinates that operate together with their own agencies in radical interrelatedness in situated, broad and distributed social collectives across vast scales. Non-human agency within curating practice I argue extends beyond museum spaces. Curating encompasses earthly and planetary processes and is refigured to become what I call "eco-curating"[13] to include all manner of entities such as chemicals, minerals, and cabling, oftentimes bonded with human curatorial agencies, as discussed in Chapter 2, with fire and the photosynthetic examined in Chapter 3, and viral agencies explored in Chapter 6. Eco-curating therefore becomes processes of planetary becoming flowing through and across museum spaces and curatorial practices. Curating would thus not be confined to the liminal spaces of museum buildings or contexts, although this has never been the case anyway.

While it is not possible to escape our own anthropocentricism, or to truly curate from a non-anthropocentric position, it is still conceivable to represent, compose, and work with other things and processes in experimental ways. This is the position from which curating practice must engage and take shape if we are to sustain planetary habitability.

This monograph takes its inspiration from Donna Haraway's 2016 monograph *Staying with the Trouble: Making Kin in the Chthulucene*. Through the development of new concepts and theories that draw inspiration from the critical posthumanities, the ecological turn, new materialisms, planetary praxis, empirical research, and practical examples, my work encourages new forms of scholarship and curatorial praxis that engage thoughtful thoughts, experimentation, and agential refiguration. Its specific goal is to promote embodied and embedded curatorial practices that involve relational and more-than-human ways of thinking and acting in the world through museum collections, exhibitions, and institutional practices. Furthermore, this text is uniquely concerned with concretely reconstituting human-centred concepts and practices to develop new museological methods that signal an enmeshed and more-than-human approach to the make-up and composition of the world. New concepts and figurations of the human, agency, life, non-life, and processes are developed in this book that suggest new understandings of what counts as thinking and acting in planetary circumstances.

International approaches to progressing real-world changes in scholarship and curatorial practice are also critically engaged with and extended in this book. The text offers inspiration for museological scholars to conduct research and for museum professionals to curate all manner of activities from a different ecological reference point so they can promote a world good enough for all things to live and thrive in radical coexistence. It acknowledges that for this to occur novel, more-than-human curatorial visions, methods, conceptual frameworks, policies, and museologies are required.

Curating for planetary habitability, mobilised through a more-than-human lens and accompanying practices, involves reworking divisions of the human and the non-human, and the injustices and exclusions that result from these limited modes of being. This text considers how the prevailing genre of the human, and of the social activated through curatorial practice, has not only excluded women, marginalised Indigenous peoples, LGBTQ, and those of colour but also other-than-human entities oftentimes rendered invisible despite empirical relations of embeddedness that attest otherwise. We have always been more-than-human and other-than-human, but such a figuration has become intensified, more visible, and more compelling.

Each of the chapters of this book aims to foster alternative ways of curating, that is, thinking, living, thriving, and enacting policy with –our radical coexistence and by fostering mindful ethical responsibility towards non-human others on which planetary survival is contingent. Through my novel refiguration of curating as eco-curating, I also examine the emergence of different curating agencies and bonded humanisms, many of which either contribute to or are borne out as planetary crises.

The arguments in the book build on my trajectory of scholarship which I term ecologising experimentations. Inspired by Bruno Latour's ecologising thesis, I developed this concept further in 2008 as an experimental practice and since have used it across domains of museum theory and practice including climate change narratives, institutional forms, networked collections and documentation, digital cultural heritage, and contemporary ecological museologies.[14] I take up Rosi Braidotti's challenge to debate what it means to be human[15] and therefore what it means to be human in more-than-human and other-than-human curatorial practice.

All the refigurations and experimental material enactments I develop in the monograph are directed to Haraway's call to become less destructive, but more response-able, attuned, capable of surprise, and able to practice the arts of living and dying well in multispecies sympoiesis.[16] But as I contend also in and with planetary processes.

There are many examples of human–non-human relations that First Nations peoples have developed over millennia, all of which are more caring and respectful of the non-human world. While my attention is directed to reworking modern curatorial practices, in this book I also consider how First Nations knowledge practices might work within museum contexts in fractiversal constellations and with other more-than-human frameworks.

This book is an occasion to re-think the concept of sustainable, or better futures. Sustainable or desired futures, as currently framed in global and museum policy frameworks, enact humanist futures. At the same time, I re-think heritage, museum curatorial temporal dispositions, and the notion of museum pasts and futures to support such orientations. These temporal frames are often presented as a continuum from the past towards a forward-thinking orientation but are not so focused on presentism. To this end, I frame novel curatorial practices of presentism, of staying with or working with all manner of entities and more-than and other-than-human worlds in the present. The promotion of presentism and ontological design in curatorial practice encompasses biodiversity, species, and materials research and also includes western economic and social lives as ecological capitalisms.

I draw inspiration from what media theorist Jennifer Gabrys calls the planetary[17] by offering both alternative configurations of planetary habitability and futures for curatorial practice. I work with new concepts such as eco-curating, practices of thingness, ecological composition, and practice as eco-curatorial. Thingness and eco-curating as concepts and praxis are not just confined to museums, or even the living and terrestrial, but rather extend beyond these earthly concerns to engage curating with and in vast sprawling webs of planetary interconnectedness. This is because the material implications of curatorial actions extend far beyond museum spaces. Curators, other museum professions involved in all manner of curatorial projects, and audiences as subjects are refigured as and in planetary processes, and as domains of influencing. Curating, in an eco-sense, can never be fully determined due to the complex nature of the agencies involved, the inconsistencies in interactions, and the indeterminacy of calculations made through machine learning, and atomic or biophysical level interactions in respect to the chemical, the agencies of fire, the photosynthetic or the viral, just to name a few examples. Eco-curatorial actions never perform the same set of interactions and therefore are never identical and replicable. Therefore, curating and creating for a habitable world is one of first acknowledging and working with our radical interrelatedness and, what Haraway calls, "thick presences"[18] in uncertainty, rather than how we are different and superior.

While the focus of this book is the refiguration of modern forms of curatorial practice in history, natural history, science, and technology contexts, it also has applications to art curation and beyond. The aim here is to consider how museums might change how they represent and therefore might contribute to changing the world. Such refigurations represent a shift from an attempt to understand the world to trying to change it, from re-enforcing established modern visions and knowledge practices to critiquing and transforming them into processual methods and knowledges that sustain.

Throughout the book, I posit that museums have a new activist role directed to changing established assumptions, values, principles, and practices and taking on new responsibilities. I suggest they invert capitalist predilections, gesture towards and make explicit curatorial practice as acts of recuperation, care, and speculative world-making. Questions explored throughout the book are numerous. They include: how can museums reorientate curatorial practices to institute recuperative measures and expand economic practice as regenerative? How can institutions cultivate attunement with others in a shared world across domains of practice and in such a way as to render others capable in new types of capacity-rendering relations? Museums can encourage what Haraway[19] suggests in regard to other domains and sectors; to call forth and trigger new ways of thinking, acting, and caring as opposed to re-enforcing the stark, impersonal, uneven, and exploitative relations of nature and culture binaries in which museums are currently complicit.

Building on earlier work I have done, I propose that interpretive and documentary practices of curatorial cuts and cutting are a suitable curatorial interpretive and documentation method.[20] Cuts have always been central to curatorial practice because any curatorial act is a cut into planetary processes. There are many different ways of making curating cuts into the vast sprawling webs of thingness, inhabitations, and their relations of connection. Cuts can be made to make visible modern anthropocentric/patriarchal frameworks typical of museum curation. Other cuts, through the lens of the Anthropocene, can reveal a curatorial disposition focused on a geo-centric orientation. Some examples are Rosi Braidotti's posthuman convergence of human and non-human entities, Jane Bennett's mesh of the vitalism of matter, Timothy Morton's biological mesh, and Haraway's compost of multispecies sympoiesis. These are all cuts and figurations of entanglement directed to promote attunement. For Haraway, the weaving together of diverse biological entities is a means of creating intimate connections to foster biological love, care, respect, empathy, bonding, and partial recuperation. In each of the chapters, I propose different regenerative inhabitations through mechanisms such as policy, exhibitions and collections, and museum engagement and through the use of new humanisms in curatorial practice as both fractiversal and multiversal communitarian design practices.

In taking account of uncertainty in compositional design, we can't know everything, or begin to know everything, about the world around us and the beings and things that inhabit it. Nor can we know how we might compose with entities in radical intimacy, in openness, and in immanence.[21] To address uncertainty, curatorial practice must remain open to process in emergence.

Foremost, curatorial practice is experimental and thus directed towards animating novel approaches, which stimulate thoughts, new conceptual frames, methods and guide developments in the field. Central to this is the framing of curating as an empirical practice focused on enlivening eco-curating in the field.

In eco-curating practice, new notions of community emerge and comprise entities that are more-than-human and other-than-human. Earthly processes, such as the atmosphere, biosphere, oceans, and ice, become critical stakeholders; they become kin in relationships and demand respect and care.[22]

Like modernity and modern thought, curatorial practice is not a homogeneous process, a set of universal procedures or the outcome of one-world thinking. In undertaking a refiguration of curatorial practice, I seek to design curatorial thoughts and methods for living with others in situations of complex relationality and involving processes mediated by new notions of community as a broader suite of coordinates in situated contexts. Such curating practices would aim to be refigured from a new positionality of what it means to be human, decentre human agency, and seek to acknowledge the unknown and unknowable. In each of the chapters of this book, I suggest ways we might re-organise curatorial practices as forms of communitarianism and as communitarian design practice.

Capitalism is embedded in all aspects of museum practice and operations and is also driving what are considered desirable futures. Typically, capitalism is strengthened and rendered normative through museums. Museums have typically presented non-capitalist economies as pre-modern or anthropological and they are rendered in the past tense. Neo-capitalism is presented overtly and subtly as the preferred economic practice for empowering human agency and capacity to control and extract. Museums though, I argue, have the potential to perform other economic worlds inspired by the community economies network. One way they can do this is by using experimental design concepts to showcase economic practices of difference, and as diverse economies that open out economic activity to the possibility of alternatives[23] that support more-than-human flourishing. To do this what is required is a critical analysis of economy in museums and how it is performed through museum practice, and then a concerted effort to engender other worlds, economic values, principles, and decision-making attuned to radical connectivity.

In Chapter 2 of this book, I draw on arguments made in my 2021 monograph, *The Future of Digital Data, Heritage and Curation In a More-Than-Human World* (inspired by environmental and critical posthumanities, digital humanities, and media ecology) to introduce the key conceptual frameworks and methods used throughout the book. I discuss how these can be used to curate digital cultural heritage in planetary circumstances and in the face of rapid technological and social transformations both beyond the archive and out in the world. Trump's tweets are used as examples to illustrate the materiality of digital cultural heritage as being ecological, of the earth, more-than-and-other-than-human, and stories of deep time and planetary crises and futures yet to come, and in which the planetary becomes a new figuration. To achieve this, I advance the notion of online digitally born heritage as a new figuration as distributed ecological compositions, and I suggest that curation, curatorial agency, and users are part of more-than-human eco-curating processes and agencies with variabilities in duration, and as operating both within the archive and out in the world.[24] In a conventional sense, curating is no longer authorial, nor a series of actions by humans or automated systems. Rather it is the result of the vitalities of more-than and other-than-human coordinates emerging as situated forms of eco-systemic curating.

In order to re-think modern human-centric curatorial practice and to reinvent it so that it becomes curation for planetary habitability and recuperation, curators and scholars require a deeper understanding of humans' environmental impacts, including the material trails of waste, the inequalities embedded in regimes of power and the disadvantages that arise from data production. This can be achieved by striving for a greater attunement to all manner of humans and non-human constituencies that are embedded in data production and consumption. Ecological and embodied awareness, rather than a humanist vision of life-making processes through data production, is mobilised and foregrounded in this chapter in order to engender ethical accountability and care.

The material traces of ecological compositions, I argue, will emerge as heritage of the atomic kind, and be fossilised in stone as anticipated forms of human, more-than-human, and other-than-human rituals of memorialisation that are yet to manifest. These emerging technofossils will attune scholarly and professional practice to the other temporalities latent in digital cultural heritage compositions, that is, temporalities beyond the eco-curating processes of the contemporary present to that of biogeochemical processes of deep time in which digital cultural heritage is becoming stone.[25] Within this vein, I argue that Trump's tweets are examples of heritage that stretch over extended time frames, and comprise multiple intersecting durations from the temporalities of climate change and environmental exhaustion, to those of techno-capitalist regimes of accumulation and exploitation that challenge linear historical time of modern heritage frameworks.[26] I conclude Chapter 2 by stating that the digital cultural heritage legacies that will be significant for future generations are not only what museums and cultural institutions seek to make and contain as the artifactual, but what is passed on as dynamic ecological compositions and their wastes.[27]

In Chapter 3, "Collections and curating human–non-human climates," I investigate how heritage collections can be curated as material agents for the cultural and social change on which the "de-carbonisation" of modern societies is reliant. I first work with a melted green plastic bucket recovered from the Black Saturday bushfires in Victoria, Australia in February 2009. This bucket is held in the collection of Museum Victoria and I explore how it is a material register of entanglements: the convergence of extreme weather events and an embodiment of climatic catastrophe in the form of wildfire. The second example is the Boulton and Watt "Lap" rotative steam engine, built in 1788 and held the Science Museum, London collection. This, I argue, is an exemplar of the Industrial Revolution that embodies a profound climate history of emissions.

Rather than viewing these collections as obsolescent technologies, as non-sentient, compliant, lacking their own agency, and only enlivened when receiving interpretive input from curators and educators in respect to their climate or Anthropocene histories, I argue that they are agential and vibrant with carbon materialities, and demonstrative of moments in the trajectory of rising atmospheric fossil fuel emissions and their consequences. Drawing on my earlier work I demonstrate how collections can be put to work according to, what I call, new types of cultural mitigation instruments mobilised through the figuration of "climatic ecological compositions" and the folding out and analysis of their material form, material, and cultural histories, processes of production and so forth. To do this I use the Boulton and Watt "Lap" Engine case study to consider the ratios of carbon isotopes embedded within their "DNA." These two examples, the bucket and the Engine act as book ends in a climate story spanning 230 years, and hemispheres apart. One is implicated in the development of fossil fuel combustion that led to the climate crisis in Birmingham, England in 1788, and the other is the unintended consequences of climatic events in the hot and dry landscapes of Australia and wildfire. I also illustrate how climate collections can serve as new forms of cultural carbon mitigation strategies when considered through the development of new figurations of modern populations as dynamic biochemical fossil fuel life forms enmeshed in the photosynthetic register of their wooden fabrics. A ^{13}C carbon isotope analysis of their wooden components and contexts reveals these human–non-human climatic enmeshments and the changes in the composition of the atmosphere over time as illustrative of rising emissions. They can thus be mobilised for these purposes by acting against the belief that modern populations are dis-embedded, hubristic sovereign individuals, and by the exercise of strong human agency can solve the climate crisis. This figuration has the potential to expand worldviews and changing beliefs in respect to their relations with the non-human world to one of profound enmeshments from a position in which new relations of responsibility, acts of care, and respectful design might emerge.

In an uncertain world facing socio-ecological collapse, museums of all types and forms around the world are mobilising for climate action and implementing global policy strategies to promote and enact more sustainable futures. These mandates are no longer on the periphery of museum concern but have become agendas and policy frameworks that drive museum strategies. However, it must be said that typcially museums tend to adopt global policy frameworks and implement them without a critical regard for what they mean and their material consequences. The discursive material and policy components of institutions – from infrastructures to legal regimes, to collecting practices and interpretive strategies – materially influence human and non-human communities in profound ways. In Chapter 4, "Museums, climate policy frameworks and the problem of humanist-driven solutions," I explore museums' engagement with climate change mitigation strategies, namely, the PA and its education policy, "the Action for Climate Empowerment" (ACE). I interrogate the philosophical premises and value systems underlaying these policy frameworks and question the futures these instruments might materially enact when implemented through museums. To do this, I critically examine the modern forms of knowledge, human agency, economy, and culture that are given authority through museums but have led to the devastating consequences of the climate crisis. These same conative frames continue to be embedded in dominant policy frameworks and museum public-facing programs.

Following on from the arguments in the preceding chapter, in Chapter 5, "Communitarian design and eco-curatorial methods: Eco-curating climate change in attunement," I explore how the PA and the Action for Climate Empowerment might be re-worked and expanded beyond their modern frameworks. At the same time, I highlight other "ecologizing" knowledge practices that might be put to work to curate climate for planetary habitability within museum spaces.

Modern populations, their lives, and histories have always been inextricably ecological because all things shape other things. And yet many societies in the West have lost their way. Modern societies are not truly thinking and feeling ecologically, and in an enmeshed sense, because they view humans as the dominant governing system. Nor do western societies, cultural infrastructures, and industrial and political systems that govern interactions with the non-human worldview these multifarious entities and things as cohabitants in a shared world. If they did ethical forms of conjoint action could be promoted that, according to Haraway, would have the potential to render non-humans capable.[28]

To this end, I illustrate how ecological thoughts of various kinds are proliferating in the creative arts, humanities, social sciences, and economic practice. I explore the potential of more-than-human knowledge practices to be inducted into curatorial practice. The approaches I bring forth and develop in Chapter 5 promote values, thoughts, and practices fully aware of how humans are connected with other beings, things, and earthly processes and how an acknowledgement of these connections has the potential to fuel curatorial inspiration.[29] Rather than institutions judiciously following the principles of the humanist global policy documents discussed in Chapter 5, I argue they can also become active agents in formulating policy and context-specific and attuned climate research by engaging diverse knowledge practices. In doing so, I reframe audiences as trans-corporeal, as embedded and embodied subjects who are cognizant of the more-than-human and therefore equipped to understand material responsibility. And while museums have been complicit in modern forms of thinking and acting through curatorial work that have been destructive, as institutions they have an important role to play in reframing knowledge, modifying disciplinary actions and research regimes, and changing the very concept of curating to become attuned eco-curating crafting practices operating "in-worlds" as laboratories for experimental communitarian design practice.

The Covid-19 virus has, in a profound and forceful way, highlighted the ecological intimacy of lifeworlds, earthly processes, and economic, social, and political systems. In Chapter 6, "Viral museologies: Curating human-species-viral worlds in sympoiesis ," I promote the idea of viral museologies as opposed to humanist museologies as a different reference point for engagement in which viruses become a significant curatorial agent. While museums portray themselves as vulnerable victims of the pandemic, I contend that institutions are complicit in the creation and spread of these new viral kin because in many instances they have promulgated thoughts and acts of dis-embeddedness and hubristic un-mindedness. Here, the porosity and vulnerability of human bodies are highlighted as levers for arguing against the notion of the sovereign individual, a conceptualisation that often forms the basis of museum programs. Museums therefore become known as radical viral inhabitations and potential spreading venues and events.[30] Through this optic I conduct an ecologising experimentation of planetary worlding in the context of the Delta and then Omicron variants and, in doing so, draw on the arguments presented in my chapter "Viral agencies and curating worldly life differently in museum spaces," which appears in the collection, *The Posthuman Pandemic*.[31]

Drawing inspiration from multispecies research, and particularly the work of Eben Kirksey, Stefan Helmreich, Celia Lowe, and Donna Haraway, as well as my own work on more-than-human curating, it becomes possible to outline new types of profoundly intimate communities and eco-curating processes that can be made visible in museum spaces. In this context, curatorial practice, museum space, exhibitions, collections, audiences, curators, and all manner of non-human stakeholders emerge as trans-corporeal entities in multiple material, biological, energetic, cultured, and discursive co-constitutive enmeshments knitted together genetically, chemically, elementally, energetically but also virally.[32] During the Covid-19 pandemic the Anthropocene became the Virosphere,[33] an unimaginatively vast and potent agent shaping political, economic, and social forces outside human capacities to control, including profound injustices and violence within geopolitical systems which themselves have the potential to become new topics for museum engagement, and social, political, and agential reform.

All manner of institutions, including the International Council of Museums and other professional bodies, are adopting Agenda 2030 and the SDGs, a set of 17 goals developed to frame global initiatives for sustainable futures. Through an expanding network of museum institutions and alliances, the SDGs are going viral across the sector.

While aspirations enshrined in the goals – such as ending hunger, alleviating poverty and violence, attaining gender equality, peace, justice, good health and well-being, clean water, and sanitation are admirable, the goals are actually profound in their failure to invoke relational interdependence with others and the non-human world in respectful, recuperative ways. Museums' un-reflexive ways of invoking the SDGs literally means they have the potential to threaten and undermine planetary habitability. In Chapter 7, "Curating sustaining practices in more-than-human worlds," I unpack the dominant thoughts, frames, and systems of value inherent in museums' engagement with the SDGs and explore the types of futures these frameworks might enact.

Sustainability, I argue, must become habitability practices of care and respect attuned to the rhythms and vitalism of the non-human world through an understanding of their enmeshments with human designs and the sustenance of these relations in respectful ways. A consideration of the non-human world as flourishing and not just a resource to meet human ends, and the exercise of practices whereby what is taken is only what is required rather than being dictated by an exploitive logic are two tenants that are fundamental to many First Nations communities. Sustainability, within museum frameworks and representational practices, therefore, is reframed as sustaining frameworks that are always in transformation and are viewed as fractiversal in being comprised of a broad range of habitability and economic living practices in specific contexts rather than adhering to one agenda. As diverse economies pioneers, Katherine Gibson and Julie Graham concur changing our understanding of the world and economy is to change the world, if only partially and locally.[34]

Founded in the 1970s and gathering momentum in the 1980s, the "new museology" fostered a post-structuralist sociological approach to museum philosophy and practice. Concerned with foregrounding the social roles and purposes of museums and deconstructing the politics of representation in exhibitions and collecting practices, these paradigms acted as an alternative approach to the old museological focus on methods and procedures.[35] While institutions have in the last three decades placed a greater emphasis on cultural pluralism, practices of cultural representation, community development, and education, these tropes continue to uphold the assumption that the human is the centre of all things and that human societies are naturally divided into a finite number of non-interacting cultural categories. Difference, in the context of "new museology," is equated as a plurality and a disparity of worldviews.

The discussion in the final chapter is directed towards an investigation of the "new museology"[36] and practices centred on the human subject and capitalist accumulation, and on reframing these processes as a relational and more-than-human ontological formulation. The more-than-human museologies I propose move beyond the current focus on human agency, subjectivities, and social constructivism and instead become open to and enmeshed with the full agency of the world and its more-than-human hybrids and entities from technology, to the digital, to plants, animals, climatic systems, human bodies, emotions, diverse ontologies, and so forth. Non-human entities become curating agencies in new forms as bonded humanisms with fire, the photosynthetic, and the viral. The figure of the human and strong human agency becomes one of humanness as extended and distributed processes of culturing, crafting, and influencing on broader planetary process with other agential processes.

"Community" in more-than-human museological figurations not only includes human and human others but also the non-human and earthly processes. "Engagement" is extended beyond the sole focus on human interactions with others and things to include entanglements with an array of coordinates. "Identity," "agency," and "history" become the emergent consequences of more-than-human and other-than-human coordinates within these new social collectives. "Subjectivity," "expertise," and "difference" become an array of affecting, entangling agencies, and affordances as eco-curating processes. Things, commonly called objects or collections, no longer act as stand-ins for human subjectivities, but rather become situated ecological compositions, comprising diverse material agencies and more-than-human and non-human domains of influencing in and across planetary processes. Museum or heritage time becomes the co-mingling of the durations and temporalities of many agencies.

Rather than seeing cultural diversity as plural, cultural expressions set against the backdrop of a one worldview and one nature configuration of life will be viewed as fractiversal and of comprising a diversity of natures and entities requiring the negotiation of different realities to co-produce shared worlds through all manner of practices and narratives within and beyond museum spaces.[37] Museum curating and the museologies that inform such practices will become ecological, embodied, and embedded forms of more-than-human communitarian design and experimentation.

Enjoy the monograph! I hope you find it inspirational as a re-narration and refiguration of curatorial practice and scholarly museologies for ethical, communitarian possible world-making.

Notes

1 Zoltán Boldizsár Simon and Marek Tamm, "Historical Futures," *History and Theory* 60, no. 1 (March 2021): 3–22, 3.
2 Simon and Tamm, "Historical Futures," 4.
3 Fiona R. Cameron, "Ecologizing Experimentations: A Method and Manifesto for Composing a Post-Humanist Museum," in *Climate Change and Museum Futures*, eds. Fiona R. Cameron and Brett Neilson, Routledge Research in Museum Studies (New York: Routledge, 2014): 16–33.
4 Saul Newman and Tihomir Topuzovski, eds. "Introduction," in *The Posthuman Pandemic* (London: Bloomsbury Academic, 2021): 1–26, 2.
5 Cameron, Fiona R. "Theorizing More-Than-Human Collectives for Climate Change Action in Museums." *L'Internationale Online Climate Change Special Issue* to coincide with the UNFCCC meeting in Paris, November 30–December 11, 2015, accessed August 3, 2022, http://www.internationaleonline.org/research/politics_of_life_and_death/48_theorising_more_than_human_collectives_for_climate_change_action_in_museums.
6 Severin Fowles, "The Perfect Subject (Postcolonial Object Studies)," *Journal of Material Culture* 21, no. 1 (2016): 9–27, 11.
7 Cameron, "Ecologizing Experiments," 18–21.
8 Nick Axel, Daniel A. Barber, Nikolaus Hirsch and Anton Vidokle, "Accumulation," *e-flux Architecture*, accessed September 19, 2022, https://www.e-flux.com/architecture/accumulation/.
9 Jeffrey Scott Marchand, "Non-human Agency," in *Posthuman Glossary*, eds. Rosi Braidotti and Maria Hlavajova (London and New York: Bloomsbury Academic, 2018): 292–95, 295.
10 Cameron, "Ecologizing Experimentations," 21.
11 Genevieve Lloyd, *Routledge Philosophy GuideBook to Spinoza and the Ethics*, Routledge Philosophy GuideBooks (New York: Routledge, 1996), 40; Kyle Powys Whyte, "Indigenous Climate Change Studies: Indigenizing Futures, Decolonizing the Anthropocene," *English Language Notes* 55, no. 1–2 (Fall 2017): 153–62.

12 Gilles Deleuze and Felix Guattari, *A Thousand Plateaus: Capitalism and Schizophrenia* (London: Continuum, 2008); Donna Haraway, "A Manifesto for Cyborgs: Science, Technology, and Socialist Feminism in the 1980s," *Socialist Review* 5, no. 2 (1985): 65–107; Donna Haraway, *The Companion Species Manifesto: Dogs, People and Significant Otherness* (Chicago, IL: Prickly Paradigm Press, 2003); Donna Haraway, *When Species Meet* (Minneapolis: University of Minnesota Press, 2007); Donna Haraway, *Staying with the Trouble: Making Kin in the Chthulucene* (Durham, NC and London: Duke University Press, 2016); Cary Wolfe, *What Is Posthumanism?* (Minneapolis: University of Minnesota Press, 2010); Pramod K. Nayar, *Posthumanism* (Cambridge: Polity, 2014); Manuel De Landa, *A New Philosophy of Society: Assemblage Theory and Social Complexity* (London and New York: Bloomsbury Academic, 2019); Jane Bennett, *Vibrant Matter: A Political Ecology of Things* (Durham, NC: Duke University Press, 2010); Timothy Morton, *The Ecological Thought* (Cambridge, MA: Harvard University Press, 2010); Bruno Latour, *We Have Never Been Modern* (New York: Harvester Wheatsheaf, 1993); Graham Harvey, *Animism: Respecting the Living World* (New York: Columbia University Press, 2006). Rosi Braidotti, *Posthuman Knowledge* (Cambridge: Polity Press, 2019).

13 Fiona R. Cameron, "Theorizing Digitizations in Global Computational Infrastructures," in *The International Handbook of New Digital Practices in Galleries, Libraries, Archives, Museums and Heritage Sites*, eds. Hannah Lewi, Wally Smith, Steve Cooke, and Dirk vom Lehn (London: Routledge, 2019): 64–66; Fiona R. Cameron, *The Future of Digital Data, Heritage and Curation in a More-Than-Human World* (London: Routledge, 2021): 9.

14 Cameron, *The Future of Digital Data,* 10; Cameron, "Theorizing Digitizations in Global Computational Infrastructures," 55–67; Fiona R. Cameron, "Stirring Up Trouble: Museums as Provocateurs and Change Agents in Polycentric Alliances for Climate Change Action," in *Addressing the Challenges in Communicating Climate Change Across Various Audiences*, eds. Walter Leal Filho, Bettina Lackner, and Henry McGhie (New York: Springer, 2019), 647–73; Fiona R. Cameron, "Posthuman Museum Practices," in *Posthuman Glossary*, eds. Rosi Braidotti and Maria Hlavajova (London and New York: Bloomsbury Academic, 2018), 349–51; Fiona R. Cameron, "The Liquid Museum: New Ontologies for a Climate Changed World," in *Museum Theory: An Expanded Field*, eds. Andrea Witcomb and Kylie Message (Oxford: Blackwell Publishing, 2015), 345–62; Fiona Cameron and Sarah Mengler, "Transvisuality, Geopolitics and Cultural Heritage in Global Flows: The Israeli–Palestinian Conflict and the Death of the Virtual Terrorist," in *Transvisuality: Dimensioning the Visual*, eds. Tore Kristensen, Anders Michelsen, and Frauke Wiegand (Liverpool: Liverpool University Press, 2015), 59–72; Cameron, "Theorizing More-Than-Human Collectives for Climate Change Action in Museums"; Fiona R. Cameron, "From 'Dead Things' to Immutable, Combinable Mobiles: H.D. Skinner, the Otago Museum and University and the Governance of Maori Populations," *History and Anthropology* Special Issue, 25, no. 2 (March 2014): 208–26; Fiona R. Cameron, "We Are on Nature's Side? Experimental Work in Rewriting Narratives of Climate Change for Museum Exhibitions," in *Climate Change and Museum Futures*, eds. Fiona R. Cameron and Brett Neilson (New York: Routledge Museum Research Series, 2014), 51–77; Cameron, "Ecologizing Experimentations," 16–33; Fiona R. Cameron, "Climate Change, Agencies and the Museum and Science Centre Sector," *Museum Management and Curatorship* 27, no. 4 (October 2012): 317–39; Fiona R. Cameron, "Liquid Governmentalities, Liquid Museums and the Climate Crisis," in *Hot Topics, Public Culture, Museums*, eds. Fiona Cameron and Lynda Kelly (Newcastle on Tyne: Cambridge Scholars, 2010), 112–28; Fiona R. Cameron, "Object-orientated Democracies: Conceptualising Museum Collections in Networks," *Museum Management and Curatorship* 23, no. 3 (2008): 229–43.

15 Braidotti, *Posthuman Knowledge*, 10.
16 Haraway, *Staying with the Trouble*, 101.
17 Jennifer Gabrys, "Becoming Planetary," *e-flux Architecture*, accessed September 19, 2022, https://www.e-flux.com/architecture/accumulation/217051/becoming-planetary/.
18 Haraway, *Staying with the Trouble*, 16.
19 Haraway, *Staying with the Trouble*, 16.
20 Cameron, "Theorizing Digitalizations," 66; Cameron, *The Future of Digital Data*, 9.
21 Morton, *The Ecological Thought*; Haraway, *Staying with the Trouble*, 16.
22 Cameron, "Climate Change, Agencies and the Museum and Science Centre Sector," 332.
23 Julie Katherine Gibson-Graham, "Diverse Economies: Performative Practices for 'Other Worlds'," *Progress in Human Geography* 32, no. 5 (2008): 614.
24 Cameron, *The Future of Digital Data*, 129.
25 Cameron, *The Future of Digital Data*, 256.
26 Cameron, *The Future of Digital Data*, 256.
27 Cameron, *The Future of Digital Data*, 284.
28 Haraway, *Staying with the Trouble*, 126.
29 Cameron, "Theorizing More-Than-Human Collectives."
30 Fiona R. Cameron, "Viral Agencies and Curating Worldly Life Differently in Museum Spaces," in *The Posthuman Pandemic*, eds. Saul Newman and Tihomir Topuzovski (London: Bloomsbury Academic, 2021): 219–46, 220.
31 Cameron, "Viral Agencies."
32 Cameron, "Viral Agencies," 220.
33 Carl Zimmer, "Welcome to the Virosphere," *New York Times*, March 24, 2020, accessed February 26, 2023, https://www.nytimes.com/2020/03/24/science/viruses-coranavirus-biology.html.
34 Gibson-Graham, "Diverse Economies," 629.
35 Duncan F. Cameron (1972) "The Museum, a Temple or the Forum," *Curator* 4, no. 1: 11–24; Peter Vergo, ed., *The New Museology* (London: Reaktion Books, 1989); Tomislav Šola, *Essays on Museums and their Theory: Towards the Cybernetic Museum* (Helsinki: Finnish Museums Association, 1997).
36 Peter Vergo, *The New Museology* (London: Reaktion, 1989).
37 Cameron, "Ecological Experimentations."

References

Axel, Nick, Daniel A. Barber, Nikolaus Hirsch, and Anton Vidokle. "Accumulation." *e-flux Architecture*, accessed September 19, 2022, https://www.e-flux.com/architecture/accumulation/.

Baumlin, James, S. "From Postmodern to Posthumanism: Theorizing Ethos in an Age of Pandemic." *Humanities* 9, no. 2 (2020): 9–46.

Bennett, Jane. *Vibrant Matter: A Political Ecology of Things*. Durham, NC: Duke University Press, 2010.

Braidotti, Rosi. *The Posthuman*. UK, Cambridge: Polity Press, 2013.

Braidotti, Rosi. *Posthuman Knowledge*. UK, Cambridge: Polity Press, 2019.

Cameron, Duncan F. "The Museum, a Temple or the Forum." *Curator* 4, no. 1 (1972): 11–24.

Cameron, Fiona R. "Object-Orientated Democracies: Conceptualising Museum Collections in Networks." *Museum Management and Curatorship* 23, no. 3 (2008): 229–43.

Cameron, Fiona R. "Liquid Governmentalities, Liquid Museums and the Climate Crisis." In *Hot Topics, Public Culture, Museums*, edited by Fiona Cameron and Lynda Kelly, 112–28. Newcastle on Tyne: Cambridge Scholars, 2010.

Cameron, Fiona R. "Climate Change, Agencies and the Museum and Science Centre Sector," *Museum Management and Curatorship* 27, no. 4 (October 2012): 317–39.

Cameron, Fiona R. "Ecologizing Experimentations: A Method and Manifesto for Composing a Post-Humanist Museum." In *Climate Change and Museum Futures*, edited by Fiona R. Cameron and Brett Neilson, 16–33. Routledge Research in Museum Studies. New York: Routledge, 2014.

Cameron, Fiona R. "From 'Dead Things' to Immutable, Combinable Mobiles: H.D. Skinner, the Otago Museum and University and the Governance of Maori Populations." *History and Anthropology* Special Issue, 25, no. 2 (March 2014): 208–26.

Cameron, Fiona R. "We Are on Nature's Side? Experimental Work in Rewriting Narratives of Climate Change for Museum Exhibitions." In *Climate Change and Museum Futures*, edited by Fiona R. Cameron and Brett Neilson, 51–77. New York: Routledge Museum Research Series, 2014.

Cameron, Fiona R. "The Liquid Museum: New Ontologies for a Climate Changed World." In *Museum Theory: An Expanded Field*, edited by Andrea Witcomb and Kylie Message, 345–62. Oxford: Blackwell Publishing, 2015.

Cameron, Fiona R. "Posthuman Museum Practices." In *Posthuman Glossary*, edited by Rosi Braidotti and Maria Hlavajova, 349–51. London and New York: Bloomsbury Academic, 2018.

Cameron, Fiona R. "Stirring Up Trouble: Museums as Provocateurs and Change Agents in Polycentric Alliances for Climate Change Action." In *Addressing the Challenges in Communicating Climate Change Across Various Audiences*, edited by Walter Leal Filho, Bettina Lackner, and Henry McGhie, 647–73. New York: Springer, 2019.

Cameron, Fiona R. "Theorizing Digitizations in Global Computational Infrastructures." In *The International Handbook of New Digital Practices in Galleries, Libraries, Archives, Museums and Heritage Sites*, edited by Hannah Lewi, Wally Smith, Steve Cooke, and Dirk vom Lehn, 55–67. London: Routledge, 2019.

Cameron, Fiona R. *The Future of Digital Data, Heritage and Curation in a More-Than-Human World*. London: Routledge, 2021.

Cameron, Fiona R. "Viral Agencies and Curating Worldly Life Differently in Museum Spaces." In *The Posthuman Pandemic*, edited by Saul Newman and Tihomir Topuzovski, 219–46. London: Bloomsbury Academic, 2021.

Cameron, Fiona R. "Theorizing More-Than-Human Collectives for Climate Change Action in Museums." *L'Internationale Online Climate Change Special Issue* to coincide with the UNFCCC meeting in Paris, November 30 – December 11, 2015, accessed August 3, 2022, http://www.internationaleonline.org/research/politics_of_life_and_death/48_theorising_more_than_human_collectives_for_climate_change_action_in_museums.

Cameron, Fiona R. and Sarah Mengler. "Transvisuality, Geopolitics and Cultural Heritage in Global Flows: The Israeli–Palestinian Conflict and the Death of the Virtual Terrorist." In *Transvisuality: Dimensioning the Visual*, edited by Tore Kristensen, Anders Michelsen, and Frauke Wiegand, 59–72. Liverpool: Liverpool University Press, 2015.

Colebrook, Claire. "Not Symbiosis, Not Now: Why Anthropogenic Change Is Not Really Human." *Oxford Literary Review* 34, no. 2 (2012): 185–209.

De Landa, Manuel. *A New Philosophy of Society: Assemblage Theory and Social Complexity.* London and New York: Bloomsbury Academic, 2019.

Deleuze, Gilles and Felix Guattari. *A Thousand Plateaus: Capitalism and Schizophrenia.* London: Continuum, 2008.

Fowles, Severin. "The Perfect Subject (Postcolonial Object Studies)." *Journal of Material Culture* 21, no. 1 (2016): 9–27.

Gabrys, Jennifer. "Becoming Planetary." *e-flux Architecture*, accessed September 19, 2022, https://www.e-flux.com/architecture/accumulation/217051/becoming-planetary/.

Gibson-Graham, Julie Katherine. "Diverse Economies: Performative Practices for 'Other Worlds'." *Progress in Human Geography* 32, no. 5 (2008): 613–32.

Haraway, Donna. "A Manifesto for Cyborgs: Science, Technology, and Socialist Feminism in the 1980s." *Socialist Review* 5, no. 2 (1985): 65–107.

Haraway, Donna. *The Companion Species Manifesto: Dogs, People and Significant Otherness.* Chicago, IL: Prickly Paradigm Press, 2003.

Haraway, Donna. *When Species Meet.* Minneapolis: University of Minnesota Press, 2007.

Haraway, Donna. *Staying with the Trouble: Making Kin in the Chthulucene.* Durham, NC and London: Duke University Press, 2016.

Harvey, Graham. *Animism: Respecting the Living World.* New York: Columbia University Press, 2006.

Latour, Bruno. *We Have Never Been Modern.* New York: Harvester Wheatsheaf, 1993.

Lloyd, Genevieve. *Routledge Philosophy GuideBook to Spinoza and the Ethics.* Routledge Philosophy GuideBooks. New York: Routledge, 1996.

Marchand, Jeffrey Scott. "Non-human Agency." In *Posthuman Glossary*, edited by Rosi Braidotti and Maria Hlavajova, 292–95. London and New York: Bloomsbury Academic, 2018.

Morton, Timothy. *The Ecological Thought.* Cambridge, MA: Harvard University Press, 2010.

Nayar, Pramod K. *Posthumanism.* Cambridge: Polity, 2014.

Newman, Saul and Tihomir Topuzovski, eds. *The Posthuman Pandemic.* London: Bloomsbury Academic, 2021.

Simon, Boldizsár Zoltán and Marek Tamm. "Historical Futures." *History and Theory* 60 no. 1 (March 2021): 3–22.

Šola, Tomislav. *Essays on Museums and their Theory: Towards the Cybernetic Museum.* Helsinki: Finnish Museums Association, 1997.

Vergo, Peter, ed. *The New Museology.* London: Reaktion Books, 1989.

Whyte, Kyle Powys. "Indigenous Climate Change Studies: Indigenizing Futures, Decolonizing the Anthropocene." *English Language Notes* 55, no. 1–2 (Fall 2017): 153–62.

Witcomb, Andrea and Kylie Message, eds. *Museum Theory: An Expanded Field.* Oxford: Blackwell, 2015.

Wolfe, Cary. *What Is Posthumanism?* Minneapolis: University of Minnesota Press, 2010.

Zimmer, Carl. "Welcome to the Virosphere." *New York Times*, March 24, 2020, https://www.nytimes.com/2020/03/24/science/viruses-coranavirus-biology.html.

2

TECHNOSPHERIC HERITAGE

Curating more-than-digital heritages in and for planetary durations

Introduction

Waking around 5.30 am each morning, former US President Donald Trump turned on his television and tuned into CNN Fox News in the White House master bedroom, often tweeting from his personal account handle @realDonald Trump or his presidential account, inspired or infuriated by what he saw and heard on the news report. All of Trump's tweets, from the mundane to the intimating, the outrageous, to the subversive – including those that were complicit in inciting violence and the occupation of the Capitol on January 6th, 2021 – have been designated as historical data by the US Presidential Records Act (PAR; 1978). They must therefore be saved in perpetuity and henceforth, within a normative humanist heritage framework, become digitally re-born heritage.[1]

Furthermore, with the development of AI and automation, all manner of entities now produce data, from social bots on social media sites to sex bots in intimate relationships, autonomous weapons, embedded medical devices, automated systems in banks, and even bacteria in bio art experiments. All this data, from Trump's tweets to the scripts of sex bots, represent the emergence of new heritage forms and the data producers of present and future life.[2]

As a sub-set of heritage, digital cultural heritage is currently framed as societal data worth saving and passing onto future generations as two distinct forms: ***digitally born*** and ***digitisations***. Like other types of heritage, digital cultural heritage is human-centred, founded on modern ideas, informed and formed by concerns of endangerment and loss, mobilised through the lens of obsolescence, and driven by the demands of capital accumulation. The arguments in this chapter posit that all these concepts and practices are limiting

DOI: 10.4324/9781315212067-2

and inadequate for a world characterised by crises such as climate change, environmental destruction, data profusion, technological innovation, and the long-term concerns of planetary habitability.

Within a social media and heritage framework and while Trump's tweets are considered a practice and technic of life, they are also much more. Inspired by the environmental and critical posthumanities, digital humanities, and media ecology, I refigure digitally born heritage using Trump's tweets as examples as material, as ecological, as being of the earth, as more-than-and-other-than-human, and as stories of deep time and planetary crises and futures yet to come. To achieve this, I advance the notion of online digitally born heritage as a new figuration, as distributed ecological compositions, and I refigure curation, curatorial agency, and users as more-than-human eco-curating processes in duration, within the archive and out in the world.[3] In light of the emergence of AI, automation the human, and the case of Trump and his tweets, the human user as a central figure is not foregone but rather refigured as humanness, as traces through and across the ecological composition.[4]

Furthermore, I pull apart human-centred understandings of digital cultural heritage, based on human actors, narratives, and object-centred forms of materiality, sociality, and the accompanying social, cultural, and technical frames of interpretation, as well as recent accounts that focus on digital materiality. In the context of a radical expansion of the types and forms of data, infrastructures and technologies and their human, more-than-human, and non-human aspects, digital cultural heritage becomes more-than-humanist forms of social or cultural expression, statements of identity, or digital material substrates.[5] I first experiment with new ideas about human subjects, objects, and agency through the lens of ontological, new materialisms, posthuman and ecological modes of thinking and action. In light of this, I figuratively theorise digital cultural heritage as unruly ecological compositions, and curating as radical, ecosystemic processes involving the action of many different types of coordinates, and the vitality of their interrelatedness in their unfolding.[6] I then illustrate how these concepts might be applied empirically in digital cultural heritage contexts working with Trump's tweets.

A more expanded version of the concepts presented in this chapter, including more-than-human and non-human heritage forms, can be found in my monograph, *The Future of Digital Data, Heritage and Curation in a More-Than-Human World* published in 2021 by Routledge.

Digital cultural heritage as unruly ecological compositions

Over the last 30 or more years, there has been a fundamental societal shift in the way we think about, produce, work with, and engage with digital data and media. This change is aligned with technical innovation, the desire to extend human capabilities and creative capacities, to produce new commodities, and use data for bureaucratic and monetary purposes. There is, however, a gap between the emerging realities of digital life and experience and the theories and practices of digital cultural heritage. In the early days, and prior to the Internet, digital cultural heritage was viewed as a singular, isolated computational object comprising data – first as punch cards, then code saved on a carrier such as a floppy disc, CD, or DVD. Digital cultural heritage was thought about according to a limited range of coordinates, such as software and hardware, and routinely as a screen-shot image.

With the development of many internets, multiple superfast wifi connections, Web 2.0, the semantic web, extensive social networks, sophisticated search queries, personalisation and mobile applications, and the emergence of different types of mobile devices, digital platforms, and media types – such as Facebook, Twitter, Instagram, and Pinterest – there are now more complex forms of data production and connectivity.[7] As technological capitalism has grown, and the digital economy has expanded amidst the complexity and messiness of planetary computational infrastructures, the rise of automation and algorithmic governance, energetic systems, cloud computing and storage, more complex platform, financial, governance, and surveillance structures begin to cross sectors and scales. The destructive forces of data in its making and disposal have accordingly come to the fore. The mining of vast quantities of rare earths and metals (cobalt, gallium, indium, lithium, platinum, aluminium, tin, copper, palladium, and silver) used in batteries, hard drives, displays, memory chips, electronic components alongside gold and solvents used in connectors, contacts, and wire bonding to enable rapid Internet connections have led to toxic electronic waste dumps, conflicts, resource depletion, environmental destruction across multiple life spans, and to a range of human rights violations.[8] As a consequence of coal and petroleum mining, fracking for energy, and carbon pollution through the burning of fossil fuels to drive the digital economy, digital practices have contributed to climate change. Furthermore, the emergence of the Technosphere (a subsystem of the Anthropocene)[9] has been brought about by the evolution of technics, large-scale technology, and media production a set of circumstances and practices that according to media theorist Jussi Parikka is a question of geology.[10]

Smart phones, for example, media theorist Jussi Parikka explains, comprise an array of geological extracts drawing from a wide range of the planet's resources that are supported by a multiplicity of infrastructures.[11] The geological components in digital media devices include cobalt from Africa, zinc from Alaska, extracted and refined into indium in Canada, as well as other metals and minerals originating from Belgium, Russia, Peru, China, Congo, South Africa, and Malaysia. Here, the non-human world, previously described as Nature, is enrolled in digital technical projects and infrastructures.[12]

In a digital economy, directed towards scientific understanding and economic exploitation of all living and non-living matter under the imperative of capital accumulation, these more-than-digital ecologies create new materials, and contaminating conditions in the ruins of digital consumption and obsolescence.[13] At the same time, these developments are also reorganising social, geographical, economic, and political life as new types of ecologically implicated conditions while blurring distinctions between humans, other species, and earthly elements and processes.[14]

As a result, an ever-expanding array of coordinates and subjects enters the field of things that the heritage industry calls digital cultural heritage, both digitally born and digitisations often mediated by AI, algorithmic culture, and machine learning. Such intimate collaborations between digital media systems, technology, organic life, and geological matter incorporate a whole host of new coordinates and forms of data and systems into the digital cultural heritage realm.[15] The expansion of digital capitalism led to the financialisation of new entities such as big data and to the rise of new forms of labour, both of which produce data. Non-human energy sources – from the sun, water, and wind – are enrolled in the digital cultural heritage composition to green energetic systems. The advent of nanotechnology, synthetic and more intelligent materials, bit innovation (for smarter, faster computing), 3D printing, personalised and predictive medicine, robotics, computer vision, extended reality, stem cell research, AI image generators, and Chat GBT Open AI all become sources for more-than-human data and knowledge production.

The hyperbolic speed and the multi-scaled digital transformation of existing capabilities, and the development of new ones wrought by the Covid-19 pandemic (directed to allowing businesses, work, and life to continue in the face of distancing measures), changed the constitution of and led to the intensification of potential more-than-human heritage.[16] The rapid increase in more sophisticated forms of more-than-human heritage is enabled through not only the rapid increase in the production of data and the consumption of media but also more dynamic sophisticated augmented reality, robotic teaching, and collaboration tools and platforms. This included, for example, new and improved cloud technology, telehealth, genomic capabilities, 3D printed surgical masks, Covid-19 tracing applications, entertainment, such a cloud raves and online museums and heritage sites, alongside core technology supply chains, robotics, and drones for cleaning and delivery all supported through fast 5G networks.

Digital cultural heritage in the making is at the same time brought down to earth leading to even more devastating and irreversible environmental destruction as digital life intensifies. Alongside this, the radical profusion and re-distribution of the agencies of digital cultural heritage as new relativities of scale, intensity and form alongside the uncertainties of their distribution, together with the creation of new job, market conditions as the products of digital capitalism and the social and economic inequalities that derive from these processes and will continue to exert their influence on the expansion of technospheric heritage.[17]

While it is acknowledged that computational infrastructures, media, and data have always operated as a human-technical ecosystem, I expand on the notion of the ecological to construct a more radical and extensive account of digital cultural heritage within these ecological formations.

Digitisations, and the born-digital, can no longer be considered solely as singularly bounded objects on an interface, such as PDFs, DVDs, or as bits and bit streams. Nor can they be seen as immaterial because of the almost instantaneous speed and lightness with which they appear and quickly disappear from computer screens. They can no longer be considered solely as technical systems. Rather they must be imagined as borderless, radically interconnected, distributed, interoperable, self-organising, and often unpredictable and unknowable in their extent and effects.[18]

Accordingly, digital data and, by implication, digital heritage can no longer be thought of as originating from a singular source similar to the way we erroneously make reference to material objects in heritage practice. Digital cultural heritage is made up of different elements from many geographical regions. Through the extraction, harnessing, and culturing of the vital immanent qualities of metals, minerals, and chemicals – in computational design and infrastructures originating from many different geographical and spatial locations – digital cultural heritage is drawn into vast, deep, and complex ecologies of an inter-galactic scale including the earth's strata and life itself.[19] Similarly, material objects never originate from one point of origin, because they are made up of many different elements and materials and are the outcome of processes often rooted in human and other-than-human histories of deep time.

Given this, digital cultural heritage objects originate from multiple geographical locations and arise from many influences. That is, from the authors of their software and the designers of hardware; from the materials from which they are made to the global logistics and supply chains of materials and electronic waste and the repurposing of scrap metals in their obsolescent afterlife; from their unique domain addresses; from text files, code, and electrical signals; from the highly mobile particles that comprise their energetic circuits of interoperability; and from user inputs and interpretations. Identifying, sourcing, and capturing the so-called essence of a digital cultural heritage object is thus an impossible task because of the radical interconnectedness and interoperability of all these coordinates and processes, and the addition of new inputs, functionalities, and capabilities as the so-called object circulates across vast scales which make it possible and active. But it is not possible to capture a whole object, as digital preservation erroneously seeks to do, because it is just too complex, dynamic, and emergent.

In addition to all this, digital cultural heritage objects, both born-digital and digitisations, are made operable through the machine mediated by its own set of interactions and the processing of information – through user inputs and narratives, through algorithms, through energetic fields, through smart materials, chemical and material agencies, and through vast planetary computational systems. Digital cultural heritage on the Internet therefore becomes open to, and takes part in the world as it circulates across planetary scales in which these processes come together at multiple points and on many devices simultaneously.[20] Curating, as I will describe later, must take account of these dispersed points of emergence and convergence.

Digital cultural heritage as data and digital media is therefore also made possible by, and implicated in, the rich embodied, enmeshed, and layered network of agencies. These agencies comprise the various concrete, geomaterial, energetic, physical, social, technical, machinic, and physiological forces described by media theorists such as Jussi Parikka, Friedrich Kittler, Wolfgang Ernst, Matthew Fuller, and Anna Reading,[21] all of which operate across the interdependent layers of global computational infrastructures.

Media theorists Anna Reading and Tania Notley explain how digital hardware and software become contingent on the capitalist-driven, self-interested, exploitative labour practices, and global supply chains through which these objects are made and remade.[22] Digital cultural heritage therefore becomes implicated in destructive mining practices and the processing of metals, and the chemical and raw minerals used in the making of hardware and software in Africa, the USA, China, and Malaysia. Digital cultural heritage also comprises supply chains and exploitative practices in which their components are made and recycled, and transported. Through the energy flows and infrastructures in which digital cultural heritage is produced, made operable, and circulated, digital data and its infrastructures also become implicated in the damage caused to the atmospheric envelope through carbon-burning practices used to make electricity.[23] Taking the ecological complexity of digital cultural heritage into account, I theorise and conceptualise new renderings of this type of heritage in a way that goes beyond the conventions of what digital cultural heritage as code and as a list of formal characteristics and descriptors of their technical features and functions.

Therefore,—born-digital collections and digitisations can no longer be considered as objects in a conventional sense. Rather they are new types of ecological compositions (even though they appear as coherent things), comprising what I call "thingness" (the multi-scaled, extensive, radically interoperable webs of heterogeneous coordinates, forces, and agencies that comprise them).[24] The variable coordinates that make up digital cultural heritage are so extensive that they interpenetrate human and non-human life itself. Digital cultural heritage, therefore, becomes an extended and dispersed spatial and temporal composition made up of diverse, conjoined, and interacting vital elements. An ecological mode of thought draws our attention to environmental crises in which digital data and digital media and therefore as I argue, heritage are implicated.[25] As a result ecological thinking, and the expansive range of coordinates that comprise these productions, collapses the human centredness of digital cultural heritage. Instead, humans and digital cultural heritage operate within an ever-expanding field of coordinates and processes. Rather than being thought of as completed historical entities, their histories are emergent.

To do this work, I have been inspired in part by linguist Bill Brown's "thingly turn."[26] The thingly turn is a method used to explore object relations in an expanded field, and not only in terms of their end function.[27] Digital cultural heritage objects are viewed as technical productions of human cultural expression and practice but are not necessarily knowable in these finite terms. Standard object classifications, technical descriptors for digital preservation, the role of digital data as informational, and virtual objects – or bits – offer certainty. They are the type of objects that we know, that we expect to encounter, and what we can grasp conceptually within our own frameworks of understanding. Such objects find their rightful place in heritage schemas. But as ecological compositions, they are full of complexities and ambiguities. Objects are stable and able to be objectified. Digital cultural heritage, on the other hand, comprises objects that "are" of a humanist heritage-centred world, but which as ecological compositions create a world of their own in collaboration with us.[28]

We, as a result, get caught up in their ecological complexity in multifarious ways. As ecological compositions, they become integrated into our material encounters with the world, from the energetic and conductive capacities of earthly chemicals and metals, for example, to environmental destruction and algorithmic manipulation.

Ecological compositions are no longer knowable in a conventional disciplinary and heritage sense. They operate in excess of what we call an object and become different types of dynamic and unique compositions, often appearing simultaneously in different locations and on many devices.[29] The distinction between objects and ecological compositions can be characterised as a desire to curtail, manage, and classify data within the archive as opposed to their unfolding in the world and operation in experience.

Thinking about digital cultural heritage as I propose as ecological compositions and as "thingness" is reminiscent of German philosopher Martin Heidegger's proposition that our being in the world is of a thingly nature, because things are all relationally connected and function together. While objects, with their seemingly hard boundaries, are conceived as separate from us, things are not. But Heidegger's and Brown's things and thingness operate as signifiers. They are still things that have substance, properties, and traits, but unlike objects are formed through sense perceptions independent of the representational systems we impose on them. Thingly things, however, can't account for the inexplicable energy, vitality, and life forces constituted by the huge variety of more-than-human, non-human, and machinic agencies that comprise ecological compositions, or their radical interoperability as congealing forces and the resulting affects. Nor do Heidegger's and Brown's things and thingness account for the different bundles of ecological, technical, and human forms of sensing and calculation that emerge through the force fields generated by their coordinates.[30]

The recruitment of digital data to concepts such as digital cultural heritage still requires the object. When the bounded object or thing is called into question, thingness becomes instead the disclosure of and folding out of its coordinates, agential and multifarious durations, temporalities, and processes. My thingness concept and method have the capacity to account for the infinite variety of coordinates present that describe the cosmos of physical, social, sensing, and practical experience that digital heritage inhabits, mobilises, or is mobilised by it.[31] Furthermore drawing inspiration from French philosopher Etienne Souriau's notion of thingness, I put this concept to work to account not only for the extraordinary, unexpected occurrences, or appearance of things perceptible by the senses [32] but also to formulate thingness in a different way to conceptualise the hidden and automated processes generated by an ecological composition's coordinates that are not necessarily discernable. The utility of thingness also draws attention to the plasticity of thingness coordinates as they interact, their multiple intensities, tensions, different forms and states, and the intended and unintended consequences of their interrelatedness.

Here the coordinates that comprise digital cultural heritage are not just networks of discrete things, of digital code, of graphical interfaces, of file formats, computers, narratives, and disciplinary perspectives. Their coordinates also include, for example, bits and software, geolocations, algorithms, programming languages, machines, infrastructures, cultural, political, spiritual, semiotic and heritage values, human bodies, narratives and inputs, techno-capitalism, smart devices, materials, human and neural aesthetics, minerals, chemicals, eco-cognitive processes, energies of many kinds, mathematical equations, machine calculations, and so forth.[33] Accordingly, thingness progresses a renewed consideration of the coordinates hidden from view or rendered invisible, such as computational cognition.

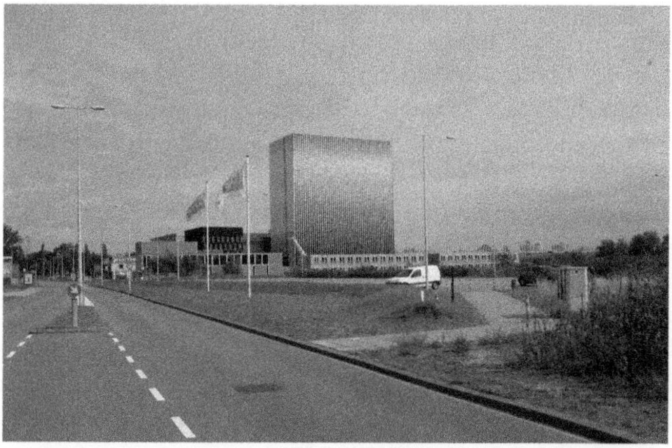

FIGURE 2.1 Data centre, Amsterdam, 2018. Photo credit: Ned Rossiter.

Political theorist Jane Bennett's work draws attention to the vibrancy of matter, its conative agencies, and affects in every-day objects.[34] The object, as an ecological composition, is mobilised through acts of composing by its aggregating bundles of coordinates, made possible through its distributed performativity. Its appearance as a coherent image emerges through the registers of sensing and influencing born out of an indefinable variety of conjoined material, mineral, chemical, discursive, energetic, social, human, non-human, natural, and cultural forces.[35] As an ecological composition, digital cultural heritage becomes a field of relatedness, embodied movement, and multiple temporal processes and represents a predilection for the assertion of its coordinates' own agencies and affective power in non-dialectical ways. Ecological compositions are therefore unique manifestations born out of their ecological processing. Because digital cultural heritage as ecological compositions is non-identical, they resist our attempts to recruit them to conventional heritage concepts, naming, and descriptive protocols. Accordingly, I suspend the notion that digital cultural heritage is, as UNESCO likes us to believe, identical in technical composition, meaning, and substance.[36]

Detailing the properties and the form of the coordinates that make up digital cultural heritage therefore occurs in the context of particular and unique ecological circumstances and through an onto-ethnographic reading of how they relate to each other. No two ecological compositions, with their web of sprawling coordinates (thingness), are identical in form, substance, or meaning because each has its own unique coordinates and histories of relatedness.

As ecological compositions, the coordinates that digital cultural heritage comprises are difficult to categorise as distinctly human, conceptual or ideational, technological, biological, or mineral, because they are radically interoperable. None have an independent existence; none are capable of being extracted into discrete things or objects.[37]

Thingness also opens up a space to consider diverse renderings of digital data as machinic, as narrative, and as other-than-human. Its state of concrescence as a composition (appearing or sensed as a solid thing) originates from many locations and through many processes, often simultaneously. This is because thingness is distributed across multiple domains as a result of the emergent consequences of the interrelatedness of its unique coordinates acting in the world, and across vast scalar, temporal, geographical, markets, and technical domains.[38] The ecological composition curates and distributes itself across vast distances and conjoins with multiple points that act as switches and connectors with other compositions. Thingness and its extensive character are also made possible by multiple user inputs, the vast network of data storage facilities, the circulation through data centres, and the electrical impulses that travel across and through wireless networks and expansive grids of undersea cables.[39] Much digital cultural heritage therefore becomes planetary scaled, distributed and dispersed, and made and remade through its processing.[40]

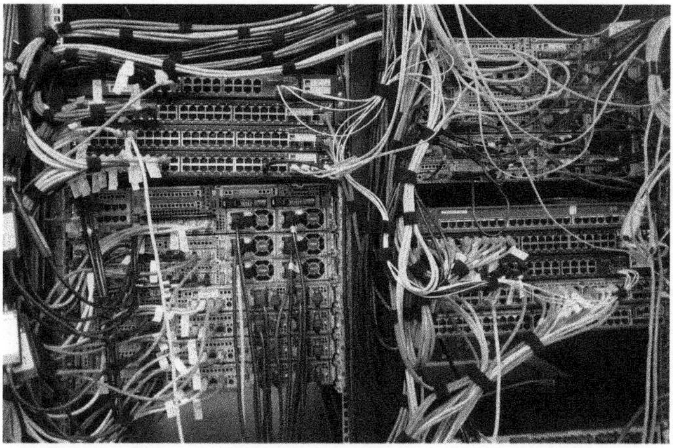

FIGURE 2.2 Networked servers in a data centre, Chile, 2017. Photo credit: Ned Rossiter.

Thingness is post-relational because there is not a simple relational or material equivalence between all its coordinates, for example, matter as quantum physicist Karen Barad suggests.[41] That is, because digital cultural heritage, as ecological compositions, incorporates different types of embodied vitalities – from social and scientific knowledge practices to materialities, databases, platform logics, bitstreams, and embedded metadata – they can't be explained simply in terms of a common substance such as matter.[42]

Ecological compositions also encompass diverse modes of existence that incorporate a wide range of culturally inflected coordinates and meanings, which is the case with the digitisations of Māori material culture (taonga, "treasured possessions") or Malagan funerary masks originating from New Ireland in Papua New Guinea, both of which are embodiments of their ancestors.

The rendering of digital cultural heritage as ecological compositions and as thingness are also post-object concepts[43] because digitisations and born-digital cultural heritage objects, in this frame, are more dispersed and open-ended than conventional notions of the digital object. As ecological compositions, they are unique and emergent consequences of the interrelatedness of a series of different types of coordinates and their affordances comprising, for example, their unique materialities, meanings, aesthetics, politics, and technicities, including the inputs of curators and users that operate across scales. Those compositions that reside on internets lack any inherent framing and are so dispersed, non-identical, and dynamic that they can no longer be thought of as a coherent object or a thing, even though they might appear so on the interface.[44]

This post-object status also emerges out of an inter-operative mode of sensing. Code, for example, is not a simple or pre-determined set of logical instructions, as might appear on a computer screen. Rather, it is performative and continually produced within and through thingness. For instance, computational processes are an operative mode of thinking that has no underlying organisational principles.[45] The ecological composition is more attuned to the notion of digital cultural heritage as an emergent gathering that concretely ties together heterogeneous coordinates, but at the same time has no underlying organisational principles or object-like structure.

The concepts of ecological composition and thingness that I developed in my previous publications to refigure digital cultural heritage acknowledge these more-than-digital heritages as programmable things in multifarious ways. These concepts capture their dispersed, non-identical character, their extended, multiple provenances, and locational politics. At the same time, they draw our attention to their composition as made up of heterogeneous interdependent coordinates, such as diverse human meanings and narratives, their multiple networked and energetic modalities, their emergence as multivalent compositions, their trans-medial, non-linear character, and their changing forms and combinations.

Eco-curating and curatorial agency as domains of influencing

The born-digital and digitisations, comprising thingness and their gatherings as complex ecological (temporal-technical-material-political-energetic-anthropological) compositions, operate as a lively federation of coordinates. As entangled entities, they exhibit multifarious temporal and spatial forms through their affordances (physical capacities to act and in relation to and with each other), and as a result they are constantly forming and reforming. Here digital heritage is not separate from us; rather, due to its radical interoperability, all these things, including human curatorial practices, function in relation to each other.[46]

But in the museum and digital heritage field, digital curating is generally viewed as human acts and more recently human–machine agency. Digital curatorial tasks are directed to data capture, to processes that seek to add value to digital data by assigning administrative, descriptive, structural, and technical archival metadata to them, and the storing of digital objects as software, hardware, and bits. The curatorial interpretation of digital data conforms to the notion of the informational object and its social constructivist and representational frames of understanding.

In the art field, curating and the term "curatorial agency" arose as concepts to progress a critical rethinking of curating contemporary art in culture and society as a series of acts directed towards changing the world.[47] No longer merely viewed as an author or presenter of pre-existing artistic concepts, the curator here is an active social agent who contributes to cross-referential understandings of art between different artistic, cultural, ethnic, class, and gender groups for the betterment of society.[48]

It is acknowledged that curatorial production has expanded with the development of internets. Curation has moved from the single object to processes in dynamic networks and is distributed through multiple agents such as technical networks and software.[49] In this frame, the politics of curating becomes systemic, at once human and technical.

Through four publications and in my 2021 monograph, *The Future of Digital Data, Heritage and Curation in a More-Than-Human World* I developed a new figuration of curatorial agency and curating processes as eco-curatorial.[50] As new types of ecological compositions, curating and curatorial agency (denoting acting in the world through selecting, organising, and influencing) take on a different meaning. Curating is no longer authorial in a conventional sense, as a series of actions by humans or automated systems, it is also the result of the vitalities of more-than-and other-than-human coordinates together emerging as forms of eco-systemic curating.[51] Here the production of curatorial knowledge and the act of curatorial authorship by humans operate in collaboration with all manner of vital coordinates, forming intermeshed alliances with them as a new type of curatorial eco-logic.

Curatorial agency emerges as eco-curatorial agencies.[52] Curating, enacted through diverse eco-curatorial agencies, is often indeterminate acts of immanent or interdependent processing in formation made actionable through the interrelatedness and interoperability between, and performed by, a wide range of its coordinates. That is, from software – programming languages, mathematical equations, machine learning, and automated processes embedded in algorithms and bots – to networks, infrastructures, and calculative storage processes, to elemental chemicals embedded in computational capacities. From bitstreams and data to data centres, to planetary computational infrastructures

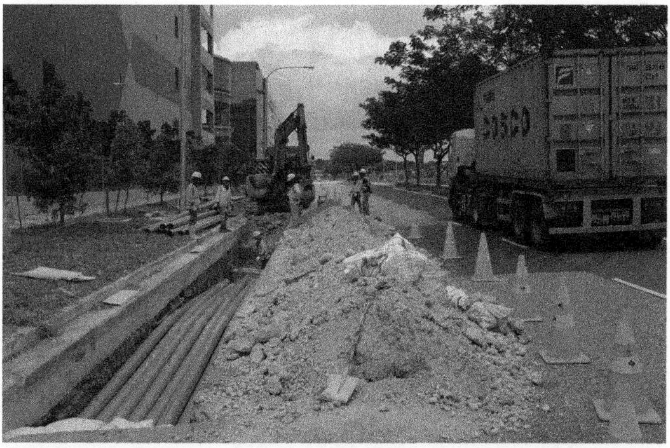

FIGURE 2.3 Excavated trench and laying network cabling for the internet, Singapore, 2017. Photo credit: Ned Rossiter.

and the electromagnetic forms of transmission that pass-through data centres, and to the rare earth minerals that act as conductors of electricity. All these things produce their own eco-curatorial aggregations, and different patterns of material, form, and performative affects. Their generative capacities also constitute extended spatial-temporal structures and material durational processes of self-assembling.

Ecological compositions, in their complexity and through the continuous and emergent interrelatedness between all their coordinates and processes of computational abstraction, cannot be separated from their biophysical complexity. Computational design today is founded in physical, biological, and chemical processing, in an assessment of their aggregate material properties, and in the ways all these things can be harnessed to constitute more effective design solutions.[53] For example, a computer chip has 60 different chemical elements,[54] all of which have their own agencies that alter and enter different types of intensive aggregated relationships. The compression of all these affordances and agencies into chips, for example, are used as calculative entities and are able to eco-curate at nanosecond speeds. As a consequence of these computational design solutions, curating is also performed via the physical, chemical, and biological processing embedded in its thingness. Together, all these eco-systemic processes contribute to the formation of an ecological composition's concrescence (the visual rendering of itself as a coming together of multifarious elements as a composition on the interface, as opposed to a vision of itself as a solid form).

The extensive, broadened range of coordinates made possible through new computational design, from data processing to algorithms, to infrastructural expansion, as examples, has distributed curating on a much larger and more

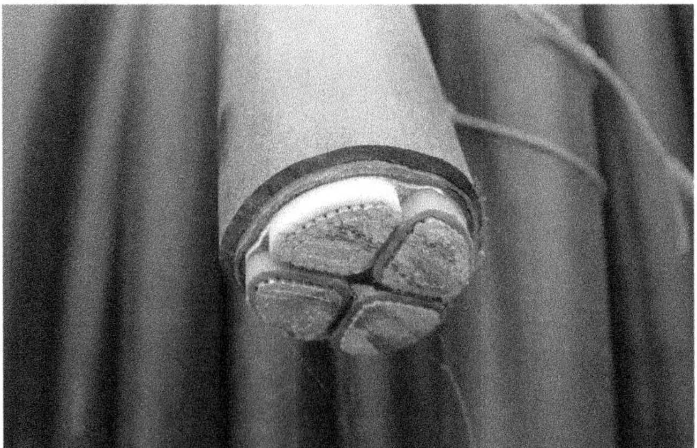

FIGURE 2.4 Electrical cable, data centre, Malaysia, 2017. Photo credit: Ned Rossiter.

expansive scale than has ever been seen before.[55] Such ecological webs and curating processes gesture towards, and make possible, the extended reach and influencing capabilities of human actors across vast scales. I will discuss this in terms of Donald Trump's tweets in the following section.

Furthermore, new eco-curating agencies are constantly enrolled in the emerging and mutating ecological composition as it moves, extends, and composes itself. New user inputs and algorithmic processing, the expansion of infrastructures through the laying of cables, the development of platforms, the enlargement of cloud storage, the ecological composition's appearance on multiple interfaces, devices, and screens, and the ever-multiplying practices of copying and distribution by all manner of agents, all involve thousands of different eco-curatorial processes.[56]

As ecological compositions, digital cultural heritage also self-curates in multifarious ways. These self-curating processes occur through, for example, the behavioural capacities of machine learning and calculations made possible by the biophysical properties of the ecological compositions' coordinates, as well as through the contingency of environmental factors and the curatorial and user interactions in which they are enmeshed in, across and through its extended composition.[57]

Curating, in an eco-sense, can never be fully determined due to the nature of its complexity, the inconsistencies in interactions, and the indeterminacy of calculations made through machine learning. Eco-curatorial actions never perform the same set of interactions and therefore are never identical and cannot be replicated.

As a result of eco-curatorial acts, new types of eco-cognitive processes emerge. Just like curatorial agency, cognition is not just human.[58] The collective capacities of vastly different forces, coordinates, and affects harnessed through computational design – such as the multiple and indeterminate potentialities, capacities, dynamics, and contingencies of material and com-putational physical, biological, chemical, and energetic behaviour, described by media theorist Luciana Parisi[59] (including human intent and desire) – form part of distributed, aligned eco-cognitive processes, and affects.[60] The human user still has their own agency and their own unique cognitive capacities when authoring or interacting with data, but these agencies and cognitive processes operate alongside and through all its other coordinates.

These multifarious eco-cognitive, thinking, and sensing processes emerge as multiple milieus, localities, temporal and spatial generative emergent patterns, and forms. In doing so, such processes open digital cultural heritage up to the indeterminate and unknown, or indeed the unknowable. Intelligence is not just human or machine. Intelligence is reworked variously as automated reason, as the behaviour of materials in computational design,[61] as the writ-ing of code and algorithmic processes alongside their capacity to change, self-organise things, and direct action in certain ways.

Ecological forms of intelligence, power, and meaning-making emerge as affective domains of influencing.[62] For example, as media theorist Elena Esposito explains, the actions of the intelligent human feed on user profiles and contributions and exploit them, and the system returns or produces affects often in a unrecognisable form or in a surprising way.[63] Searches on Google for the movie *300 Rise of the Empire*, portraying the battle of Thermopylae between the Persians and Spartans in 480BC, for example, return a link to the Project 300 website. While it may be known how some elemental materials behave because they are used in computational design, other agencies and cognitive processes remain elusive. Therefore, eco-curating processes, eco-cognition, and resulting eco-intellectual affects generally will stay indeterminate and to a degree contingent.

Digital cultural heritage is therefore no longer solely a question of digital materiality. As ecological compositions, the analogue, born-digital, and digitisations are called into being in self-differencing ways that cannot be explained simply in terms of differences in materiality. And in a vital frame, heritage more generally emerges in plural and self-differencing forms, thereby replacing formal divisions based on substances such as digital, intangible, and material heritage.[64] By viewing heritage as ecological compositions, and as self-organising ontological processes, they disrupt these formal heritage distinctions both in the way they are described and how they might be valued. Here, each born-digital object or digitisation is characterised by distinctive ecological coordinates and circumstances, rather than being seen as bounded domains of things, whether materially or conceptually.

Diversifying the range of coordinates that comprise digital thingness, and therefore the ecological composition, exceeds what we currently view as heritage. As ecological compositions, digital cultural heritage operates in excess of human system meanings, habits, or projects, or indeed exceeds what we have the ability to comprehend or know. As a knowledge and curatorial practice, the ecological composition itself, its collection and documentation, will always be impartial, contingent, multiple, and never completed.[65]

The aim here is not to treat concepts, the ecological composition, and thingness as better reflections of reality or more objective ways to standardise digital collections, as many new approaches seek to do. Rather these concepts operate as both a mode of thought and empirical experimental practice.

Donald Trump's tweets as ecological compositions

In the introduction to this chapter, I discussed the tweeting activity of former President Donald Trump as a new form of heritage in the making. Trump's use of Twitter raised his profile in the 2016 US presidential primaries, enabling him to secure the Republican nomination and the presidency.

Trump's tweets, as official records of his presidency, are considered digitally born heritage because they are viewed as historical data that must be saved in perpetuity by virtue of the PRA of 1978.[66] The Act stipulates that documents of administrative, historical, informational, or evidentiary value cannot be disposed of or destroyed without the written permission of the archivist because they must be kept for future reference and for political accountability.[67] This includes records created digitally from posts made on Facebook, Twitter, and YouTube, and electronic communication such as email.

Debates regarding the standing of Trump's tweets as heritage focus on whether original or altered tweets should become the archived record. One hundred days into his presidency, Trump had put out 517 tweets and deleted 11 of them.

> After a misspelled statement from Mr Trump's private Twitter account was altered on Saturday (local time) and later deleted, archives spokeswoman Miriam Kleiman said presidential tweets, like all electronic communications "created or received" by the president or his staff, were considered presidential records... Mr Trump wrote on his private Twitter account: "I am honered (sic) to serve you, the great American People, as your 45th President of the United States." The tweet was later corrected to "honored", the correct American spelling, and then later removed entirely.[68]

During Trump's presidency, his administration was required to keep all tweets, including those from his official White House and private Twitter accounts.[69] In doing so, Trump also preserved his own digital data. Within the context of the Act's jurisdiction, Trump's tweets became elevated to the status of heritage even before they were interred in the archive. They were historical data in the making.

Trump's tweets most succinctly demonstrated the operation of born-digital cultural heritage as vast, sprawling ecological compositions made up of multi-scaled and meta-territorial domains of influencing of infinite reach and depth. Through their interoperability, Trump's tweets appeared as interfacial images simultaneously in multiple locations through digital devices, and in doing so incited real-world, material effects.

Trump's ambition to assert his authority through data regimes was based on the premise that millions of people are connected by smartphones and computers, receiving and generating large quantities of digital data. His tweets are viewed as digital speech acts intended to capture hearts and minds, claim sovereignty, and mobilise power through digital networks. Trump became centre stage on which the world became subject to his politics through the distribution of his opinions and agendas across large distances and scales.

Accordingly, Trump's tweets were intended to play an important role in his efforts to restructure global fields of power and knowledge, assert influence, and drive action according to his own agendas. With an estimated 88 million supporters (an estimated half of whom are simply social bots or the tweets of click farmers), Trump sought to bolster his support through automated means, and through the exploitation of low-paid workers.

Trump's tweets were economic, political, and social investments directed towards national and global sovereignty in which social bots, angry US and Mexican citizens, heads of state, North Korean ballistic missiles and Trump supporters gathered and took strong positions. They were Trump's attempts to refashion relationships between people and states activated through an interconnected mesh of people, technologies, smartphones, data, algorithms, words, and so forth.[70]

The political potential of Trump's tweets was realised at the very moment they are made machine and human readable, appearing in their concrescence as visible "objects" on the interface and therefore able to accrue power.[71] In doing so, Trump's politics, mediated through his tweets, sought to interrupt relations between the USA and other nation-states, such as Mexico, Iran, and North Korea and challenge the outcome of the 2021 Presidential election. Actions included attempts to build a wall between the USA and Mexico to stop illegal immigration, re-investment in American manufacturing, the closure of borders to migrants from Muslim countries, and attempts to put a stop to North Korea's nuclear weapon program.

While Twitter is thought of as a field of data production, Trump's tweets were far more than socio-technical arrangements of humans and machines. Nor were they merely representative of his acts of governing or solely the outcome of what media theorists Evelyn Ruppert, Engin Isin, and Didier Bigo call data politics in which speech acts are merely constituted as an object of power and influence.[72] As ecological compositions, his tweets were used to articulate political questions, acts of authority, and defiance. Trump's tweets were not only his own curatorial acts and opinions, they were made possible through processes of eco-systemic curating involving many vital coordinates with their own capacities for involvement in political action, which they in turn enabled.[73]

Through their interoperability, Trump's tweets appear as interfacial images simultaneously in multiple locations through digital devices, and in doing so incited real-world, material effects such as the storming of the Capitol on January 6th, 2021 as lawmakers met to certify Joe Biden's victory. Trump sent a tweet at 01:42 local time on December 19, 2020, telling his supporters:

> Peter Navarro releases 36 page report alleging election fraud more than sufficient to swing victory to Trump. A great report Peter. Statistically impossible to have lost the 2020 Election. Big protest in D.C. on January 6th. Be there, will be wild.[74]

The post, "electrified and galvanised his supporters," to converge on Washington DC on January 6th, 2021 because they believed Trump's claims that the election had been stolen from him.

Trump's tweet emerged from many different coordinates and their interoperability. As a digital data set, it comprised textual characters, usernames, timestamps, unique identifiers, links to URLs on the web, metadata, APIs (application programming interfaces), abstracted patterns of zeros and ones that shuttle between storage, and the display of that data as his speech acts appeared on technical devices such as computers and smartphones.

But such compositions are also political, historical, and material of a qualitatively different kind. They comprise technological coordinates through the relatedness of infrastructure such as equipments, cables, routers, data servers, data farms, wireless communications, switches, devices such as computers and smartphones, wireless communications bounded together by a binary machine code of zeros and ones, and automated systems such as algorithms and social bots. In addition, they included human coordinates, such as Trump himself, designers, programmers, engineers, citizens, nation-states, right-wing supporters, and archivists.[75]

Media and its waste, Parikka convincingly argues, becomes entangled in its earthly computing nature through its material substrates. This includes the exploitation of the resources of the Earth's crust, through the mining of earth minerals and chemicals used in the production of all manner of computing devices and their associated geopolitics, the search and acquisition of energy sources to drive the economy and the grim labour existent in the factories of computational production[76] and their afterlife.

Trump's tweets also comprised natural coordinates such as the terrain central to the event – the Capitol building in the Mall, Washington DC – and a geophysical politic. His tweet of December 19th, 2020 to his supporters including far-right extremist group Proud Boy are earthly computational entities and comprised natural coordinates that were themselves inherently political in quite unexpected ways. These included the silicon, ferrous metals, zinc, and copper extracted from the ground used in computer hardware, software, media devices, and infrastructures through which his tweets were received by Proud Boy and used to conduct electrical currents. The consequences of environmental destruction were wrought through many avenues. These included mining practices across the globe from the Congo to Brazil, geopolitical resource conflicts, the complicity between the materials in Trump's iphone and that of Proud Boy, the working conditions, and the health problems of workers who made it at Foxconn, Zhengzhou, China to the contaminating conditions in US e-waste dumps in China where devices in their afterlife leak toxic elements into the soil. Trump's tweet was also implicated in the politics of the global logistical supply chains that mobilised all these things, the burning of oil and coal in US electricity plants producing energy to distribute his tweets, cool data centres through which they circulated, and their resulting emissions.[79]

Furthermore, enmeshed in his tweet of December 19th, 2020 was electoral policy coordinates including electoral voting processes, citizen votes, and proclaimed so-called electoral fraud alongside the desire to incite a coup to remain in power and a refusal to concede the election to Biden. Economic coordinates, central to the ecological composition, were also made visible through reference to Trump's desire to "make America great again." Most importantly, his tweets comprised ideological coordinates materialising as crowds gathering at the Capitol building, chanting, "Fight for Trump," violence, the storming of the building, broken glass, Confederate flags and death. Hence, Trump's tweets as ecological compositions are part ideological, part textual, part political, part policy, part technological, part economic, part mineral, part computer code, part program, part elemental, and part algorithm.

Recalling the North Korean standoff by Trump in respect to the North Korean nuclear program his tweets, through eco-curating processes and the domains of influencing that emerged as a result created what Bratton calls geoscapes,[78] contested terrains that are resoundingly fractious and potentially dangerous. His tweet of December 22, 2016, threatening to expand US nuclear capabilities unless North Korea curbed its own nuclear program, prompted the emergence of a particular set of relations, events, and a milieu characterised by heightened tensions between the USA and North Korea, which could have led to war but thankfully did not.

This geoscape was in part formed through the circulation of Trump's plans, projections, claims, ideologies, and aspirations and made actionable through more-than-and other-than-human domains of influencing.[79] These actions were on the one hand planned by Trump to further his own political aspirations, but on the other hand had unintended consequences. Chillingly, the use of intimidation and fear became a power of affect within and across these domains of influencing.

Trump's statement was written into data, made actionable through computer programs and algorithms, search engine results, as socio-cultural-technical-material computational environments and infrastructures and the minerals, metals, and chemicals embedded in the media that composed them made possible by designers and engineers. Accordingly, these eco-curating acts emerged as different registers of action, as the affectual, the calculative, the automatic, the non-conscious, and the conceptual, as a series of interdependencies inciting real-world material affects.

Machine and non-human curating also produced different types of eco-cognitive processes and politics outside Trump's control because all these things curate.[80] Trump was no longer the only author of his tweets and indeed never was. Rather, communication becomes radicalised through multiple cognitive agents where a different communicative and relational-sense culture emerged, not just through technical and network processes, or infrastructures and machine learning, but through the eco-curating interrelatedness and the ensuing actions of all its coordinates.

Eco-cognitive processes, enacted by thingness coordinates such as neural networks, machines, algorithms, automations, and artificial intelligence, operate together at indeterminate speeds changing sequences of operations and at the same time processing data through calculation, recording, and reflecting. In doing so, they multiplied their own contingencies producing their own forms of calculative intelligence and structures. All of these processes represented the eco-extension of cognition into the world across multiple locations and scales, at times simultaneously.[81] These processes are never completed, they are vast, indeterminable, uneven, and sometimes chaotic.

New forms of digital communication and machinic automated thoughts and memory emerge based on the quantification of data from which new capacities for power and manipulation arise. Trump harnessed these capacities to promote his politics. The subsequent chains of events illustrated how domains of influencing operate in directing real-world consequences.

In a New Year's Day speech, and in response to Trump's tweet of December 23rd, 2016, North Korean President Kim Jong Un warned that North Korea was preparing to launch an intercontinental ballistic missile (ICBM). He asserted that the nation would continue to build its nuclear capability for "pre-emptive strikes" unless the US ended naval exercises with South Korea. Perceived as a threat to Washington D.C. and the Federal government, Kim Jong Un's remarks prompted an angry response from Trump on January 2, 2017 with an implied threat that North Korea will be prevented from developing a nuclear weapon capable of reaching US soil.[82]

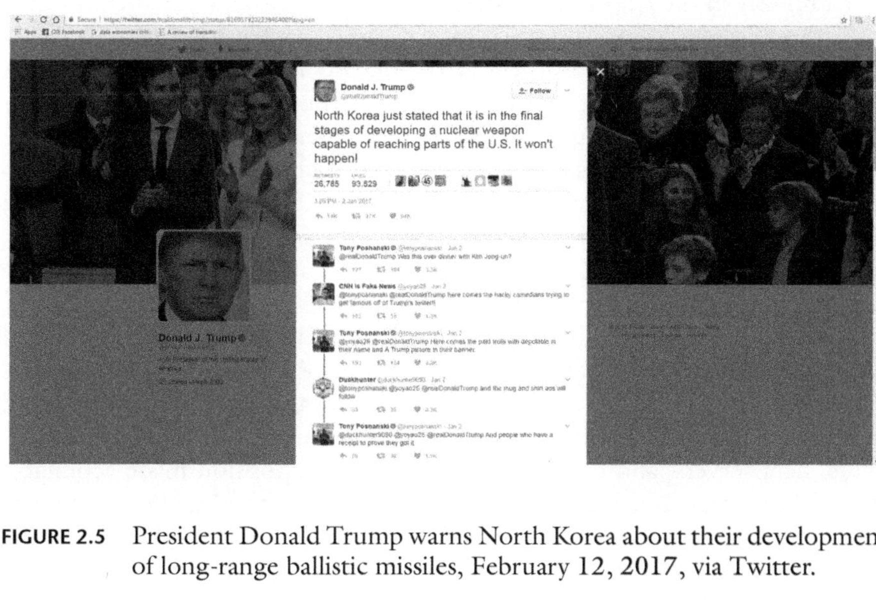

FIGURE 2.5 President Donald Trump warns North Korea about their development of long-range ballistic missiles, February 12, 2017, via Twitter.

Following this, the North Korean state media announced on Sunday February 12, 2017 that it had successfully completed the launch of a new ballistic missile:

> The Pukguksong-2, previously unpublicized part of its arsenal which North Korean state media described as a medium long-range ballistic missile, was test fired on Sunday under the supervision of North Korean leader Kim Jong Un, according to North Korea's Korean Central News Agency, KCNA.... The agency described the missile as a "Korean style new type strategic weapon system."[83]

A US official reported that the missile launched from North Pyongan province travelled 500 km (310 miles) before landing in the Sea of Japan.[84]

Crucially, this event occurred at the very moment Trump was entertaining former Japanese prime minister, Vale Shinzo Abe, at his Palm Beach club, Mar-a-Lago. The bizarre sequence of events that followed was reported by Eliot Weinberger in the *London Review of Books*:

> Interrupted by news of a North Korean missile launch, Trump chose to display his presidential power to the wowed and tweeting club regulars by holding an emergency security meeting at the restaurant table. Aides used their cell phones as flashlights to read classified documents while waiters served entrees over the papers. One member even posted a selfie ("Wow!") with, he wrote, "Rick", the guy who follows the president with the "football" – the briefcase carrying the nuclear codes. Luckily, the missile landed harmlessly in the ocean before dessert.[85]

At the same time, the investor, actor Richard DeAgazio, posted pictures from Mar-a-Lago showing President Donald Trump and his entourage conducting their security meeting over wine and dessert.

Many different registers of narrative, calculation, sensing, and action emerged as multi-directional and scaled domains of influencing, including those arising from the non-linguistic and machinic, the human, more-than-human, and other-than-human such as chemicals' and minerals' affective capacities.[86] Through eco-systemic curating processes, their effects, often self-organising, became folded into meaning-making and the real-world consequences that subsequently emerged. These include the words of Trump on December 22, 2016 as textual characters, digital data emerging through computer display, and their mediation through quartz and silicon ferrous metal, electrical currents, data servers, and cables. The ecological composition made actionable drew in material things and events such as nuclear bombs, the building of nuclear capability, an enraged North Korea, the words of President Kim Jong Un on Sunday January 1, 2017, the launch of an ICBM and North Korean military bases, North Pyongan province, metal and split atoms, the time and date, Sunday February 12, 2017, the Sea of Japan, and the meeting of Trump and Abe at the beach club, Mar-a-Lago.

Trump's tweets, although earmarked as heritage, continued to be produced and operated as live data emergent in the world. This challenged the very foundation of digital cultural heritage production because they were not necessarily historical in a conventional sense, indicative of prior events. They were not interred and locked away in the archive. His tweets did not remain in an original state, they gathered responses and were distributed far and wide.

It is not possible to view, know, or map Trump's tweets as ecological compositions in their entirety. We can't describe or map their full complexity. We do not yet know enough about these automated cognitive processes and how they operate. Perhaps we will never unpack them. This is because the subjectivities of machine learning, or "soft thought," are not simply a case of the execution of instructions in a logical fashion but emerge as a result of any given tweet's own algorithmic modalities in which sequences of instructions change on the basis of the way a machine orders data.[87] But most importantly, working with the concept of the ecological composition as a new way of looking at digital cultural heritages opens an opportunity to look at these things as active and as processes out in the world.[88]

Twitter permanently suspended Trump's tweet accounts due to the risk that they might further incite acts of violence following the storming of the Capitol.[89] However, on November 21st, 2022, the new CEO of Twitter Elon Musk reinstated Donald Trump's Twitter account following a poll on the platform in which 52% of the more than 15 million Twitter users responded in favour of Trump's return.[89] Many of whom were most likely pro-Trump bots.

Trump's tweets, interpretive, and documentation practices

By considering Trump's tweets as more-than-human and ecological, the use of heritage descriptive schemas and classifications, as distinct media types and genres directed towards upholding the original, are foregone in favour of engagement with them on their own terms as multiple dynamic interlocking, technical, cultural, social, narrative, biological, geological, economic, and political processes of re-mediation.[90] Documentation then becomes onto-choreographic, requiring the mapping of the tweets' coordinates, the aggregations they form by coming together and moving apart in different locations, and the ways in which they affect and are affecting by themselves and with others in their interrelatedness depending on the forces that create those intensities at any specific point. In some instances, human narratives might be foregrounded, and at other times more-than-human and other-than-human coordinates, agencies, and affects might take precedence.

Therefore, it can't be assumed that all so-called ecological compositions are awaiting representation through artefactual production. In some contexts and circumstances, digital data is best not interred in the archive. In these cases, heritage institutions could host specific ecological compositions that are still addressed and used in the world. Heritage institutions could work collaboratively to support other entities, people, and organisations to keep ecological compositions alive in the world without collecting them themselves. One example where this approach has been taken is with Trump's tweets. The National Archive instructed the White House to save all his tweets in their original form. These tweets are therefore simultaneously out in the world and saved in the archive. Here institutions can operate an inventory of data seen as worth keeping active and ongoing, similar to a heritage register of buildings.[91] Collecting, according to intent, can be regarded as non-linear, iterative, and interactive.

The notion of the cut or incision is inspired by Barad's agential cut as a material-conceptual device for sampling material processes.[92] Drawing inspiration from Barad, media theorists Sarah Kember and Joanna Zylinska conceptualise photographs as visual expressions of a moment in time, and as cuts into the duration of life.[93] According to these theorists, cutting takes two forms: "as something that *is*, thought or perceived to be and as something that is *thought to be taking place*."[94] So here cutting Trump's tweets involves both collecting for representation – that is, what the ecological composition *is* deemed to signify as a moment in time, as a fixed thing or in its active state, or both – and as the sampling of what it is or *what is thought to be occurring* in its emergence in the world.[95]

Accordingly, the documentation of Trump's tweets, within the archive and out in the world as interpretive and documentary cuts, could involve the capturing of his words, the sequence of responses, a snapshot of what the tweet looked like on his Twitter account, and the collection of the original code. But as an ecological composition, the tweet is no longer solely documented as a static textual record in the archive. It can also be documented in dynamic ways as an ethnographical reading from the field, as multiple and overlapping domains of influencing involving observations of its history, in terms of its emergence through multiple locations, through the effects of its distribution, through the dynamic social collectives and extended networks to which it is subject, and through all manner of data analytics such as visualisations of all these things. Trump's tweets, as an embodiment of his thoughts, feelings, actions, and politics, become a situated and relational form of expression, as part of a unique and dynamic more-than-human convergence of coordinates.[96]

Trump's tweet of December 22, 2016, and the threat made to North Korea to curb their nuclear program, for example, are interoperable compositions of thingness such as textual, ideological, policy, technological, economic, mineral, computer code, programs, elemental, and algorithmic coordinates. Such documentation possibilities include, for example, a consideration of what is enacted through the many locations into which his tweet crosses, what situations and events emerge in the world – such as the launch of ballistic missiles in North Korea, protests in the US regarding presidential politics – and the different responses and registers of meaning, value and significance that his tweets hold for people and the responses to them made through other media. In terms of the more-than-and other-than-human coordinates that comprise the tweet, documentation could involve the agencies of its coordinates, from the bit-streams to file formats, to the material and chemical properties embedded in its material substrates that enable the tweet's distribution, and the human, AI, and automated interactions and narratives that appear through these processes and their multifarious effects.

Cutting here operates as a mechanism for prioritising documentary editing of the streaming of content, the flows, frictions, accidents, and serendipitous occurrences. Documentation also privileges the different registers of action and cognition activated by material, technological, organic, elemental, or human narratives and their machinic distribution and movement across platforms utilising onto-choreographic procedures. Within the wider realm of computational affordances and machinic jurisdictions, these latter inputs are not always known in advance. Having said all this, it is not possible to produce a comprehensive record of the ecological composition due to its complexity.

The documentation of the December 22nd, 2016 in the museum and the worldly archive as an ethnographic form of sampling could involve: first an inventory of the coordinates that Trump's tweet comprises, inhabits, mobilises, or is mobilised by as an extended composition in the field; and second their multiple agencies, sequences, interoperabilities, and spatial and temporal dimensions; followed by an onto-cartographic mapping of its eco-curating processes; and finally a consideration of the constellations of the different kinds of relations, and machinic patterns of distribution that arise alongside their intended and unintended consequences of its emergent histories and effects.

Central to this is the documentation of the material histories, and the multiple provenances of Trump's tweets as a form of geomaterial documentation.[97] This could involve, as a new form of empiricism, a laboratory analysis of the origin and composition of the material substrates and their histories in his media devices, hardware, and infrastructures that supported their circulation and their links to global logistical supply chains, climate change, environmental exhaustion, and its contaminating conditions.

These forms of documentation support the analysis and description of Trump's tweets as technical-sociological-political-material-geological and temporal compositions. Here such performative accounts shift from knowing the object as artefactual to a compositional engagement with the world through its multi-agential, material, and ecological rendering. Eco-curating processes are not always known or visible to humans, such as machine learning, and it can also be difficult to describe these processes in words such as code, a file format, or the visualisation of a neural network. Cuts are made by autoediting systems and platforms. Here social bots could be trained to gather and describe content and undertake onto-cartographic work in respect to their collection and documentation. Dynamic cuts of content could also record the interactivity of web pages and social media posts, similar to the Rhizome.org's Webrecorder. The historical emergence of the duration of any given ecological composition, for example, Trump's tweets still operating out in the world, could also be viewed and documented through time machine curating tools. Platforms could be developed for aggregating digitally-born resources in the cloud, for keeping ecological compositions active, documenting cuts, aggregating elements for their re-release for historical enactment, for hosting open-source data, and/or for connecting them to Github versioning systems. All these approaches, mediated by eco-curating processes for example those directed by AI and machine learning, signal processual rather than object-directed interpretations of them.

Due to the limitations of traditional forms of documentation, this process also requires a new vocabulary, but not as a form of identification. Nouns are routinely used to identify and describe stand-alone bounded objects. Yet here nouns, adjectives, and verbs can be used differently – not to describe or categorise things as conventional heritage description dictates, but rather to express coordinates, and their interrelatedness, qualities, alignments, and processes. So here nouns are used to describe all manner of the ecological composition's coordinates, not just the forms of them, but also the ideas encapsulated within them, the events in which they are embedded, and the multifarious cultural frameworks of value they enact. Adjectives are used to describe their qualities. And most importantly verbs and syntax express in language the congealing of agency and process, their deeply interconnected matrices of their thingness for example, bits, technologies, infrastructures, people, the networks of aligned, entangled interests, relational connections, and the intensities of their interrelatedness and effects as a nuanced account of a Trump tweet. Data visualisations could offer other forms of documentation alongside AI-generated images of text. Data-based analysis of Trump's tweets in the political realm seeks to examine how his tweets continued to influence US politics. Techniques, such as text-based analysis of over 24,000 tweets, seek to understand his mind, priorities, and thoughts through the use of words and their frequency, and these analyses are illustrated through data visualisations. Algorithmic and AI documentation in collecting and sampling processes could include Chat GBT interpretive narratives, live geovisualisations of data traces and responses to his tweets, social networks of re-tweets; frequency of tweets, common phrases, timelines of sentiment, subject and word clouds and patterns; and other types of data analytics. Documentation, therefore, involves a series of sampling processes to update interpretive cuts made through multiple entry points.[98]

Documentation also becomes future-orientated. This practice no longer solely operates in the past tense, or as a retrospective practice, but is emergent and focuses on the ongoing effects of Trump's tweets. Documentary updates trace their emergence. These emergent and future-orientated dispositions challenge the very notion of unchanging and human forms of documentary heritage.

Humanist notions of digital data and heritage interpretation co-exist with other coordinates radically different from a humanist sense of meaning and classification. That is because the congealing of collective memory and interpretive agency comprises many different registers of sensing and action, including those arising from the non-linguistic and machinic, the human, more-than-human and non-human, such as chemicals and minerals all of which form affective capacities and whose affects are often self-organising. These eco-curating processes emerge as different registers of cognition and action. Trump's tweets, for example, move in their own direction through automated processes of data quantifications. While representation and meaning-making are still important considerations in the interpretation of Trump's tweets as heritage, what also emerges through eco-curating processes is a more expanded understanding of digital memory as the affectual, calculative, sensing, automatic, and the non-conscious all operating as domains of influencing.[99] Therefore, it no longer makes sense of digital cultural heritage through narratives of the human, because human meaning-making and action are folded together with other eco-cognitive affects. Machine learning, and decisions made by artificial intelligence and algorithms – in combination with all the other coordinates embedded in eco-curating processes – signal the end of the notion of human memory, reasoning and truth as the central features of heritage collecting, documenting, and interpretive work. Therefore, cultural meaning in a traditional humanist sense is displaced.

Accordingly, Trump's tweets are interpretive communities comprising all manner of human and non-human coordinates. To explore this, I draw inspiration from Rosi Braidotti's posthuman project to re-define community and subjectivity, displacing human-centric notions of self to include biological, technical, and affective influences.[100] Opinions that appear on the interface are a product of the collective. Trump's tweets for example are a collaboration between all these coordinates, and not just between Trump and his human participants.

These new collectives illustrated through Trump's tweets as forms of participation, unsettle museum practices of representation, documentation, and normative forms of identity making. How can we understand the notion of community and communication across private social networks and public domains which enable and produce new forms of cognition and sensing? Community is no longer an individual human or human collective subject but rather comprises the radical interconnectedness of a wide range of coordinates that are involved in eco-curating processes in which human actors such as Trump are just one element. Therefore, what emerges is the formation of hybrid human and non-human socialities and communities where ecosystemic forms of agency gesture towards not only new community forms but also new forms of sociality that interact in heritage and museum spaces. By extending the notion of community, I also extend the notion of societal data and the entities actively involved in potential digital cultural heritage making, documentation, and interpretation to include human coordinates such as social bots, automated and algorithmic systems, global infrastructures, chemical and mineral substrates, and human users active in click farming. Clearly, we too must relinquish our position as the central figure in documentation and interpretation. Subjectivity becomes pluralised where human cognition is congealed with other coordinates, merging as intra-actions, effects, and emergence.[101] Rather than defining data according to humanist narratives and representational meanings, the world instead becomes bound up in the ecological composition.

Trump's tweets as technospheric more-than-human heritages

Trump's tweets when made visible as ecological compositions are complicit in carbon emissions and environmental exhaustion. His iphone produced 60 kg of emissions in its manufacture. Trump's tweets and their circulation are entrenched in the stockpiles of electronic waste such as computers, mobile phones, printers, keyboards, processors, and the toxic residues that issue from them. They are complicit in earthly forces and environmental destruction through mining for phone and infrastructure components, the burning of oil and coal in the USA to energise cooling data centres, and the circulation of tweets through cables and data centres towards personal devices.

Transported around the globe in their afterlife, often to China, India, or the Philippines for dumping and processing, rare metals such as gold extracted from Trump's iphones, computers, including those of his followers and social bots produce e-waste. These hardware components leak contaminants into the soil and produce toxic vapours and cause environmental and bodily contamination that can lead to cancer in the bodies of labourers who do the work. As a consequence, Trump's tweets are materially implicated in climate change and in Technosphere sedimentations.[102]

Trump's tweets also become technospheric heritage through their involvement in climate denial. His statements, such as "it's freezing here, where the hell is global warming," "we have ended the war on coal," "beautiful clean coal" and, in response to the bushfire crisis in California in September 2020, "it will get cooler, just you wait," are deeply implicated in atmospheric durations and rising greenhouse gas emissions as both an expression of policy and their material consequences of extraction, exploitation, and environmental destruction. His tweets, and the denial that issue from them, sought to protect the US fossil fuel-based economy, save the energy source from disappearing from the US energy mix, bury emissions underground through carbon capture and storage, and dismantle climate policy.[103]

Drawing inspiration from Jussi Parikka's analysis of geomedia, digital cultural heritage is also complicit in the atmospheric durations of climate change, and also those of geological times, through its critical media resources and infrastructures. When thinking about digital cultural heritage as ecological compositions, it becomes evident that the materials from which data is sourced from mining and geomaterials through the penetration and excavation of rare earth minerals. The stratification of digital cultural heritage and its aesthetic is made explicit through the interpenetration of the lithosphere through increasingly deeper cuts into the geological layers as deposits are exhausted and new drilling technologies emerge to extract more.[104]

Media is, according to Parikka, complicit in contested planetary politics of cloud storage, in land and sea territories as part of hidden military infrastructures of transmission and acts of surveillance.[105] Digital cultural heritage is power and enacts violence. Trump's tweets, as official records of his presidency, constitute legally effective military orders, no matter how confusing or irrational they may appear to activate military infrastructures. On April 22nd, 2020 Trump issued a tweet, "I have instructed the United States Navy to shoot down and destroy all Iranian gunboats if they harass our ships at sea." This tweet was issued a week after 11 vessels from Iran's Islamic Revolutionary Guard Corps Navy (IRGCN) came close to US ships in the Persian Gulf.[106]

Conclusion: Trumps tweets and eco-curating for planetary habitability

Fundamentally, therefore if we are to curate for planetary habitability we require a deeper understanding of digital cultural heritage and their environmental impacts, the material trails of waste, and the inequalities and regimes of power and disadvantage that arise from data production. This can be foregrounded through a greater attunement to all manner of humans and non-human constituencies that are embedded in data production and consumption. Ecological and embodied awareness, rather than a humanist vision of life-making processes through data production, can be put to work to engender ethical accountability and care.

Analyses founded on an ecological compositional model of digital cultural heritage, as well as Trump's tweets, can be used to raise questions about how we currently live on the Earth, ways we can live differently, and the recuperative processes and actions required to help curb the future impacts of these problems. Here interpreting digital cultural heritage as ecological compositions, and tracing the cartographic movements, points of contact, and effects of its coordinates, opens up reflective sensibilities and modes of acting.[107] By doing this we can take account of the interrelatedness of human actions and the wider non-human world in relation to data production, examine shifts in the chemical composition of the atmosphere, and the increasing toxicity of the earth to make visible the very existence of a common world and the figurations and alliances in which all humans are embedded. Using such information, we can gain the understanding of what is at stake to negotiate these processes if we are to curate in ways that ensure earthly survival.

Trump's tweets as forms of digital cultural heritage are also likely to become what I call technospheric waste. That is, human and more-than-human artefacts or technofossils buried in landfills, contained or preserved, to appear later in the archaeological and geological record as indicators of the Technosphere. At the same time, the emergence of technofossils, as a stratigraphic imprint of the new geological epoch of the Anthropocene embedded in the Earth's crust, brings with it, as cultural theorist Ben Dibley describes, a new type of lithic intimacy in the future becoming.[108] Because digital data is now embedded in almost everything western societies make and use, technofossil waste is likely to comprise a significant proportion of the burgeoning quantities of human-made materials, unknown or rare, in the other-than-human world – plastics, silicon, metals such as aluminium, titanium, and uranium, all of which returns to the earth in a cultured form that does not break down and is not able to be recycled or naturally processed by the biosphere.

The visible and hidden infrastructures of Trump's tweets as compositions have the potential to become technofossils. The making of technospheric heritage, and its trace fossils, is most likely to be from the hard parts of digital technologies, decommissioned electricity plants and data centre buildings, subterranean cables and the rare minerals that comprise them, the metal, wires, plastic, CDs, and DVDs, and other types of materials that are in the process of decomposing.[109] At the same time, these materials are contaminating the Earth's air, waterways, and oceans as they become buried in the Earth's crust.

What museums and other cultural institutions have overlooked or forgotten to collect for the digital heritage record can be recovered in the future, albeit as fragmentary, from the records decomposing sediments of the archaeological record as new types of legacies. These remains will become future fossils of the digital cultural heritage regime and of technological capitalism. Examples include the plastic disks that held data recovered from the Ground Zero rubble, now on exhibition at the 9/11 Memorial and Museum in Manhattan, and the remains of dead satellites that circle the Earth or shoot off into space, no longer producing data.

Other material traces of these ecological compositions will emerge as heritage of the atomic kind, fossilised in stone as anticipated forms of human, more-than-human, and other-than-human rituals of memorialisation yet to manifest. In doing so, emerging technofossils alert us to other temporalities latent in digital cultural heritage compositions, temporalities beyond the eco-curating processes of the contemporary present to that of other series of biogeochemical and eco-curating processes of deep time in which they are becoming stone.

Mapping technospheric heritage, and its circulations through an ecological compositional figuration, reveals a new emergent form of residue accretion. As ecological compositions, digital cultural heritage produces more intensive, deeper alignments of energy, capital accumulation and geology, environmental waste, and exhaustion.

Technospheric heritage makes explicit the more-than-human entangled histories and the emerging social relations between the exploitation of the earth's strata, environmental exhaustion, climate change, obsolescent design, and digital rubbish.[110] At the same time, technospheric heritage is complicit with digital capitalism, wealth accumulation of escalating proportions but with multifarious and unknown impacts.

This mapping exercise as detailed above also reveals not only the material and exhaustive elements embedded in Trump's tweets as technospheric more-than-human heritage but also the multiplicity of agencies, temporalities, and meanings entangled within them beyond those revealed through humanist forms. Accordingly, genealogies of digital cultural heritage, and heritage-like practices in a technospheric and ecological form, also reveal how Trump's tweets as heritage is stretched over long-time frames, comprises multiple intersecting durations from the temporalities of climate change and those of environmental exhaustion, to the temporalities of techno-capitalist regimes of accumulation and exploitation and the specific linear historical time of modern heritage frameworks.[111]

The digital cultural heritage legacies that will be significant for future generations are not only what museums and cultural institutions seek to make and contain as the artefactual, but what is passed on as dynamic ecological compositions and their wastes in duration.

Notes

1 Fiona R. Cameron, *The Future of Digital Data, Heritage and Curation in a More-Than-Human World.* (Abington: Routledge, 2021), 1.
2 Cameron, *The Future of Digital Data,* 227–257.
3 Cameron, *The Future of Digital Data,* 129.
4 Cameron, *The Future of Digital Data,* 3–14.
5 Cameron, *The Future of Digital Data,* 3–14.
6 Cameron, *The Future of Digital Data,* 139–143.
7 Cameron, *The Future of Digital Data,* 131–132.
8 For a discussion of globital digital memory and its material political economies see Anna Reading and Tanya Notley, "The Materiality of Globital Memory: Bringing the Cloud to Earth," *Continuum: The Journal of Media and Cultural Studies* 29, no. 4 (July 2015): 511–521; Jussi Parikka, *A Geology of Media.* (Minneapolis, London: University of Minnesota Press, 2015).

9 For a discussion of the Technosphere as a concept see Peter Haff, "Technology as a Geological Phenomenon: Implications for Human Well-Being," in *A Stratigraphical Basis for the Anthropocene?* eds. C. N. Waters, J. Zalasiewicz, M. Williams, Michael Ellis and Andrea Snelling (London: Geological Society London Special Publications, May 2014), volume 395, accessed February 25, 2023, http://dx.doi.org/10.1144/SP395.
10 Parikka, *A Geology of Media*, viii.
11 Parrika, *A Geology of Media*, 46.
12 Gilbert Simondon states that all modes of existence and being become subject to the technical, and Nature is enrolled to support the technical 1989. *Technical Mentality*, accessed February 25, 2023, https://www.parrhesiajournal.org/parrhesia07/parrhesia07_simondon2.pdf.
13 For a discussion of digital rubbish and its material contingencies see Jennifer Gabrys, "Digital Rubbish," in *Posthuman Glossary*, eds. Rosi Braidotti and Maria Hlavajova (London: Bloomsbury Academic, 2018): 108–109.
14 Benjamin Bratton, *The Stack: On Software and Sovereignty* (Cambridge, MA: The MIT Press, 2015); Jussi Parrika, "Mutating Media Ecologies," *Continent-Journal* 4, no. 2 (2015): 24–32; Rosi Braidotti, *Posthuman Knowledge* (Cambridge: Polity Press, 2019): 96.
15 Cameron, *The Future of Digital Data,* 132.
16 Cameron, *The Future of Digital Data,* 132.
17 Cameron, *The Future of Digital Data,* 134.
18 Cameron, *The Future of Digital Data,* 134.
19 Cameron, *The Future of Digital Data,* 134.
20 Cameron, *The Future of Digital Data,* 151.
21 For a discussion of the multifarious material, mineral, technical, social, and capitalist forces operating in the global computational infrastructures and digital economies see Parrika, "Mutating Media Ecologies"; Jussi Parikka, *A Geology of Media.* (Minneapolis, London: University of Minnesota Press, 2015), viii; Reading and Notley, "The Materiality of Globital Memory."; Friedrich Kittler, *Gramophone, Film, Typewriter* (Redwood City, California: Stanford University Press, 1999); Wolfgang Ernst, *Digital Memory and the Archive* (Minneapolis: University of Minnesota Press, 2012); Matthew Fuller, *Media Ecologies: Materialist Energies in Art and Technoculture* (Cambridge, MA and London: The MIT Press, 2005).
22 For a discussion of global supply chains and the mining of minerals see Parrika "Mutating Media Ecologies"; Reading and Notley, "The Materiality of Globital Memory."
23 Cameron, *The Future of Digital Data,* 136.
24 Cameron, *The Future of Digital Data,* 136.
25 Erich Hörl, "Introduction to General Ecology," in *General Ecology: The New Ecological Paradigm*, eds. Erich Hörl with James Burton (London and New York: Bloomsbury, 2017): 1–74, 9; Cameron, *The Future of Digital Data.*
26 Bill Brown, "Thing Theory," *Critical Inquiry* 28 (Autumn 2001): 1–22.
27 Brown, "Thing Theory," 1–22.
28 Cameron, *The Future of Digital Data,* 134.
29 Cameron, *The Future of Digital Data,* 136
30 Cameron, *The Future of Digital Data,* 137.
31 Cameron, *The Future of Digital Data,* 137.
32 Bruno Latour, "Reflections on Etienne Souriau's *Les Différents Modes d'Existence,*" 32, accessed February 25, 2023, http://www.bruno-latour.fr/sites/default/files/downloads/98-SOURIAU-GRAHAM-GB.pdf.
33 Cameron, *The Future of Digital Data,* 137.

34 Jane Bennett, *Vibrant Matter: A Political Ecology of Things* (Durham, N.C.: Duke University Press, 2010), 208, for a discussion of the conative capacities and affects of matter.
35 Fiona Cameron, "Posthuman Museum Practices," in *Posthuman Glossary*, eds. Rosi Braidotti and Maria Hlavajova (London: Bloomsbury Academic, 2018), 349–351.
36 Cameron, *The Future of Digital Data,* 137.
37 Cameron, "Posthuman Museum Practices," 352.
38 Cameron, *The Future of Digital Data,* 137.
39 Cameron, "Theorizing Digitizations in Global Computational Infrastructures," 55–67.
40 Cameron, *The Future of Digital Data,* 137.
41 While the notion of relationality operates within a posthumanist notion of performativity –one that incorporates important material and discursive, social and scientific, human and non-human, and natural and cultural actants – it implies there is a relational equivalence between things such as the interaction of all things as matter. This is Karen Barad's approach in *Meeting the Universe Halfway: Quantum Physics and the Entanglement of Matter and Meaning* (Durham, N.C.: Duke University Press, 2007), 808.
42 Cameron, *The Future of Digital Data,* 138.
43 Cameron, "Theorizing Digitizations in Global Computational Infrastructures," 65.
44 Cameron, *The Future of Digital Data,* 139.
45 For a discussion of operative modes of thought and sensing see Luciana Parisi and Stamatia Portanova, "Soft Thought (in Architecture and Choreography)," in *Computational Culture: A Journal of Software Studies* 1 (November 2011), accessed February 25, 2023, http://computationalculture.net/article/soft-thought.
46 Cameron, *The Future of Digital Data,* 140.
47 Influenced by Alfred Gell, *Art and Agency: An Anthropological Theory* (Oxford: Clarendon Press, 1998); Suzana Milevska "Becoming-Curator," in *The Curatorial: A Philosophy of Curating*, ed. Jean-Paul Martinon (London: Bloomsbury, 2013): 65–72; Cameron, "Theorizing Digitizations in Global Computational Infrastructures," 65.
48 Milevska, "Becoming-Curator," 69.
49 Joasia Krysa, *Curating Immateriality: The Work of the Curator in the Age of Network Systems* (Brooklyn, N.Y.: Autonomedia, 2006): 7; Fiona R. Cameron and Sarah Mengler. "Transvisuality, Geopolitics and Cultural Heritage in Global Flows: The Israeli–Palestinian Conflict and the Death of the Virtual Terrorist," in *Transvisuality: Dimensioning the Visual*, eds. Tore Kristensen, Anders Michelsen, and Frauke Wiegand, (Liverpool: Liverpool University Press, 2015): 59–72.
50 Cameron, "Posthuman Museum Practices,"; Cameron, *The Future of Digital Data,* 140; Cameron, "Theorizing Digitizations in Global Computational Infrastructures."
51 Cameron, "Theorizing Digitizations in Global Computational Infrastructures."
52 Cameron, *The Future of Digital Data,* 140.
53 For a discussion of material computation see Parisi and Portanova, "Soft Thought," 83.
Hörl, "Introduction to General Ecology," 29; for a discussion of network agents see Elena Esposito, "An Ecology of Differences; Communication, the Web, and the Question of Borders," in *General Ecology: The New Ecological Paradigm*, eds. Erich Hörl with James Burton (London and New York: Bloomsbury, 2017): 289.
54 Hörl, "Introduction to General Ecology," 29 for a discussion of network agents see Esposito, "An Ecology of Differences," 289.
55 Cameron, "Theorizing Digitizations in Global Computational Infrastructures," 56.

56 Cameron, *The Future of Digital Data,* 142.
57 Cameron, *The Future of Digital Data,* 142.
58 Cameron, *The Future of Digital Data,* 142.
59 Hörl, "Introduction to General Ecology," 22. Hörl summarises Parisi's argument in regard to computation and materials as new forms of eco-rationality in design.
60 Cameron, *The Future of Digital Data,* 143.
61 For a discussion of the intelligence of materials as new types of ecological power in computer design see Luciana Parisi, "Computational Logic and Ecological Rationality," in *General Ecology: The New Ecological Paradigm*, eds. Erich Hörl with James Burton (London and New York: Bloomsbury, 2017): 75–100, 75.
62 Cameron, *The Future of Digital Data,* 142.
63 Esposito, "An Ecology of Differences," 293.
64 Cameron, *The Future of Digital Data,* 143.
65 Cameron, *The Future of Digital Data,* 143.
66 Presidential Records Act (PRA) of 1978.
67 "The COVFEFE Act Would Preserve Donald Trump's Tweets as Presidential Records," *The Washington Post*, June 13, 2017, accessed February 25, 2023, https://nationalpost.com/news/world/the-covfefe-act-would-preserve-donald-trumps-tweets-as-presidential-records; Presidential Records Act (PRA) of 1978."
68 "Donald Trump's Tweets are Presidential Records, but What Happens When They Get Deleted or Altered?" *ABC News*, January 24, 2017, accessed February 25, 2023, http://www.abc.net.au/news/2017-01-24/donald-trump-tweets-presidential-records-deletions-edits/8206920.
69 Shontavia Johnson, "Donald Trump's Tweets are Now Presidential Records," *US News*, February 1, 2017, 3.00 pm, http://www.usnews.com/news/national-news/articles/2017-02-01/donald-trumps-tweets-are-now-presidential-records; "Trump's Tweets to be Preserved by the National Archives," *CBC News*, April 3, 2017, https://www.cbc.ca/news/world/trump-s-tweets-to-be-preserved-by-national-archives-1.4053277; Olivia Beavers, "White House Will Preserve all of Trump's Tweets," Report, *The Hill*, March 4, 2017, 09.57, 2017, accessed February 25, 2023, https://thehill.com/homenews/administration/327125-white-house-will-preserve-all-of-trumps-tweets-reports.
70 Cameron, *The Future of Digital Data,* 151.
71 Cameron, *The Future of Digital Data,* 151.
72 Evelyn Ruppert, Engin Isin and Didier Bigo, "Data Politics," *Big Data and Society* 4, no. 2 (2017): 1, accessed February 25, 2023, https://doi.org/10.1177/2053951717717749, Ruppert, Isin and Bigo, in a special issue of *Big Data and Society*, frame a new concept, "data politics," to describe the way data has been constituted as an object vested with certain powers, influence, and rationalities. Data politics asks questions about the ways in which data has become such an object of power and explores how to critically intervene in its deployment as a subject of knowledge. Data politics is concerned with the conditions of possibility of data that involve things (infrastructures of servers, devices, and cables), language (code, programming, and algorithms) and people (scientists, entrepreneurs, engineers, information technologists, designers) that together make up particular spaces of relations or worlds. Trump's tweets, I argue, exert a different type of data politics that is a result of much more complex eco-curating processes and the domains of influencing that emerge through these actions.
73 Cameron, *The Future of Digital Data,* 152.
74 Capitol Riots Timeline: What Happened on January 6, 2021? June 10, 2022, accessed February 25, 2023, https://www.bbc.com/news/world-us-canada-56004916.
75 Cameron, *The Future of Digital Data,* 152.

76 Jussi Parikka, *A Geology of Media* (Minneapolis, London: University of Minnesota Press), viii.
77 Cameron, *The Future of Digital Data,* 153.
78 Bratton, *The Stack*, 246. Bratton develops the idea of the geoscape: "A geoscape is a contested terrain of contested terrains, a shifting landscape made up of shifting landscapes: images, maps, projections and plans, irredentist land claims, borders, and jurisdictions, strata and striations, imagined worlds, macroeconomic forecasts, projected homes and homelands, addressing systems and various terra incognita. Geoscapes are a form of content that cannot *not* be designed and designed for where unplannable and unresolved territories, jurisdictions and programs are put into play."
79 Cameron, *The Future of Digital Data,* 153.
80 Cameron, "Theorizing Digitizations in Global Computational Infrastructures,"65.
81 Cameron, *The Future of Digital Data,* 154.
82 Cameron, *The Future of Digital Data,* 155.
83 Steve Almasy and Joshua Berlinger, "North Korea Calls Ballistic Missile Test-fire a Success," *CNN*, updated 0643 GMT (1443 HKT) February 13 2017, accessed February 25, 2023, http://edition.cnn.com/2017/02/11/asia/north-korea-missile/index.html.
84 Almasy et al., "North Korea Calls Ballistic Missile Test-fire a Success."
85 Eliot Weinberger, *LRB blog*, "The Month of Trump," (blog), *London Review of Books*, February 16, 2017, accessed February 25, 2023, https://www.lrb.co.uk/blog/2017/02/16/eliot-weinberger/the-month-of-trump/.
86 Cameron, *The Future of Digital Data,* 155.
87 For a discussion of soft thought as computational processes see Parisi et al., "Soft Thought," 10.
88 Cameron, *The Future of Digital Data,* 156.
89 Megan McCluskey, "What Donald Trump's Twitter Reinstatement Means for TRUTH Social Time," November 21 2022, accessed February 25, 2023, https://time.com/6234846/trump-twitter-return-truth-social.
90 Cameron, *The Future of Digital Data,* 213.
91 Cameron, *The Future of Digital Data,* 212.
92 The agential cut as a material, conceptual device was developed by Barad and adapted by Sarah Kember and Joanna Zylinska, *Life after New Media: Mediation as a Vital Process.* (Cambridge, MA: The MIT Press, 2012): 81.
93 Kember et al., *Life after New Media*, 83.
94 Kember et al., *Life after New Media*, 79. The concept of cut in photographic practice is inspired by Henri Bergson's philosophy of perception.
95 Cameron, *The Future of Digital Data,* 210.
96 Cameron, *The Future of Digital Data,* 213.
97 Cameron, *The Future of Digital Data,* 214.
98 Cameron, *The Future of Digital Data,* 215.
99 Cameron, *The Future of Digital Data,* 217.
100 Braidotti, *The Posthuman.*
101 Cameron, *The Future of Digital Data,* 218.
102 Cameron, *The Future of Digital Data,* 252.
103 Cameron, *The Future of Digital Data,* 252.
104 Cameron, *The Future of Digital Data,* 253.
105 Parikka, *A Geology of Media*, 30.
106 Reuters, Trump Instructs U.S. Navy to Destroy Iranian Gunboats "if they harass our ships at sea" April 22, 2020, accessed February 25, 2023, https://www.reuters.com/article/usa-iran-military-idUSL2N2CA0CX.
107 Cameron, "Posthuman Museum Practices," 52.

108 For a discussion of technofossils see Jeffrey Jerome Cohen, *Stone: An Ecology of the Inhuman* (Minnesota: University of Minnesota Press, 2015) and Ben Dibley, "The Technofossil: A Memento Mori," *Journal of Contemporary Archaeology* 5, no. 1 (2018): 44–52.
109 Cameron, *The Future of Digital Data*, 255.
110 Cameron, *The Future of Digital Data*, 256.
111 Cameron, *The Future of Digital Data*, 265.

References

Almasy, Steve and Joshua Berlinger. "North Korea Calls Ballistic Missile Test-fire a Success." *CNN*, updated 0643 GMT (1443 HKT) February 13, 2017, accessed February 25, 2023, http://edition.cnn.com/2017/02/11/asia/north-korea-missile/index.html.

"As Trump Orders Wall, Mexico's President Considers Canceling U.S. Trip." *New York Times*, January 25, 2017, accessed February 25, 2023, https://www.nytimes.com/2017/01/25/world/americas/trump-mexico-border-wall.html.

Barad, Karen. *Meeting the Universe Halfway: Quantum Physics and the Entanglement of Matter and Meaning*. Durham, N.C.: Duke University Press, 2007.

Beavers, Olivia. "White House Will Preserve All of Trump's Tweets." Report, *The Hill*, March 4, 2017, accessed February 25, 2023, https://thehill.com/homenews/administration/327125-white-house-will-preserve-all-of-trumps-tweets-reports.

Bennett, Jane. *Vibrant Matter: A Political Ecology of Things*. Durham, N.C.: Duke University Press, 2010.

Braidotti, Rosi. *The Posthuman*. London: Polity Press, 2013.

Braidotti, Rosi. *Posthuman Knowledge*. Cambridge: Polity Press, 2019.

Bratton, Benjamin. *The Stack: On Software and Sovereignty*. Cambridge, MA: The MIT Press, 2015.

Brown, Bill. "Thing Theory." *Critical Inquiry* 28 (Autumn 2001): 1–22.

Cameron, Fiona. "Posthuman Museum Practices." In *Posthuman Glossary*, edited by Rosi Braidotti and Maria Hlavajova, 349–351. London: Bloomsbury Academic, 2018.

Cameron, Fiona. "Theorizing Digitizations in Global Computational Infrastructures." In *The International Handbook of New Digital Practices in Galleries, Libraries, Archives, Museums and Heritage Sites*, edited by Hannah Lewi, Wally Smith, Steve Cooke, and Dirk vom Lehn, 55–67. London: Routledge, 2019.

Cameron, Fiona, R. *The Future of Digital Data, Heritage and Curation in a More-Than-Human World*. Abington: Routledge, 2021.

Cameron, Fiona R. and Sarah Mengler. "Transvisuality, Geopolitics and Cultural Heritage in Global Flows: The Israeli–Palestinian Conflict and the Death of the Virtual Terrorist." In *Transvisuality: Dimensioning the Visual*, edited by Tore Kristensen, Anders Michelsen, and Frauke Wiegand, 59–72. Liverpool: Liverpool University Press, 2015.

Cohen, Jeffrey Jerome. *Stone: An Ecology of the Inhuman*. Minnesota: University of Minnesota Press, 2015.

"The COVFEFE Act Would Preserve Donald Trump's Tweets as Presidential Records." *The Washington Post*, June 13, 2017, accessed February 25, 2023, https://nationalpost.com/news/world/the-covfefe-act-would-preserve-donald-trumps-tweets-as-presidential-records.

Dibley, Ben. "The Technofossil: A Memento Mori." *Journal of Contemporary Archaeology* 5, no. 1 (2018): 44–52.

"Donald Trump's Tweets are Presidential Records, but What Happens When They Get Deleted or Altered?" *ABC News*, January 24, 2017, accessed February 25, 2023, http://www.abc.net.au/news/2017-01-24/donald-trump-tweets-presidential-records-deletions-edits/8206920.

Ernst, Wolfgang. *Digital Memory and the Archive.* Minneapolis: University of Minnesota Press, 2012.

Esposito, Elena. "An Ecology of Differences; Communication, the Web, and the Question of Borders." In *General Ecology: The New Ecological Paradigm*, edited by Erich Hörl with James Burton, 283–301. London and New York: Bloomsbury, 2017.

Fuller, Matthew. *Media Ecologies: Materialist Energies in Art and Technoculture.* Cambridge, MA and London: The MIT Press, 2005.

Gabrys, Jennifer. "Digital Rubbish." In *Posthuman Glossary*, edited by Rosi Braidotti and Maria Hlavajova, 108–109. London: Bloomsbury Academic, 2018.

Gell, Alfred. *Art and Agency: An Anthropological Theory.* Oxford: Clarendon Press, 1998.

Haff, Peter. "Technology and Human Purpose: The Problem of Solids Transport on the Earth's Surface." *Earth System Dynamics Discussions* 3, no. 1 (November 2012): 149–156.

Haff, Peter. "Technology as a Geological Phenomenon: Implications for Human Well-Being." In *A Stratigraphical Basis for the Anthropocene?*, edited by C. N. Waters, J. Zalasiewicz, M. Williams, Michael Ellis and Andrea Snelling, volume 395, no. 1. London: Geological Society London Special Publications, May 2014, accessed February 25, 2023, http://dx.doi.org/10.1144/SP395.

Heidegger, Martin. "The Thing." In *Poetry, Language, Thought*, translated by Albert Hofstader, 161–180. London and New York: Harper Perennial, 1975.

Hörl, Erich. "Introduction to General Ecology." In *General Ecology: The New Ecological Paradigm*, edited by Erich Hörl with James Burton, 1–74. London and New York: Bloomsbury, 2017.

H.P. "Human Artefacts as Technofossils: Picking Over the Traces." *The Economist*, April 3, 2014, accessed February 25, 2023, http://www.economist.com/blogs/babbage/2014/04/human-artefacts-technofossils.

Johnson, Shontavia. "Donald Trump's Tweets are Now Presidential Records." *US News*, February 1, 2017, accessed February 25, 2023, http://www.usnews.com/news/national-news/articles/2017-02-01/donald-trumps-tweets-are-now-presidential-records.

Kember Sarah and Joanna Zylinska, *Life after New Media: Mediation as a Vital Process.* Cambridge, MA: The MIT Press, c. 2012.

Kittler, Friedrich. *Gramophone, Film, Typewriter.* Redwood City, California: Stanford University Press, 1999.

Krysa, Joasia. *Curating Immateriality: The Work of the Curator in the Age of Network Systems.* Brooklyn, N.Y.: Autonomedia, 2006.

Latour, Bruno. "Reflections on Etienne Souriau's *Les Différents Modes d'Existence*," accessed February 26, 2023, http://www.bruno-latour.fr/sites/default/files/downloads/98-SOURIAU-GRAHAM-GB.pdf.

McCluskey, Megan, "What Donald Trump's Twitter Reinstatement Means for TRUTH Social Time," November 21 2022, accessed February 25, 2023, https://time.com/6234846/trump-twitter-return-truth-social.

Milevska, Suzana. "Becoming-Curator." In *The Curatorial: A Philosophy of Curating*, edited by Jean-Paul Martinon, 65–72. London: Bloomsbury, 2013.

Parisi, Luciana. *Contagious Architecture: Computation, Aesthetics, and Space.* Cambridge, MA: The MIT Press, 2013.

Parisi, Luciana. "Computational Logic and Ecological Rationality." In *General Ecology: The New Ecological Paradigm*, edited by Erich Hörl with James Burton, 75–100. London and New York: Bloomsbury, 2017.

Parisi, Luciana, and Stamatia Portanova. "Soft Thought (in Architecture and Choreography)." *Computational Culture: A Journal of Software Studies* 1 (November 2011), accessed February 25, 2023, http://computationalculture.net/article/soft-thought.

Parikka, Jussi. *A Geology of Media.* Minneapolis, London: University of Minnesota Press, 2015.

Parrika, Jussi. "Mutating Media Ecologies." *Continent-Journal* 4, no.2 (2015): 24–32.

Presidential Records Act (PRA) of 1978. *National Archives*, accessed February 25, 2023, https://www.archives.gov/presidential-libraries/laws/1978-act.html.

Reading, Anna and Tanya Notley. "The Materiality of Globital Memory: Bringing the Cloud to Earth." *Continuum: The Journal of Media and Cultural Studies* 29, no. 4 (July 2015): 511–521.

Reuters, Trump Instructs U.S. Navy to Destroy Iranian Gunboats "if they harass our ships at sea," April 22, 2020, accessed February 25, 2023, https://www.reuters.com/article/usa-iran-military-idUSL2N2CA0CX

Ruppert, Evelyn, Engin Isin, and Didier Bigo. "Data Politics." *Big Data and Society* 4, no. 2 (2017), accessed February 25, 2023, https://doi.org/10.1177/2053951717717749.

Scolari, Carlos A. "Media Ecology: Exploring the Metaphor to Expand the Theory." *Communication Theory* 22, no. 2 (May 2012): 204–225.

Simondon, Gilbert, *Technical Mentality*, accessed February 24, 2023, https://www.parrhesiajournal.org/parrhesia07/parrhesia07_simondon2.pdf.

Tauberg, Michael. "Analysing Trump's Tweets: A Data-Based Analysis of Trump's Language on Twitter," October 31, 2018, accessed February 24, 2023, *The Startup*, https://medium.com/swlh/analyzing-trumps-tweets-5368528d2c90.

"Trump's Tweets to Be Preserved by the National Archives." *CBC News*, April 3, 2017, accessed February 24, 2023, https://www.cbc.ca/news/world/trump-s-tweets-to-be-preserved-by-national-archives-1.4053277.

Trump Twitter Archive, accessed April 21, 2020, http://www.trumptwitterarchive.com/.

Weinberger, Eliot. *LRB blog*. "The Month of Trump" (blog). *London Review of Books.* Posted February 16, 2017, accessed February 24, 2023, https://www.lrb.co.uk/blog/2017/02/16/eliot-weinberger/the-month-of-trump/.

Woolley, Samuel C. "Automating Power: Social Bot Interference in Global Politics," *First Monday*, April 4, 2016, accessed February 24, 2023, http://firstmonday.org/article/view/6161/5300.

3

COLLECTIONS AND ECO-CURATING HUMAN–NON-HUMAN CLIMATES

Introduction

In this chapter, I investigate the proposition that cultural collections can be put to work as material agents for the cultural and social change on which "de-carbonisation", climate change mitigation and adaptation are contingent. Central to this is a consideration of non-human others as stakeholders in governing projects and in policy implementation as curatorial agents. I first work with a melted green plastic bucket in the collection of Museum Victoria and explore how it is a material register of entanglements: the convergence of extreme weather events and as an embodiment of climatic catastrophe in the form of wildfire. The second example is that of the Boulton and Watt "Lap" rotative beam engine. Built in 1788 it is an early example of the development of rotative motion and was used in Boulton and Watt's Soho Manufactory Birmingham to drive 43 metal polishing machines.[1] The Engine is now on permanent exhibition in the Energy Hall in the Science Museum, London and only one of three surviving examples from the 1780s. It is an iconic engine of the Industrial Revolution, a technology that embodies a profound climate history of emissions.

Much recent scientific analyses of biological collections, from herbaria, to animal species, and seed and tissue banks, document species responses to climate change as time series data.[2] Climate-sensitive biological specimens are used in climate change research to determine the spread, diversity, and extinction of species due to climate variability and, through comparison with fossils, changes in species body size.[3] Climate variability over time can also be recorded through carbon isotope sampling of plant collections.[4] Biological collections record the impact of warming as one written into the DNA of species, which

DOI: 10.4324/9781315212067-3

is transmitted through and as changes over time. Botanical collections record changes in carbon isotopes as a register of atmospheric biochemical change. Likewise, a history of climate change emissions is bonded through the photosynthetic into the wooden componentry of cultural collections.

Cultural collections are no longer obsolescent, nonliving, non-sentient, and only vital when subject to interpretive input from collection managers and curators in respect to their climate or Anthropocene histories. These collections are vibrant with carbon materialities, and demonstrative of atmospheric fossil-fuel emissions and their consequences. They can be put to work as new types of cultural mitigation strategies through an analysis of their material form and composition and, in respect to the Boulton and Watt "Lap" Engine case study and related cultural collections containing wood, ratios of carbon isotopes within their wood "DNA." As a result, cultural collections can act as catalysts for the cultural and social change required for the reduction of emissions in novel, powerful ways. The two examples explored in this chapter act as book ends in a climate story spanning 230 years and are hemispheres apart; one implicated in fossil-fuel combustion that initiated the climate crisis in Birmingham, England and the other the unintended consequences of fossil-fuel combustion in the hot and dry landscapes of Australia and its vulnerability to wildfire.

In this sense heritage institutions such as museums can become important pedagogical institutions, developing novel, cultural mitigation strategies by using their cultural collections as material indices of climate history. A consideration of cultural heritage collections as material indices of climate history begins with an understanding of the enmeshment of all manner of materials, industrial, human, and non-human processes that comprise their distributed agencies bonded into the fabrics and circumstances in with they are embedded. For the purposes of this discussion, these collection items, the bucket, and Engine are comprised, respectively, from elements that are made with plastic and from petrochemicals and the fossilised plants used to manufacture them. Their componentry is made from the metals and minerals extracted from the earth's crust. The Engine for example comprises elements including those made of wood that register the atmospheric conditions when the tree was growing. Both items are through the materials, manufacturing processes, and cultural practices through which they are made and used are representative of high emissions sectors and, respectively, subject to the eco-curating processes which comprise the element of fire and the agential capacities of photosynthesis.

As both emitters and sinks, steam machines and instruments of atmospheric observation to everyday cultural objects such as wooden washing machines have the potential to bolster international carbon reduction ambition. At the same time, these items bring climate change back to the here and now, embodied in ourselves, our homes, and our everyday lives, in our built environment, in our heritage – the very things we seek to pass onto future generations. They are also a reminder of the colonising processes of modernity in which heritage institutions such as museums are complicit, and in which carbon dioxide (CO_2) emissions and their after-effects including the wildfire register in material culture and in the organic fabric of multispecies communities reside. This CO_2 register, and that of fire embedded in collections, not only produces new knowledge about collections and atmospheric change but as proposed in this chapter, a new understanding of what it means to be human in a climate-changed world.

Collections and socio-ecological crises

Cultural heritage collections housed in museums range from the extraordinary – such as the 1,000-year-old Viking ships, Gokstad, Tune, and Oseberg in the Vikingskipshuset (Viking Ship Museum, Oslo)[5] – to the everyday, like the matches, pencils, and toothbrushes found in the eclectic collection of mass-produced items housed in the Museum of Everyday Life, Glover, Vermont. Astonishing or banal, such curated objects are routinely put to work to invoke cultural memory and affective responses, telling human stories of modern technological achievement and economic modernisation, of cultural history and of the social, ideological, and cultural constructions of the human subject such as identity, cultural significance, historical events, and social biographies.

Routinely in exhibitions on climate science, such as "Atmosphere" at the Science Museum in London (2010-) the foremost exhibition on climate science and ecological modernisation strategies[6] (discussed in detail in Chapter 4) collections stand for and represent the effects of climate change on Nature. Each object in the exhibition is passive and put to work in the service of science and scientists. Objects act as undeniable documentary evidence of warming trends and therefore are a ballast, or stabilising force, for affirming modern discourses of climatic change. A section of a tree ring, for example, operates in modern climate change "speak" as a climate diary where variability over the tree's life is etched into their tree-grown rings. Likewise ice, particles of dust, plants, and animal skeletons act as documentary records of past warming. In doing so scientists and curators speak for the non-human, rendering "things" as passive, having no agency of their own except when put into service to support these modern ends.[7] The non-human animals in the "Atmosphere" exhibition, for example, cows, are seen as unminded bodies subject to biopolitical compliance. Here the human task is to manipulate cows' diets to reduce greenhouse gases.

In recent years, technological objects have also been deployed to represent the histories of industrialisation and colonialism and their complicity in the destructive planetary forces now associated with the Anthropocene, the damaging consequences of modern human-centred modes of existence.

For example, emblematic objects of the early Industrial Revolution have been repositioned as carbon emitters integral to global fossil-fuelled capitalist economies.[8] These objects include, for example, the Boulton and Watt rotative beam engine built in 1786 to pump water and later to grind barley for the Barclay & Perkins Brewery in Southwark, London[9], and J. Edward Earnshaw and Co's Trestle steam machine (1862) that powered Nuremburg's Municipal Gasworks.[10] Both these objects are now viewed as engines of the Anthropocene. The Boulton and Watt's rotative beam engine is an iconic object permanently exhibited in the National Museum of Scotland, Edinburgh. The Trestle steam machine took centre stage in the Deutsches Museum exhibition, "Welcome to the Anthropocene: The Earth in Our Hands" (2014–2018).

Everyday consumer items have also been curated as objects implicated in the climate crisis. For example, cultural objects recovered from the 2009 Black Saturday bushfires in Victoria, Australia, held in Museum Victoria's collection – such as a burned-out Holden 48–215 sedan, part of Chris Lee's vintage car collection housed on his property in Koornalla, and Bill Putt's melted green plastic bucket used to save his house from the inferno – are curated to memorialise the event and recount stories of and as expressions of human empathy, emotion, loss and survival. These contrasting curatorial approaches – documenting climate change as scientific natures, monumentalising the engines that inaugurate the Anthropocene, or memorialising mundane objects caught in environmental disaster – nevertheless share a common orientation. They put human experiences – whether it be "great" men, the scientists, inventors, and entrepreneurs, or ordinary, everyday individuals – at the centre of their story telling.[11] These approaches go some way to acknowledging the modes in which the Anthropocene, the climate crisis, and human and non-human worlds are profoundly entangled. However, technological objects like those at the Deutsches Museum and the National Museum of Scotland, Edinburgh, which are now rearticulated as harbingers of the Anthropocene, or those of Museum Victoria, now cast as the bearers of loss in the aftermath of wildfire, work predominately at a discursive or cultural level. From the epochal to the event, from the advent of a human-made geological era, to the personal catastrophe of wildfire, these curatorial strategies seek to re-animate museum objects in relation to modernity's "unintended consequences" – foremost that of the climate crisis.[12] This is so in the sense that such cultural objects are at once the material expressions of carbon economies and cultures that have generated them.[13] However, there is a materiality to such objects that is beyond the representational, one that inscribes them in a more-than-human world, which goes largely unacknowledged in these strategies.[14] Such approaches share an orientation that belies the ways the climate crisis is an emergency borne out of a failure to acknowledge western societies' deep enmeshment with the non-human world.[15]

Cultural heritage collections can convey new stories that articulate how cultural objects might serve as material expressions of climate change – ones that do not rely on the sentimental tropes and anthropocentrism on which such exhibits are contingent for their resonances. Rather new narratives account for these objects as ones dynamically embedded in climate change processes.[16]

Fire curating in eco-curating processes

The Black Saturday bushfires of February 7, 2009 was one of Australia's most devastating natural disasters. It lead to the death of 173 people, 11,000 farm animals and more than 1 million native species, the loss of 3,400 homes, the razing of 430,000 hectares and the destruction of communities and infrastructure; unprecedented environmental impacts ensued. The artefacts, stories, and images collected by Museum Victoria recorded the impact of the bushfires on communities and the emergency response, and illustrated the role of bushfires in the historical landscape, both in the past and in scientific contexts.[17] This collection was much more than just products of the human imagination directed to illustrate the impacts of fire on human and biodiverse populations and environments, but also revealed the emergent consequences and effects of contingent and heterogeneous enactments, and eco-curatorial performances comprising an array of human and non-human coordinates in planetary enmeshments. In Rosi Braidotti and Maria Hlavajova's seminal posthuman glossary published in 2018 (Bloomsbury Academic), I narrated the story of Bill Putt's melted plastic bucket used to save his house from the inferno.[18] Here, drawing on my contribution in the glossary, I recite the story of Bill and his bucket again in which fire becomes the central curatorial agent as its material figuration to explicate the relations between climate change and the emergence of extreme weather events such as wildfire.

Bill Putt's green melted plastic bucket is an example of how cultural collections might act as mechanisms to re-orientate and reattach ourselves to a more-than-human world different from the human subject we are ensconced.[19] The green bucket is framed through human-centred curatorial and documentation descriptions of Bill Putt's accounts of survival, and the bucket is used to recall and recite his final attempts to save his house, "Rosewood at Strathewen," from the impending furnace. The bucket, however, is no longer solely about Bill, and his attempts to save his house – it is also profoundly elemental in which fire its affordances as and in eco-curating processes have the capacity to transform and destroy. It is also a material illustration of the enmeshment of humans, non-humans, and elemental earthly processes operating together in more-than-human eco-curatorial collaborations such that the curatorial becomes ecologically orientated.

The collections record describes the object as a "green plastic garden bucket, melted on one side and missing the handle. The spout is still visible.

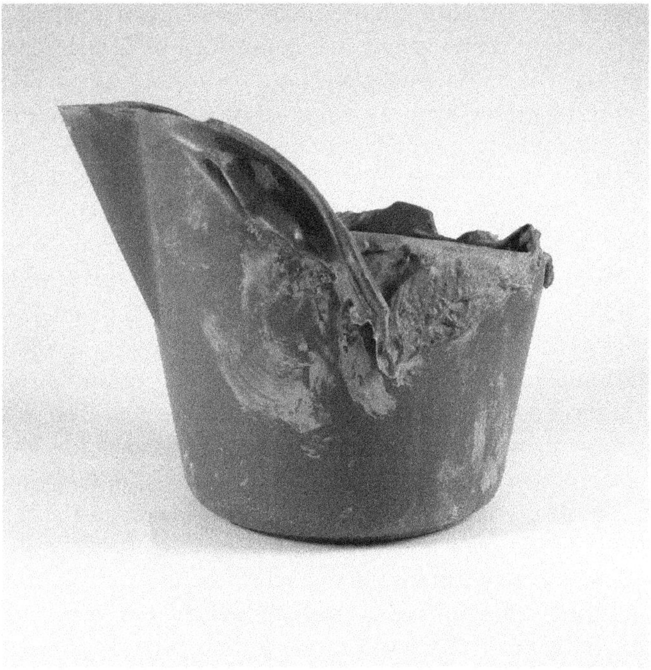

FIGURE 3.1 Bill Putt's bucket – Strathewen, February 7, 2009 (bushfire damaged). Museum Victoria collection.

Traces of dried mud adhere to its surface."[20] While the bucket's curatorial interpretation acknowledges the subjectivities of this event and its description gestures towards its materiality and form, these human subject-orientated frames promote the bucket as a static representation of itself in service of the human social.[21]

The objects and stories that were collected by Museum Victoria were acquired because they were considered likely to be of benefit to people in the future. That is, as examples of community resilience following disasters, and the adaptations and strategies required while raising awareness of the danger of bushfires in high bushfire risk peri-urban settings in Victoria.[22] The curatorial emphasis was on the human experience, that is on "objects representing the fragility of life, the depth of loss, the miracle of survival and the ferocity of fire." The collection also covers ecology, wildlife, and the interconnectedness of humans and the environment.[23] Many of the collection items illustrate the extraordinary force of the fire,[24] through which many were reduced to ashes or altered fundamentally in material composition and form. In doing so curator Dale Hallett took an interdisciplinary approach to collection development, in which the experience of bushfire and its impacts in social, cultural, environmental contexts were registered through a diverse range of acquisitions into the history and science collections. The interdisciplinary approach was also evident in collections management systems in which keywords allowed curators to log in to any collection, and transverse searches across them using terms such as "bushfire," "fire," and "burnt," retrieving collection records across all museum disciplines.[25] The fire collection comprised other items such as mineral deposits profoundly altered by bushfires many thousands of years old, fire sticks and art work from Northern Australia, and a clutch of birds that fell out of the sky during a dust storm that preceded Ash Wednesday in 1983.[26] The Leadbeater nest box, representative of endangered species, while formally located in the history collection, was interpreted together with specimens in the natural history collection in order to recite historical, cultural, and scientific forms of storytelling.[27]

Relational thinking across the collection, enacted through database searches and interpretive strategies, was directed to bringing disciplines and collections together thereby circumventing the disciplinary divide. Furthermore, interdisciplinary and ecological forms of thinking were configured in the development of the Leadbeater display as part of the fire collection through an interpretive collaboration, and an ecological thought practice, involving a historian, an endangered species biologist, a park ranger, an environmentalist, and activists.

The fire collection is represented as a quantification of "whole objects," as bounded and as representative, standing for the bushfire event, its processes of destruction, and what came after for human populations and their environments. In doing so its curatorial framing renders invisible the other material, discursive, technological, biological, elemental, and non-human aspects of the green bucket's agential and eco-curatorial processes all of which also played a starring role.

When we begin to see the world differently – as entangled with other people, animate, inanimate things, and elemental processes – human subject–object-based museum practices appear incongruous, even deceptive. Untying human subject–object-based understandings of the bucket as a bounded "object," and as "artefact," requires its re-ascription in more-than-human terms as "thingness" and as "ecological composition." As "thingness" I refer to the bucket as an extended and dispersed spatial and temporal thing made up of conative and ontologically diverse vital coordinates that include the non-human and, further to its distributed performativity, incorporating material, discursive, social, political, economic, scientific, human, non-human, natural and cultural elements.[28]

The first task in the bucket's ontological refashioning is to re-work the humanisms that hold its objecthood and interpretation in place. Disrupting these modern humanisms and the agential relations organised around Bill, his experiences and actions, that currently underlay our understanding of the green plastic bucket, necessitates the disassembling of these interrelations and their reconstitution into a metanomically diverse list that represents the bundles of elemental, material, technical, conceptual, ecological, social, and emotional components that make up its distributed composition and its provenances that cannot always be known.[29]

This assortment of conative thingness comprises not only Bill Putt, his bodily actions, and his stories, but it also brings to the fore the bucket's processes of production, its geographical location, its extended histories from the deep past to its museological ascription, and its material composition and non-human coordinates, features, and form and its situated contexts.[30]

To this end, the list comprises plastic, carbon, hydrogen, oil, coal, natural gas, crude oil, salt polymers, mud, fire, moulds, digital code, climate change, trees, greenhouse gases, temperature, wind speed, Strathewen, Victoria, 7 February 2009, a mud brick house, a failed water pump, curators, collection managers, documentation procedures and so forth. Many of these elements are invisible in its formal object description.[31] From this ontologically diverse list of conative elements, Bill and his subjectivities become just one of many coordinates that intermingle with the other components comprising its composition and their agencies.

The list also operates as a mechanism for re-ascribing each of its components as "vital" coordinates. This task re-works agency away from the human to acknowledge the bucket's multiple agencies and temporalities, including its materials, physical form, the element of fire, and multifarious discursive and affective framings. The situated human becomes embodied humanness distributed across and through the bucket, from those that design and craft the bucket into its form, to the corporations, waged labourers, and machines that mine and extract the materials and the crude oil from which it is made, to Bill, the bucket, his enmeshed interactions and his own fossil-fuel intimacies and dependencies and that of modern populations that are refiguring as atmospheric chemical compositions.

In order to attribute agencies and affordances to each of the buckets' elements – for example biological, elemental, ecological processes, technical and technological processes, human actions, thoughts, desires, crafting, and so forth – I draw inspiration from the work of Rosi Braidotti,[32] Ian Hodder,[33] Jane Bennett,[34] and my own work,[35] alongside relevant technical, scientific, and historical information.

The plastic, for example, in which the bucket is made comprises a series of tight interdependences, from its historical genealogy in the development of its plastic, to its human coordinates such as civil engineers, to its sequences of production from its component materials: carbon, oil, polymer to molten plastic, pressure, and injection moulding. By gathering together the coordinates that make up the "composition" and reassembling them in new types of relations, we can consider the performative effects of their interdependencies and affordances overall as distributive eco-curatorial agencies.

With the primary elemental curatorial agency of fire, for example, its affordances – including its physical and chemical properties – operate as a series of tight interdependences that exhibit different temporalities, clusterings, tautness, and enabling effects that cohere in certain ways around the bucket. These include temperatures in the 40s, atmospheric biochemical conditions, heating afforded by modernities' fossil-fuelled populations interacting with other affordances such as wind speed, the potential for the rapid oxidation and combustion of the Australian bush, the release of heat, light, burning embers and smoke enabling the Kilmore East fire to emerge and converge on Bill Putt's property, and the bucket. This sequence was locked into a series of other enmeshed relations with a failed water pump, the bucket's form and spout as an affordance to hold and pour water, the chemical composition of plastic combined with radiant heat and its molten effects, Bill's fear of loss, dying livestock, burning buildings and the act of running.[36]

Objects like the bucket, as a combusted ecological composition, become a material history of wildfire etched into its "DNA" through its transformed molecular composition and form. The bucket becomes a compositional lesson of extreme weather and the climate emergency. The unprecedented ferocity of the 2009 Black Saturday bushfires in Victoria saw a new "catastrophic" category emerging through the curatorial agencies of more-than-human and non-human elements.

The bucket is made of polymers, long chains of chemicals, minerals, and other elementals of the earth consisting mostly of carbon atoms with hydrogen, sulfur, and oxygen. Through the bucket's eco-curatorial burning event toxic gases like dioxins, furans, mercury, and polychlorinated biphenyls (BCPs) were released into the atmosphere and at once became an emitter through the release of black carbon (soot), thereby contributing to atmospheric climate change and air pollution. Through the eco-curating of atmospheric conditions – through modernity's fossil-fuel emissions and from the carbon and the other materials in the bucket – multiple temporalities and domains of influencing emerge.

Fire and combustion, for example, as eco-curating processes, are strongly enmeshed in global heating, in very hot days and low humidity, in fuel, terrain, ignition agents, and people, all of which supply the conditions for its spread, during which the material agencies of the bucket are activated in profound ways changing its very composition. Its compositional lesson becomes an expression of Australia's carbon economies, of climate inaction at the time and acts as a directive for Australia to cut its emissions rapidly and deeply to reduce the risk of even more extreme events. The bucket therefore becomes a new type of cultural climate mitigation strategy. It is so when it becomes the material consequence of fire, in which wildfire and the convergence of all the bucket's interdependencies are made present and become visible.

Given this, as an ecological composition, the green bucket is no longer a static thing. Rather it emerges through multiple locations and as the effect of its distribution across many diverse coordinates, as "thingness," as a part of more diverse, dynamic social collectives and extended networks of climate and wildfire in its many dimensions. Its solidity acts as a stand in for its distributed and congealed agencies that comprise it. Bill did indeed save his house and his life however he lost several outbuildings and animals to the fire.

Cultural collections as material archives of fossil-fuel emissions

Cultural collections can also be a different type of material climate historical archive, since many objects' fabric – the inorganic and organic material of which they are comprised – holds a history of rising fossil-fuel emissions and of human carbon-burning activity. The following sections consider how items held in the Science Museum, London can be rearticulated to highlight the climate histories that are embedded in their material compositions.

With its foundation in the 1851 Great Exhibition, the Museum was officially established in 1857 and comprised the Royal Society of Arts collection, surplus items from the Great Exhibition as part of the South Kensington Museum, together with the Victoria and Albert Museum collections. The technology and engineering collections became the Museum of Patents in 1858, and the Patent Office Museum in 1863. In 1883, the science, technology, and engineering collections founded the Science Museum.[37] Its collection of machinery has historically been curated as an exposition of scientific, technical, and industrial progress. The Museum therefore became a celebration of modernity and Britain's accomplishment. However, in recent years, its collection has come to represent a more ambivalent legacy. It now has other resonances – of extraction, of ecological crisis, and of colonial expansion and Indigenous dispossession. The Science Museum's engineering and technology collections have histories deeply entangled in a capitalist economy and its extractivist activities associated with the development of consumer culture, which have supported Great Britain's economic growth, and contributed disproportionately to global carbon emissions. In 1712, Thomas Newcomen invented a practical and reliable steam engine that powered the Industrial Revolution replacing other forms of energy and enabled the pumping of water from mines therefore facilitating the exploitation of the planet's mineral wealth.[38] The Museum has in its collection a Newcomen atmospheric engine built by Francis Thompson of Ashover near Chesterfield in 1791. Other examples include the oldest surviving Boulton and Watt's single-acting beam engine, Old Bess built in 1777 and used to pump water at Matthew Boulton and James Watts' Soho Manufactory. Old Bess played a critical role in the development of efficient steam power through the exploitation of the expansive properties of steam. The early steam locomotive, Stephenson's Rocket, designed by Robert Stephenson and made by Robert Stephenson & Co., Newcastle upon Tyne, England in 1829 is an example of advanced superior motive power.[39] Others such as the collection of vacuum cleaners and early incandescent electric lamps that illuminated domestic spaces first invented by Joseph Swan and then perfected by Thomas Edison with the invention of the filament[40] facilitated the lifestyles of the nation's population. All these examples are objects inseparable from the energy used to fuel, manufacture, and use them.[41] In other words, like the green plastic bucket, they are objects animated by the ignition of fossil fuels implicated in the historical emissions that we now face as climate change. They are objects with climate histories.[42] In this sense, the bucket and the Science Museum's science, technology and history collections, and many others internationally are already climate collections, but they await to be articulated as such. This is true not only in the sense that such objects are icons of the carbon-intensive cultures that produced them but also the history of climate change and extreme weather events are layered into the materials of which these objects are composed.

This chapter builds on this concern by drawing out the ways in which its collection items are implicated in carbon economies and associated cultures are also objects with climate histories as a different matter of fact. This proposition draws on background intellectual work and research used to develop the multi-authored paper, "Climate Collections and Photosynthetic, Fossil-Fuelled Atmospheres" published in the July 2023 issue of the *Environmental Humanities* journal and for this chapter. This background work was also the incubator for the Australian Research Council Linkage project, "Curating Museum Collections for Climate Change Mitigation" which is concerned with the material sampling of museum collections and their ^{13}C analysis.[43]

These publications are the result of an interdisciplinary collaboration. It engaged my own work on museum collections as climatic ecological compositions, posthumanities thinking, more-than and non-human agencies, and their potential as new cultural carbon mitigation strategies. Professor David Ellsworth's ^{13}C analysis of atmospheric conditions bonded into timber, his work on EucFACE, a world-class experiment that simulates the impacts of future carbon-dioxide-rich climates on Australia's native forests, investigations into the potential for sampling of the wooden components of cultural collections and his graph simulations of ^{13}C depletion in oak in the Midlands, the home of the Industrial Revolution and eucalypts on the Western Sydney University, Hawkesbury campus, Sydney provided the scientific rationale. In addition, Dr. Ben Dibley, a museum scholar of the Anthropocene and heritage provided context for the collections under investigation.

The insights presented here in respect to cultural objects and their climate composition are drawn from the environmental humanities and the critical posthumanities and are augmented with other knowledge practices from the biogeochemical sciences. This includes, for example, carbon dating – whose methods provide the opportunity to locate particular cultural objects in relation to the deep time of planetary climate change and atmospheric CO_2 enrichment.[44] In doing so I further develop the proposition and concepts that, articulated as such, objects as complex "climatic ecological compositions"[45] and, understood as such, can serve as "cultural carbon mitigation strategies"

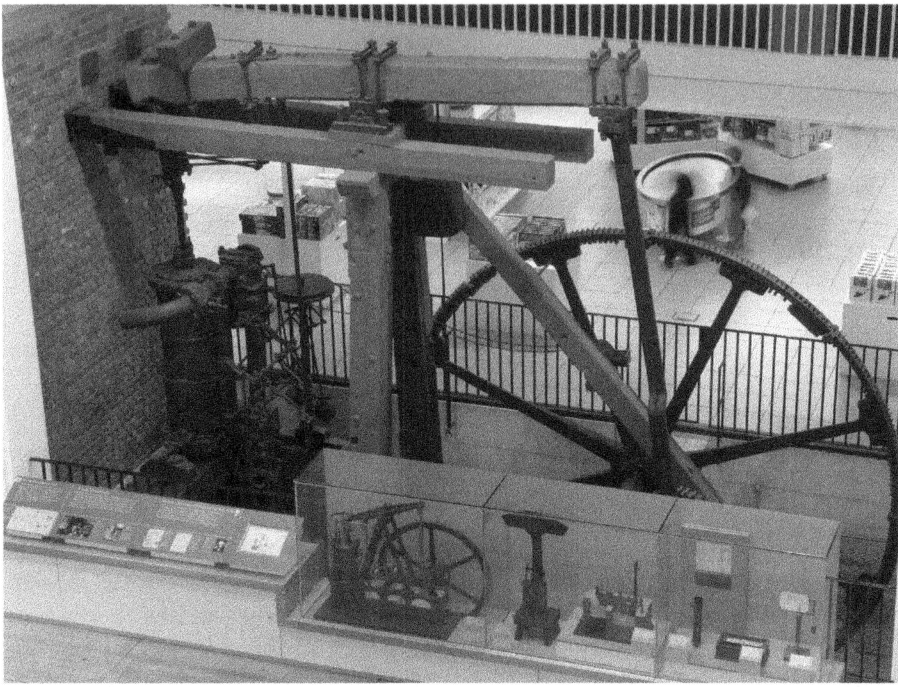

FIGURE 3.2 Boulton and Watt "Lap" rotative steam engine, Science Museum, London. © Science Museum Group.

that can complement climate mitigation policies and programs. This proposition is trialed in relation to a key collection item in the Science Museum, London the Boulton and Watt Rotative Beam Engine or "Lap" Engine, Object no 1861–46.

However, I am not here to demonise Boulton and Watt engines as grotesque embodiments of the Anthropocene. Nor am I interested in human-centred histories that cast entrepreneur Matthew Boulton and inventor James Watt as climate men. But rather I am interested in the engines as having agency, or indeed multiple agencies, that can be put to work to reinterpret them and indeed climate change action in many different ways.

Agency, climate, and collections

Recent efforts by museums, aimed at reducing the carbon footprints of their publics, have been based on science pedagogies directed towards the communication of scientific information around the levels of dangerous greenhouse gas emissions. Often deploying dramatic visual imagery of climate impacts, these strategies are used to signal the loss of place and home. To achieve this, visualisations in the form of melting ice, rising sea levels, ice cores samples, and tree growth rings bring scientific data into an affective register. Similarly, the testimonies of communities facing climate predicaments are regularly included to foreground climate change as an interconnected crisis and aim to mobilise an empathic response from visitors. These affective and ethical strategies are put into the service of raising awareness of climate change and mobilising personal commitments to pursue solutions and reduce emissions. Such approaches often repeat the anthropocentrism of climate science and policy agendas, positioning individuals and communities as potential commanders and controllers of the atmosphere.

While I pioneered scholarship on museums and climate action, and in advocating a critical posthuman turn for museum practice and theory, these fields have expanded considerably over recent years.[46] There is a growing shift in curation that aims to invoke a more entangled optic on human–non-human relations, for example, as expressed in the work of colleague Ben Dibley.[47] The exhibition, "Human Nature," held at the National Museum of World Cultures, Gothenburg, Sweden (2019–), which focused on mass consumption sought to promote the idea of interconnectedness between human consumption practices and multispecies survival.[48] And another example was the exhibition, "We are Nature: Living in the Anthropocene," held at Carnegie Museum of National History, Pittsburgh in 2017 which delivered the key message –"We are not separate from nature, *we are nature*, and our decisions affect all life on earth" – seeking to reposition humans in ecological systems through the lens of science.

These exhibitions, among others, are laudable developments. However, what is proposed here is a deepening of this ecological optic by pressing the questions of materiality further. This is achieved by adding to the interpretation of cultural objects a biochemical analysis of their material fabric. This is a distinctive and novel strategy proposed by tree ecologist, climate researcher, and colleague Professor David Ellsworth.[49] Such an approach positions such objects as distinct forms of human–non-human biochemical relations that occasion possibilities of new material and climatic attunement.

Refiguring humanness after humanism

In discussing ecological crises, Claire Colebrook suggests human populations are merely another life form on Earth.[50] Following this historians, Marek Tumm and Zoltan Simon articulate a new characterisation of what it means to be human and historical, as life forms in movement, dynamic, and indeterminate.[51] I contend that modern populations become relationally configured in particular contexts as dynamic fossil-fuelled biochemical life forms in movement that can complement climate mitigation policies and programs.[52] To develop this contention, I drew insights from the critical post-humanities that not only decentre the figure of the human but distribute and make relational the collaborating and contesting agencies that have made this figure legible.[53] Fossil-fuel-burning populations are folded into the chemical composition of the atmosphere, soils, plants, multispecies, organic materials, distant others, extreme weather events, melting ice, and different forms of material culture, including the built environment and, indeed, museum collections made visible through Ellsworth's carbon isotope methods. Furthermore, the refiguration of a western citizenry as fossil fuel, and as enmeshed agents of atmospheric change, brings to light the need to refigure not only the way we identify ourselves as human but also our subjectivity. Modern populations are fossil-fuelled biochemical agents when considering the processes of atmospheric transformation and its unintended consequences as a novel figuration complementing that of geological agents as central figures of the Anthropocene. That is modern populations become entities of relational interdependency in which the sense of ourselves, our knowledge practices, and bodily constitution become viscous, prompting the formulation of new subjects along the lines of what Stacey Alaimo calls "transorporeal."[54] For Alaimo, the transcorporeal subject describes a new figuration of the human after humanism, conscious, and attuned to materialsm and transcorporeal responsibility. A subject, for this purpose, constituted through the convergence of embodied human–non-human biochemical agents that have transformed and are transforming the biochemical composition of the atmosphere and the life forms and earthly processes it supports. This is in the sense that fossil-fuel-combusted hydrocarbons pass through and across us as we draw sustenance, carbohydrates, and glucose from the consumption of plants, which in turn fuel us, and through the emissions we breathe and rebound in tissue, in cell reproduction and in our very genetic make up powering the replication of DNA. Our bodies are made of carbon and at the same time embody a record of industrially processed hydrocarbons. We are at once subjects of a fossil-fuel capitalism and embodiments of an emission-enriched biochemical life form.

Central to the refiguration of agency and the vital bonds that bind us to others are the insights afforded by the environmental and the critical post-humanities, alongside other modes of ontological thinking. Rejecting human exceptionalism and moving beyond anthropocentrism, these critical insights return human bodies, knowledge, actions, and powers to their ecological place. These knowledge practices explicate the ways in which the situated human is knitted together. That is, not just with other biological species, in the context of evolutionary theories and the exchange of DNA, but also with all manner of things, animate and inanimate – including non-human entities, chemicals, metals, living tissue, materials, and earthly processes.[55] The way we think about our agency and the form, quality, specificity, and proximity of connections between our bodies, our lifestyles, our fossil-fuel-burning activities, and the things we own, make, and use matters fundamentally in our efforts to bolster climate mitigation ambition.[56]

In Donna Haraway's discussion of multispecies survival and human–non-human relations of partial recuperation, she argues, that, in the face of the unfolding ecological crisis, the key question is one of who, how, and in what ways are we bound together.[57] Taking the lead from Haraway, questions of cultural collections and climate mitigation are inseparable from an understanding of the biochemical agents enmeshed within collection items and of the curatorial and pedagogical strategies that make them visible and tangible. In this approach, the human is not posited as a forcing agent. Nor as one of command and control directed to stabilising the biochemical composition of the atmosphere. Rather modern populations become deeply entangled relations of human–non-human biochemical agencies; relations on which the future trajectory of planet's biosphere now rely.

Cultural collections as climatic ecological compositions

Importantly, the articulation of these agential relations in cultural collections can provide a strategy to rethink the positionality and situatedness of human populations in climate history. It serves as a mechanism and as an invitation to engage in the work of composing ethical, habitable worlds. Ones that proceed in the acknowledgement of the uneven distribution of the drivers and consequences of climate change across populations and ones that seek to redress the injustices of this distribution.[58] Here, the who, how, and in what ways we are bound together in vital biochemical relations of human–non-human interconnectedness becomes the key question.[59] This is a question central to how we understand ourselves, and the ecological relations in which we are now enmeshed. The question, and the understanding of the material entanglement on which it is contingent, warrants special attention from museums. This would provide these institutions with a distinctive way to engage with the predicament of the climate crisis and to build new publics around climate issues and action.

In providing answers to the questions raised above, they are contingent on a particular conceptual reconfiguration of cultural objects, which I call "climatic ecological compositions."[60] Through this formulation, the aim is to highlight the ways in which cultural heritage objects are not simply items of material culture – understood as bounded, discrete entities and expressions of human-centric histories – but are semio-material assemblages, which emerge in an alliance of discourses, human bodies, biological systems, earthly processes, and elemental materials.[61] This mode of analysis folds out to make visible the vital entanglements and articulations of human and non-human convergence embedded in an object's composition. It also seeks to explicate the mindless damage in which heritage collections are implicated which led to global heating in the first instance.[62] This mode of analysis offers the means to investigate the proposition that heritage collections are material expressions of the convergence of human and non-human agencies in atmospheric relations.[63] The development of cultural collection items as "climatic ecological compositions" also takes inspiration from the emerging field of chemical ethnography[64] to investigate the ways in which the description of chemical agents thickens our understanding of museum objects.

For this purpose, wooden objects, or those with wooden components, are ideal candidates for this mode of analysis. This is so given the material's significance as a substrate recording atmospheric climate change, its amenability to techniques of carbon dating, and its relative prevalence across collections as a material of manufacture.[65] Given these factors, the aim is to explicate how wooden collection items, and those with wooden components, are registers of anthropogenic atmospheric biochemical change and, in particular, how such registers bear witness to these processes as an unequivocal outcome of human fossil-fuel burning.[66] The concern is understanding how these objects register the ways in which human thoughts, acts, and bodies – which are enrolled in carbon-intensive economies, cultures, and the extractive processes of deforestation and of industrial production, such as the manufacture of cement – are intermeshed bio-chemically with the atmosphere and then through the dynamic processes of photosynthesis.

Collections as "cultural carbon mitigation strategies"

Building on this is the contention that such objects can be used to expand the repertoire of existing climate mitigation strategies. And, in doing so, they can be advanced as what I propose as their capacity to act as "cultural carbon mitigation strategies." Conventionally understood, carbon mitigators are those instruments and techniques directed at market restructure and behavioural change designed to reduce greenhouse gas emissions. The most well-known are financial instruments such as tradeable carbon credits used to compensate for carbon emissions through emission trading schemes or carbon offset programs. These instruments leverage the capacities of trees and soils to provide "ecological services" as sinks to soak up atmospheric carbon.[67] While not directly incorporated into such market mechanisms, heritage collections can be be refigured as mitigation strategies. This is not to use carbon offset in a way that it has been regularly criticised for, that is as a practice offering to rekindle capitalism's promise that we can enjoy consumerism without being too concerned about ecological crisis.[68] The analysis of tree rings used for climatic reconstruction is conventionally directed to understanding changes in surface temperature as tree rings are responsive to climate change conditions.[69] Rather here it is to use the collections as a strategy to demonstrate our co-existence, shared vulnerabilities, and futures with the non-human world. In this way, rather than relieving visitors of responsibility, such demonstrative instruments could have the potential to activate affective and cognitive responses in unique ways to effect social change and emissions reduction. As registers of anthropogenic carbon emissions linked to a history of atmospheric CO_2 levels, cultural collections can be pressed into the service of reducing and preventing more greenhouse gases being produced by expanding and changing worldviews, attitudes, and practices. That is, by highlighting an accountability to non-human worlds by making visible entangled climate histories, material situations, and consequences. These figurations have the potential to build ecological and communitarian orientations to atmospheric and more broadly planetary processes, signalling the deep bonds shared with all manner of entities – human and non-human. Modern populations become not an "I" or "We" but an "Us," signalling the deep bonds they share with all manner of non-humans as collective polities.

Biochemical convergence: ^{13}C analysis

Through the dynamic processes of photosynthesis, plants absorb carbon dioxide from the air and soil, draw in ^{12}C and ^{13}C carbon atoms through their leaves, and combine them with the sun's energy in specialist cells called chloroplast to make glucose and carbohydrates. This leads to growth and then the release of oxygen back into the air, the vital biochemical agent that sustains our survival.[70] The key technique for identifying changing atmospheric carbon dioxide concentrations in relation to plants is through measuring the relative ratios between ^{12}C and ^{13}C carbon atoms. The latter decreases over time in the composition of carbon isotopes in wood tissue.[71]

The biochemical basis of ^{13}C analysis is that wood is formed from assimilating all forms of carbon from the atmosphere through a plant's photosynthetic tissues. Plants prefer to process lighter and more common ^{12}C molecules leaving heavier and more enriched ^{13}C behind in the atmosphere. These ratios are analysed relative to geological standards and climate records. Fossil-fuel emissions are the only carbon source consistent with the isotopic fingerprint of the carbon present in today's atmosphere, comprising fossilised plant matter. Because fossil fuels such as coal, oil, and gas are made from long-dead plants and micro-organisms, they are a depleted form of ^{13}C. The decline in ^{13}C in the atmosphere over time (around 3% per year) is consistent with the burning of fossil fuels. The global atmospheric CO_2 concentrations of ^{12}C and enriched ^{13}C at the time when the carbon dioxide from the plant that made the wood was living can be calculated with some biological assumptions.[72]

The method by which such ratios are measured involves taking small samples of wood from selected items, from which carbon dioxide molecules are combusted, separated, and quantified.[73] The isotopes ^{12}C, ^{13}C, and ^{14}C exhibit different, distinctive flight registers in mass spectrometry in which ^{13}C isotopes can be counted relative to ^{12}C with fossil fuel carbon enriched as fossil-fuel-based carbon comprising fewer ^{13}C molecules.[74] Thus, carbon molecules of both ^{12}C and declining enriched ^{13}C, embedded in the plant cells, are pulled apart and extracted through a process of combustion and release, repeating the burning and emissions process on a micro-scale. A host of scientists including David Ellsworth advanced these techniques in the 1990s.[75]

The application of Ellsworth's techniques to the wooden elements of cultural heritage objects offers a novel approach to locate such items in relation to a trajectory of anthropogenic atmospheric biochemical change.

Simulated charts articulate the depletion of ^{13}C in oak and eucalypts in which values were gathered through tree sampling in Staffordshire, the Midlands in the UK, the heartland of the Industrial Revolution from the 1780s and the trees from the Driftway, Western Sydney, Australia. These values were expanded using the atmospheric $\delta^{13}C$ signal through publicly available monthly $\delta^{13}C$ values from the Mauna Loa observatory, Hawaii, compiled records of carbon isotopes in atmospheric CO_2 for historical simulations and an annual scaling of those values back to 1780. Using these historical simulations rather than the material sampling of objects themselves, I cross-reference this data with the provenance of specific Boulton and Watt rotative engines and the rise of fossil-fuel emissions in the UK and Australia. Building on this conceptual work related to the species from which the objects are manufactured and with reference to their historical contexts, these items come to provide a distinctive biochemical signature of industrial human–non-human relations and signal the unintended climate consequences of lifestyles, economies, and processes of extraction that these objects embody. Such heritage objects bear the hallmark of an anthropogenic biosphere: they are registers of the earthly biogeochemical processes that are decisively shaped by the carbon-intensive economies and cultures of consumerism in which they are implicated. Combining ^{13}C analysis with perspectives from the environmental humanities and posthumanities and more-than-human curatorial theory through a series of transdisciplinary and experimental cogenerative dialogues (cogens), we sought to entangle humans with enriched carbon atmospheric processes, posthuman curatorial practice as it pertains to collections interpretation, and tree ecology.[76] If the techniques of ^{13}C analysis locate these objects in climate history, it is the protocols of the environmental and posthumanities that describe the agency of carbon isotopes in this conceptual framing. This is an emergence that necessarily reconfigures the environmental, technical, and social narratives in which these objects have been historically inscribed as heritage collections. Explicating the climate histories layered into these items in this way greatly enhances their interpretive value and pedagogical worth. This we explore in relation to our selected heritage object.

The Boulton and Watt Lap Engine, Object no 1861-46.

The Boulton and Watt "Lap" Engine in the Science Museum collection is the oldest almost intact machine of its type in existence. Built by Watt in 1788 in Boulton and Watt's Soho Manufactory in Birmingham, the lapping engine was used to drive 43 metal polishing machines used for lapping and polishing Boulton's manufactured articles such as buckles, buttons, and silverware, and from 1797 blanking presses for the production of coinage. By the time Boulton and Watt's partnership ended in 1800, they had built 451 Engines in which 268 examples were rotative.[77] The 1788 Engine was decommissioned in 1858 and was acquired in the same year by the Science Museum, where it is exhibited as a celebration of the history of modern technological advancement. In recent years, these Engines' role in anthropogenic climate change has been acknowledged.

These are some key moments in these engines complicated modern humanist history, moments which oscillate between the specificity of this particular object and its material composition and the broader technological, economic, social, and environment processes in which the history of the Engine is implicated, and which this particular example of its engineering comes to represent. Attention is also drawn to the contours of their multiple humanisms and histories before returning to questions of its biochemical composition and its implications for this history.

The eighteenth-century Scottish engineer and inventor, James Watt, alongside his collaborator, the entrepreneur Matthew Boulton, changed economic and social history and indeed the future of the planet's biogeochemistry.[78] They did so, according to conventional historical accounts, with their epoch-making device – the steam-driven, rotative beam engine. These engines have long been both celebrated as an emblem of the "age of steam" and denigrated as an instrument of labour's subjection in the factory system. But as devices deeply embedded in an emerging global carbon economy, they are now presented as much a harbinger of climate change as it is a symbol of the Industrial Revolution. In 1784, the first Boulton and Watt rotative beam steam engine was steamed and put to work in Henry Goodwyn's Red Lion Brewery in London. It was an early candidate for Crutzen's and Stroemer's suggestion, in 2000, of the date initiating a new geological epoch called the Anthropocene.[79]

Deploying "the sun and planet gear," method to convert reciprocating motion to rotary motion, it was William Murdoch, an engineer and employee of Boulton and Watt, who developed the machine.[80] Watt patented the design in 1781. Ending an earlier era of renewable energy from wind and water, this coal-fuelled rotative steam engine among others became a more efficient method of mechanical production. Instead of using atmospheric pressure acting on top of the engine's piston, the innovation of the Boulton and Watt rotative beam engines was to use steam to condense above and below the piston. The rotative motion of these engines was applied to numerous industrial processes and was improved in the "Lap Engine" built in 1788 with the addition of a centrifugal governor that controlled its rotational speed. In doing so it supplied the means to exploit more efficiently than previously possible energy-dense fossil fuels and to more effectively exploit the labour power of workers harnessed to the routines of industrial processes and their machines.[81] In this the Engine was to have a major role in the development of rotation devices in the Industrial Revolution, driving not only the machinery of Britain's factory system, including cotton, sugar, flour, paper, textile factories, iron mills, breweries, and distilleries but also their supporting infrastructures, canals, and waterworks.[82]

When the Boulton and Watt "Lap" Engine was manufactured, it expressed an aspiration of a human future – one that Boulton and Watt felt entitled to build and shape for themselves and for others that purchased their steam engines. While an achievement in technological advancement, it also marked the beginning of an enmeshment with fossil-fuel-driven capitalist aspirations. These engines were presented as a landmark development in man's command over nature. This is evident in the excited responses to demonstrations of rotative engines in the Boulton and Watt's factory and the sense of urgency on the part of entrepreneurs to procure their own. The first enquiry to Watt came from Henry Goodwyn on April 17, 1784.[83] He sought an engine for his Red Lion Brewery and his enquiry led to the commissioning of a four-horse engine to grind malt. The Engine was trialled in early August of the same year and as Goodwyn reported in his letter of the August 9th it exceeded his expectations.[84] The second enquiry came from Samuel Whitbread on the August 26th, 1784, ordering a 10 horse-power engine.[85] This engine survives and is in the Powerhouse collection, Sydney. The correspondence in respect to these two early engines concerned economies of scale, levels of productivity, and the machine and boiler capacities.[86] The volume of coal required to drive the machines was of particular concern because coal at that time was being heavily taxed.[87]

The new invented engine incited wonder and awe. People marvelled that these engines could do the work of 24 horses, were amazed by the productivity it could achieve, and hoped it could incite moral and material improvement. In a marketing coup for both the brewer and the Engine's manufacturer, King George III and Queen Charlotte visited the Whitbread Brewery on May 24, 1787, adding royal patronage to the Engine's admirers. The Whitbread Engine became in demand for viewing by potential local and foreign buyers.[88]

The Boulton and Watt "Lap" Engine is a centrepiece of the Energy Hall formerly the Watt Hall at the Science Museum, London dedicated to Britain's engineering achievements since the Industrial Revolution. The "Lap" engine was first acquired in 1858. The Watt Hall was first opened on January 14, 1936 to celebrate the bicentenary of Watt's birth on January 19, 1736. The now Energy Hall includes three original beam engines including Watt's original experimental models from 1765 showcasing his work on the separate condenser.[89] The Hall seeks to "trace the remarkable story of steam and how it shaped the world we live in today." Interpretive labelling articulates the role of fossil-fuel-burning technology to generate steam as "the driving force behind British industry for 300 years."[90] Another feature of the Watt Hall is the original attic workshop of engineer James Watt, preserved as it was when he died in 1819. This exhibit, titled James Watt and the World venerates Watt's fossil-fuel-driven inventions as remarkable and as one of the "greatest benefactors of the human race."[91] From steam power to tea services, visitors are invited to "explore the relationship between Watt's steam engine and a new age of consumption and discover his incredible legacy."[91] The Museum aimed to reproduce the veneration that the Engine had initially produced in elite circles.

Other humanisms embodied in these engines apart from Watt's veneration as a man of modernity include what Gilbert Simondon calls a technical mentality.[92] The interpretive materials accompanying the "Lap" Engine and through its documentation, for example, were framed through its technical mentality. This concerns a disposition that views any technical artefact as a system of distinct objects performing particular but mutually dependent functions, all of which have different qualities and substances, interacting and changing one another as different phases in its emergence.[93] These include how the steam engine works, its technical development and improvement, its materials of manufacture, the physical principles under-laying its operation, the personal histories of its principal inventors, the use of models to show its inner workings, and its display using open platforms to observe the Engine.[94] This mentality has long accompanied the Engine, as is clear in the Boulton and Watt papers in respect to the "Lap" Engine. This includes details of the Engine's design and engineering, its components, and how each of these functions together, the different materials used to make each component and how each operates mechanically.[95] It was an orientation that the Science Museum in the Energy Hall endorsed without reflection. This technicist interpretation of the Engine across its history even today shares in what Rosi Braidotti calls an anthropocentric mindset, where only human qualities, needs, actions, creativity, technological achievements, and knowledge matters are significant.[96] Such understandings are at the expense of the object's historicity and ignore the violence inscribed into its operations.

The history of the Boulton and Watt "Lap" Engine is a history inseparable from that of capitalist extractivism, exploiting not only the energy latent in coal but also the labour power of a disenfranchised peasantry and their children, who formed a new urban working class, and of slaves, whose plantation labour supplied the raw materials for British factories and the cheap calories that fuelled the bodies of factory workers. The Engine's genesis and operation are inseparable from colonial plantations the factory system's raw materials were derived from and capital's profit extracted.[97] It was profited from the transatlantic slave trade that provided Britain with capital for its industrialisation. Many of the central innovations and entrepreneurial ventures were funded by such profits. The development of Boulton and Watt's rotative steam engines were no exception; the William Deacon Bank, an institution that was founded by wealthy plantation owners, supplied the finance. Subsequently between 1790 and 1800, numerous engines were exported to the sugar cane and cotton fields of the Caribbean for mills to refine the crop.[98] The modern humanisms, technical, and anthropocentric mentalities that have accompanied the Boulton and Watt "Lap" Engine not only belie the violence of extraction to which it was integral but also the ways in which its enrolment in rising CO_2 emissions and environmental destruction are implicated in what Nigel Clark calls inhuman nature and futures – at odds with human intent, action, and human rights or a sense of justice.[99]

Human–non-human histories

Boulton, Watt, and engineer William Murdoch did not invent and build the Engine alone. The Engine embodies an array of non-humans entangled with human designs. Supplementing the object's social and economic histories with a consideration of its implication in biogeochemical processes, and the more-than and non-human agencies involved the Boulton and Watt "Lap" Engine thus becomes an object to think with and act with differently – as a different type of object lesson and a distinctive type of marvelling – not one of mastery, veneration and hubris but one of ecological embeddedness. The Boulton and Watt "Lap" Engine is no longer solely a context or background to illustrate human creativity and industrial progress but is deeply embedded in the physical substrates of the biosphere and ecosphere, troubling futures, and life itself.

The Boulton and Watt "Lap" Engine has multiple radical interfaces of atmospheric relations – condenser – boiler – copper and later iron, burning fossils, plants and animals long dead. The Engine was also an inferface between different expressions of energy – between the inhuman, in its explosive release of the energy stored in coal, and the human in terms of the various forms of labour power harnessed to the machine through factory systems and plantation slavery. The Engine therefore embodies an array of non-humans entangled with human designs and forms of labour power and its exploitation.

The Engine is, however, a cultural heritage item, with the potential to not only challenge the privileging of the human-centred histories, design, actions, and value in terms of generational remittance but, most importantly, to "put humans back in their ecological place."[100] This refiguring has the potential to bring our radical co-existence with the non-human world and with earthly, including atmospheric, processes to the fore. And to make possible the re-positioning of human and technological inventions as autonomous agents and symbols of modern achievements to entities entangled in dynamic material, processes that have profound ecological consequences.[101]

While the Boulton and Watt "Lap" Engine looks like a cultural, technological, and heritage object, it is also what I call a dynamic climatic ecological composition of multiplying processes because when all its elements and histories are folded out the tangles of earthly elements such as carbon, fire, water, steam, metals, iron ore, zinc, copper, and timber are made visible.[102] Furthermore, human creativity and intent, mining, the Industrial Revolution, capitalist aspirations, social and economic inequalities, consumerism, atmospheric chemicals such as carbon dioxide, capitalist ideologies, industrial processes, the burning of fossil fuels as a practice, and design feature become material conditions. Through this conjoinment, the Boulton and Watt engines embeddedness in extractive industries, environmental destruction, the abolition of slavery, settler colonialism and First Nations dispossession, and ecospheric problems comes to the fore.

The conceptual reconfiguration of the Boulton and Watt "Lap" Engine – as a climatic ecological composition, as webs of interconnected things, as an alliance of things, people, ideologies, and elemental materials and processes rather than as bounded objects, and the folding out and making visible of the entanglements and articulations of human and non-human agencies embedded in its composition. This figuration makes tangibly visible the unminded damage in which this atmospheric machine is implicated that led to global heating in the first instance.[103] Conversely, this refiguration promotes a sense of attunement to climatic processes in which the situated human is embedded, transforms, and is transformed. It does this by making evident their enmeshment and the registers of sensing and influencing born out of the conjoined material, mineral, chemical, discursive, intentional, energetic, social, human, non-human, natural, and cultural forces that comprise them.[104] At the same time, such figurations bring to light many different distributed and interconnected processes, and their degrees and intensities from radiative forcing and their impact on extinctions, oceans acidity, biodiversity loss and colonial dispossession, and so forth.

The folding out of all these elements, their interactions, and the mapping of these emergent processes illustrate modern populations and their entanglements, shared vulnerabilities, and futures, not only with other humans but with the more-than and other-than-human world.[105] It also promotes a sense of attunement to climatic and environmentally destructive processes by making visible the relationships between all these elements, ourselves, emerging material circumstances and the forms of knowledge produced as part of these collectives. Collectives become distributed agencies and humanness becomes situated, embedded, and enmeshed thereby dislodging the taken for granted humanisms and the dominating thoughts of narcissistic power and exploitation and indeed temporal registers.

Significantly, the Engine, as a climatic ecological composition, acts as a marker of this new condition and also a novel way of thinking and acting. It also alerts curators and audiences to how human, more-than-human, and other-than-human forces operate together in the Boulton and Watt "Lap" Engine configurations as constituted and constituting forms and as domains of influencing within the extended, distributed enmeshments and the eco-curating processes that comprise it.

The emergence of planetary collections and non-human agencies

The Boulton and Watt engines are not just steam machines emitting into the atmosphere. These engines have implications in geological time, extraction, global capitalism, deforestation, large-scale agriculture, and so forth. The "Lap" Engine, for example, embodies the minerals and chemicals from many places – it is an ecological composition of planetary dimensions. Reading the Boulton and Watt "Lap" Engine according to Jussi Parikka's geoforms of media and Manuel deLanda's notion of history as deep time, of geological time, the Engine's story stretches back millions of years.[106] That is, from the formation of rocks, minerals, chemicals, the universe and earthly processes and elements, plant forms, metals that comprise its fabric, of water, vapour, steam, air pressure, of dead plants and animals turning into coal and of atmospheric durations. It comprises in itself stories of repression and resistance, exploitation, the development of steam technology, metallurgy, engineering, and so forth.[107] The Boulton and Watt "Lap" Engine's story extends not only back into the deep past, but its legacy also stretches into the far future – to many generations of planetary life yet to come.[108] Steam machines and other forms of technology comprise different planetary enmeshments as human-non-human histories and designs.

The non-human world, in the orthodox accounts of the Boulton and Watt "Lap" Engine, acts as a background and a resource for the development of society. Haraway invites us to think in thick presences, in terms of our biological relationships with others, and to consider who and how we are enfolded.[109] In the case of the Boulton and Watt "Lap" Engine, the enfolding of human designs with non-human forces and entities is central to its operation. It is more-than-human and a non-human creation. It comprises inhuman earthly elements of fire and water. Watt nurtures and directs these forces through his new steam condenser by harnessing water in its steam phase by coal-fuelled boiling, releasing carbon dioxide embodied in the remnants of non-humans long dead. His humanness is enmeshed and embedded in the Engine. Other non-humans central to the making and operation of the Engine include its timber and metals: deals of wood such as conifer[110] and oak used in the engine's beams for its strength, copper and cast iron for its conducting properties (and the latter for its strength), lead for its non-corrosive properties, and brass (a combination of copper and zinc) for its bearings due to its durability. Watt's Blotting and Calculation Book 1782 to 1783 is preserved in the Boulton and Watt collection at the Birmingham Library. It is a record of Watt's experiments on different types of more-than-human cultured materials simulated in machine contexts to determine how non-human raw materials and their agencies may be useful in designing engines and machinery. Under the guise of scientific methods and management experiments recorded included the bending and breaking tests of bars of cast iron, wrought iron, oak, and deal.[111] All of these non-human elements become cultured by the coppersmiths, ironmasters, blacksmiths, fitters, and turners through turning, boring and drilling, casting, and forging are also embodied in the Engine's design and operation. From 1795 and the establishment of an iron foundry at their Soho Manufactory, Boulton and Watt established the first engineering plant with their own smelting furnaces[112] thereby culturing their own metals on site.

Within the interpretive material accompanying the "Lap" engine in the Science Museum's Energy Hall, references are made to non-human agency as physical principles. These human–non-human entanglements are evident in the exhibition through working models and engines, animations, and interactive screens such as for example the demonstration of the physics of combustion and the expansive properties of steam and steam pressure as a more-than-human agency. The capacity of water and steam embodied within the design of the Engine was a central force and cultured innovation that further enabled capitalist expansion. Through the use of new materials and modified design features across the life of these engines, multifarious non-human agencies are harnessed in service of capitalist accumulation directed toward making the engines more efficient, productive, and powerful.

The blueprint of the engines and their parts changed over time and were presented as improvements directed to making them more efficient, productive, able to produce more power through the use of new materials and design features. These changes harnessed multifarious non-human agencies in service of capitalist accumulation.

The Boulton and Watt "Lap" Engine as a climatic ecological composition

The Engine is a "climatic ecological composition" in the sense that it is composed of the interrelations between metals belonging to the geological substrates of the planet and their extraction, the fossil-fuel combustion integral to its mechanic function, the rise of capitalist industrial production, the exploitation of land and labour, colonial expansion, Indigenous dispossession, and air pollution. But the changes to the Engine also reveal not only a story of the entanglements of human and non-human histories central to the functioning of the Engine but also an emerging climate history embodied in its wooden components. This is a result of its fossil-fuel combustion and its correlation with increases in carbon dioxide concentrations in the atmosphere. Its mechanic and combusting functions also generated energy and thus contributed to these concentrations. Significantly these engines' wooden componentry holds a baseline of rising CO_2 levels from the start of the Industrial Revolution.

The social activation of the Boulton and Watt "Lap" Engine as a "climatic ecological composition" can be done through in this case the analytical cutting into its wooden composition and contexts measuring the ^{12}C and ^{13}C ratios and examining these ratios' depletions using simulated charts. These ratios were calibrated by Ellsworth over time against the photosynthetic sequences of oak and eucalypt timbers and are dependent on the species of wood used in its construction, through adjustments made to the Engine during its life, and the timbered fabric of its historical contexts. Furthermore, such carbon ratios, as a register of biochemical change, can be linked to key moments in the history of these engines; the date in which engines were commissioned (with some degree of variation depending when the oak tree was living) and the beginning of the Industrial Revolution/Anthropocene.

In thinking of climate change as a specific milieu in which we are subject, Colebrook explains we lose sight of the degree that human life is deeply implicated in the climate's timeframes and rhythms.[113] The conventional view of modern populations in climate change processes and as a lever for activating mitigation behaviour, for example, is premised on the idea of human populations as radiative forcing agents whose carbon-burning practices arise as emissions and these accumulating greenhouse gases in the atmosphere cause heating by absorbing terrestrial infrared radiation.[114] As a result and scientifically, modernity's populations are configured as radiative forcing agents through the chemical forcing of these greenhouse gases – carbon dioxide primarily as well as methane, CFCs, ozone and N_2O – leading to heating. But as climate-forcing agents, modern populations are positioned outside the climate – it is the chemicals released in emissions that intermingle with other particles as free gases and force heating. Mitigation therefore is framed around a greenhouse gas-centric logic – the chemical constitution of the atmosphere and how it might be stabilised to maintain temperatures through strong human agency by changing our energy sources and therefore emissions while continuing on the trajectory of capital accumulation. Conventionally nature-based solutions especially those that deploy plant agency seeks to exploit their agencies to absorb emissions to achieve these ends.

Wood in the form of wooden objects and their wooden components offers another more compelling insight into human agency in climate change. Drawing on Nicolas Shapiro and Eben Kirksey's invitation to think chemically as a "chemical species,"[115] Through ^{13}C simulations and more specifically through material analysis, human agency is transformed from a consideration of modern populations as separate from the non-human world except for the intermingling of emissions and their chemical signals to one of deep enmeshment in which dynamic biochemical life lifeforms become a register of atmospheric changes bonded into wood, and thus the fuel used to drive photosynthetic processes. Therefore fossil-fuelled intimacies and lifestyles are bonded into wood. These same populations are not merely forcing agents in which the chemicals of carbon-burning promote heating effects but are more deeply enmeshed as material-conjoined fossil-fuelled biochemical life forms in dynamic planetary movement. A repertoire of embodied greenhouse gas intimacies of life itself is recorded in the atmospheric register of plants, as the timeframes and rhythms of plant–human life through photosynthesis.

Here I draw on historian Bob Johnson's notion of fossil-fuel intimacy. He argues that our erotic attachments to fossil fuels are profoundly deep and extensive, "embodied … and rooted in minutia, in the intimate quotidian rituals of the home, workplace, streets and stores."[116] As a result, modern populations are embedded and embodied in an infrastructure of mineral energy flows that assert themselves in and on the flesh, the psyche, and the horizon of life itself. When the Goodwyn engine was first steamed in August 1784, fossil-fuel combustion began what Johnson calls "a trajectory of bodies' immersion and saturation in the materiality and praxis of modernity's rituals, conditions of life under the rule of fossil capitalism."[117] The Boulton and Watt engines, as embodiments of these aspirations, and as a transfer point for the expansion of capitalism, led to the ignition of new modern fossil-fuelled biochemical life forms, ones in which combustion became a defining element of capital and its unintended consequences. At the moment, Goodwyn's engine was first fired up those that came to admire it were the unwitting witnesses to the birth and intensification of more-than-human co-constituted climates. The Boulton and Watt engines complicity in fossil capitalism, through fossil-fuel combustion and its emerging plural modernities, energetically powered standards of living and capital growth.[118]

This refiguration of western and non-western modernities, and their populations, in this mode is not directed to promoting a message of command, control, and climate stabilisation. Furthermore, individual actions of mitigation, through the modification of carbon-burning practices, are not the only figuration of human–non-human relations directed to promoting mitigation. Using this biochemical metaphor, I gesture towards unequivocal complicity with climate processes and its affects – atomic, molecular, particulate mattering, dust, dirt, soot, smoke, and toxicity in multiple processes of co-creation.[119] This figuration is made tangible not just as calculations of CO_2 as parts per million (ppm) in the atmosphere and associated registers of heating, but also as human–non-human atmospheres in plants as mutual worldings of biochemical change, registered through photosynthesis and relative material incorporations and ex-corporations of ^{12}C and ^{13}C.

Returning to the scientific ^{13}C simulations proposed by Ellsworth, this analysis using heritage collections acts as a pivot and lever shifting an understanding of ourselves from one of dis-embeddedness to one of deep enmeshment within climate processes. New mitigation strategies have the potential to arise that foreground how to live in the world differently with others including the non-human world. Trees are no longer service providers but stakeholders in new strategies.

Human–non-human fossil-fuelled biochemical climates

The Boulton and Watt "Lap" Engine, through its wooden components and through the events and contexts in which the Engine was implicated or became connected through its dispersal, registers a trajectory of emerging human–non-human climates from the start of the Industrial Revolution. Significant moments in the Boulton and Watt history of combustion, to which it contributed to rising carbon dioxide levels registered as parts per million in the air, also register the trajectory of the emergence of fossil-fuelled combusting biochemical lifeforms in dynamic movement within each of the engine's construction[120] Different types of ecological compositional cuts can be made into and across its material composition and its registers of human–non-human history and deep time, but what is interesting here is making visible modernities' combustible traces that historically distinguish humans as biochemical species and mapping their materialisms in bio-social realms. The following simulated data details analytical cuts into the Boulton and Watt rotative beam engines as human–non-human biochemical climates, or atmospheres, and their worlding with wood in which depleted ^{13}C values as parts per thousand (permille) are markers of human agency. These analytical cuts represent data gathered from live trees from which simulated values over the course of more than 220 years were mathematically calculated. Significantly, the Engine registers the multiple temporalities of the fossil-fuelled biochemical lifeforms in the photosynthetic, in the atmospheric, and through its chemical combustion.

The oak through which the engines are connected through their plant history is a human–non-human wooding bond borne out of the carbon dioxide gathered and stored to provide energy for the plant's growth.[121] This photosynthetic process produces a register of the atmosphere's composition in oak tree tissue which carbon analysis makes legible. Multiple registers of these biochemical life forms appear in different kinds and quantities in data. Data registers the beginning of combusted fossil-fuelled biochemical lifeforms, and their folding into the atmosphere as a free gas, as concentrations of carbon dioxide per million molecules of air and then into plant biological and chemical bonds: first, in oak and as $\delta^{12}C$ values and declining carbon $\delta^{13}C$ values as molecules per million; and, then, in Boulton and Watt engines during their construction.[122] Each ratio is an analytical cut into the photosynthetic wood bonding≈formulated mathematically from live tree samples and historical simulations as human–non-human climate changes.

In a letter of April 27, 1781 John Iddins, a timber merchant in New Street, Birmingham offered Matthew Boulton several oak trees suitable for making beams.[123] Until 1798 all rotative engines' framing was made of oak, including the working beam, its support, and connecting rod. The cylinder was also mounted on an oak frame. At first, these beams were plain oak logs, then beams built from several logs were introduced, but with the advent of the double-acting engine the repeated reversal of stress to which the beam was subjected made it peculiarly liable to failure.[124]

Ellsworth's chart was developed from oak trees growing in Staffordshire in the West Midlands, the heart of the Industrial Revolution around 38 miles from Boulton and Wat Soho Manufactory in Birmingham. The oak used in the construction of Boulton and Watt's engines was likely grown and sourced locally. Therefore, these simulated charts likely represent a register of carbon emissions in this location from the start of the Industrial Revolution and the scaling up of fossil-fuel combustion through industrial development over more than 230 years.

In Figure 3.3, Ellsworth demonstrates the simulated trend in the depletion of $\delta^{13}C$ values in oak wood. The decline in ^{13}C in the atmosphere is consistent with the ^{13}C composition of fossil fuels. Burning plant or fossilised plant material, such as fossil fuels, releases into the atmosphere the ^{12}C isotopes that plants absorbed through photosynthesis millions of years ago in the form of CO_2 molecules. The added ^{12}C isotopes in the atmosphere decrease the overall $^{13}C/^{12}C$ ratio.

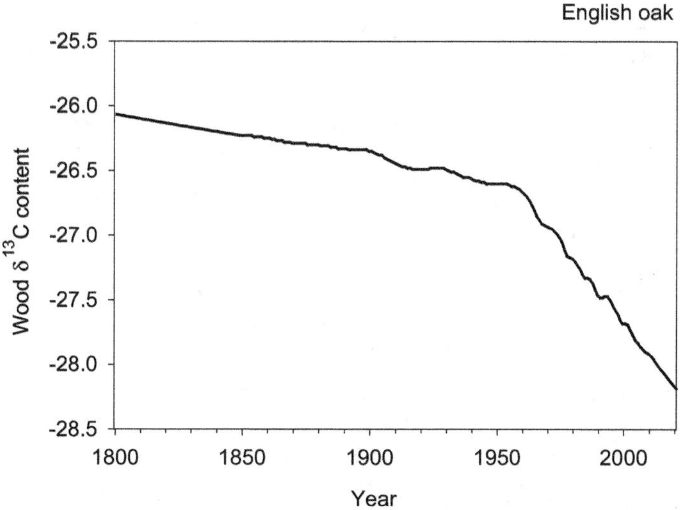

FIGURE 3.3 Simulated ^{13}C depletion in English oak from 1800 © David Ellsworth.

In James Watt's letter to Samuel Whitbread on December 7th, 1784, the process of engine construction and the materials to be used were explained. The Engine's timbers were ordered in December 1784. The Engine's working beam was made of English oak, because of its properties as durable, straight grained, and seasoned. The cape piece of the frame, which carries one end of the shaft of the flywheel, and the corner of the shaft of the flywheel were also made of English oak. The main uprights, the spring beams, and cross beams were made from fir from Danzig (present-day Gdansk, Poland) as were the side planks, cistern ends, and the spring beams. The plug frame and plug tree were made of wainscot oak.[125]

In 1784, the year that the first rotative engine was commissioned and steamed in Goodwyn's Brewery the expected $\delta^{13}C$ value was –24.32 permille in its oak structure, beam, and its wooden teeth driving the gear. At the time, there were no instruments measuring atmospheric carbon dioxide concentrations. Calculations back-estimated from the wood suggest this value to be indicative of CO_2 levels of about 293 ppm at the start of the Industrial Revolution and the construction of the first rotative beam engine commissioned by Goodwyn.

Likewise, the existing, largely unaltered "Lap" Engine built in 1788 support structure was made from English oak, held together with wrought iron straps and bolts. The flywheel was 16 feet in diameter and was fitted with 304 wooden gear teeth also made from oak and used to drive a counter-shaft which drove each individual lapping and polishing machine. When the "Lap" Engine was commissioned and put into service in 1788, the expected $\delta^{13}C$ value in its oak components was likely to have been –24.34 permille.

In 1795, the beams in engines were strengthened with oak, and $\delta^{13}C$ values of oaks in the Midlands, from which the wood was likely sourced are expected to have registered as –24.35 permille, demonstrating a slight decline in its^{13}C register signaling the first indication of the emergence of fossil-fuelled biochemical life forms and human–non-human atmospheres. By 1800, Boulton and Watt had built 451 engines in which 268 were rotative like the Lap Engine although there were also by that time many other manufacturers in the market.[126] Between 1785 and 1799, 21 engines were ordered by foreign buyers and by 1825, 86 engines.[127] The largest markets for Boulton and Watt rotative engines were the Netherlands, Spain, Germany, France, and India and as far afield as the USA, Canada, Brazil, and Russia.[128] Oak beams were used in all engines from 1786 to 1798. From 1798 cast iron was used more frequently in construction: the working beam, the connecting rod, the column under the beam, and the gearing due to its superior strength.[129] Some oak was still used in their construction. Nineteen years later in 1814, condensers were lined with oak[130] and at that time likely registered –24.42 permille of ^{13}C relative to the fossil standard.

In 1858 when the Lap Engine was decommissioned and acquired by the Science Museum – the oak trees near the Soho Manufactory would have likely registered $\delta^{13}C$ values of –24.65 permille, demonstrating a greater decline in its ^{13}C register and a likely 0.1 increase in global surface temperature.[131] In 1936 and the opening of the Watt Hall at the Science Museum $\delta^{13}C$ values were likely to be –24.70 permille and a 0.2 increase in global surface temperature.[132] This value therefore registered an intensification of emissions and of fossil-fuelled lifeforms in the photosynthetic register.

According to Ellsworth's calculations, the decline in $\delta^{13}C$ values in oaks from this location shows a sharp decline after 1950 with $\delta^{13}C$ of oak wood values of –24.75 permille further declining by 2023 to –28.9 permille with a reading of atmospheric CO_2 on May 2023 of 420ppm[133] with a likely global increase in temperature of around 1.4 degree Celsius surrounding the now archaeological remains of the manufactory and the housing estate that exists on the site. The "Lap" Engine remains a time capsule of biochemical atmospheric relations embodied in the oak trees that grew in the Midlands at the very beginning of the emergence of photosynthetic fossil-fuelled atmospheres.

In 1788, the year the "Lap" Engine was built and put into service, the First Fleet and its human cargo of convicts arrived in New South Wales, the first of 80,000 transportees drawn from Britain's underclass, Many of these transportees worked in the factories powered by Watt's steam machines. The likely $\delta^{13}C$ values in eucalyptus in 1788 was most likely –25.90 permille on First Nations Darug Country, now the area of Western Sydney close to Western Sydney University, Hawkesbury Campus.

The year the Whitbread engine arrived in Sydney in 1888 from Birmingham later to become part of the Powerhouse, Sydney collection coincided with

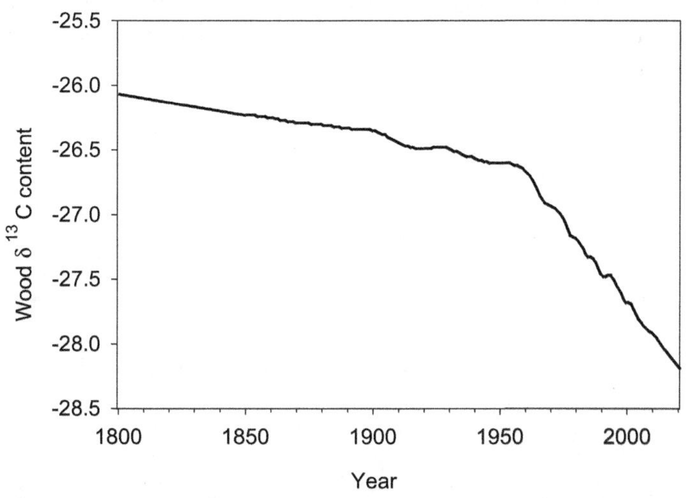

FIGURE 3.4 Simulated ^{13}C depletion in Australian eucalypts from 1800 © David Ellsworth.

the one-hundredth anniversary of the European invasion of the Australian continent. By that year, 1888, global surface temperatures had already risen by 0.15 degrees Celsius.[134] Eucalyptus trees would have likely registered –26.40. By 1988, Australia's bicentennial year registered $\delta^{13}C$ at –27.31 permille from the simulation. This translated into a CO_2 concentration of 360 ppm[135] and a global surface temperature increase of 0.7 degrees Celsius.[136] The $\delta^{13}C$ of –28.9 permille embedded in eucalypt timber and corresponding oak timber on the Soho Manufactory site equated to CO_2 concentrations of 420ppm in May 2023;[137] and a projected 1.4 degrees Celsius increase in global surface temperature.[138] As curatorial agencies, the oaks and eucalypts, whose timber contributed to the fabrication of the Engine and its historical contexts, register the emergence of fossil-fuel biochemical life forms and their combustive activities etched into trees in whcih their tissue remains as a testimony.

Over the last million years, and the intervals between ice ages and warm periods, atmospheric carbon dioxide never rose more than 300 parts per million. At the end of the last ice age, around 20,000 years ago, it was 280 ppm and today this figure has risen to around 420ppm in 2023. In the interval when the Boulton and Watt Lap Engine was commissioned in 1788 and the present day, the rise of atmospheric carbon dioxide, as our figures show, is 70% greater than when the Earth emerged from the last ice age and is increasing at a rate 100–200 times faster.[139]

Conclusion: Eco-curating human–non-human climates and cultural carbon mitigation strategies

Cultural collections as climatic ecological compositions, in respect to both the bucket and the Boulton and Watt Lap Engine, have their beginnings in the deep past and geological time and are woven into the multiple temporalities and shared histories of the human and non-human. They embody the processes and consequences of colonial expansion, Indigenous dispossession, social, gender and economic inequalities, extraction and capitalism, the destruction of culture, and environmental violence. These emerging material circumstances in which the bucket and the Engine are complicit have always been and always will be more-than-human and at the same time deeply non-human.

Both these collection items can promote ethical forms of conjoint action that acknowledge our entanglements, shared vulnerabilities, and futures, including extreme weather events, with not only other humans but with the more-than-human world. By disassembling and refiguring each as climatic ecological compositions, for example, boundaries between humans and non-humans are removed. Cultural collections become deeply connected to land, air, plants, trees, insects animals, soil, earth, elements, chemicals, and metals. They also become embedded in an expanded cosmos and multiple temporal and enfolding processes and scales, including planetary conditions requiring recuperation. The emergence of biochemical fossil-fuelled life forms not only become enmeshed in the technologies and everyday objects across cultural divides as transversal embodiments but are also integrated into museum's fabrics, exhibitions, and building structures. This material refiguration of deep enmeshment between humans and non-humans opens up a space to consider how we might curate and live with the non-human world in respectful ways that operationalise their potential to act as novel cultural carbon mitigation strategies. That is not solely in terms as a measure of one's carbon footprint as a narrow form of greenhouse gas thinking but rather opens up a suite of practices that goes to the heart of how to live differently in the world.

Cultural collections hold other multiple biochemical and material human–non-human bonds. These result from carbon-intensive cultures for example the minerals and metals extracted from the earth through mining. The toxic residue they contain has directly or indirectly contributed to climate change, environmental destruction, death, and even extinction. Furthermore, these objects comprise multiple humanisms that are at once discursive in the way they are described and the different perspectives that are brought to bear on them, but also the ways they are rendered material in respect to the raw materials that are sourced to make them and how they are crafted and the intent and thinking behind them.

This figuration of collections, as ecological compositions and cultural carbon mitigation strategies, invokes Haraway's awareness of the possibilities for thinking in different forms and in mobilizing an ethical commitment to "thinking with" –as opposed to the myth of the hubris human focused on "think for himself," or indeed in the case of the Boulton and Watt "Lap" Engine creating and enacting a world of monetary excess for himself.

The Boulton and Watt "Lap" Engine's history, for example, extends from the past into the present and future as legacies of combustion and fossil capital genealogy. The consumption of its aura, its tendency to affect as a destructive force, and to resonate with the notion of human agency as geological are just three examples of its interpretive possibilities. It is also combusting and worlding as future climates and warmings in which our forbearers, and more recently we, are biochemical lifeforms registered through ^{13}C data in its composition as timbered human natures that can be made visible through sensing and combusting samples to release carbon isotopes.

As ecological compositions, both the bucket and the Engine are thick with human–non-human climates and diverse temporalities. This includes the minerals of extractive industrial capitalism and the legacy of fire and plant curation and its register of human–non-human atmospheric relations. In respect to the Engine, fossil-fuel biochemical lifeforms are energetic and equate with correlations between an increase in CO_2 concentrations and the rise in temperature from 1784.

Multiple fossil-fuel biochemical lifeforms and plant histories are embedded in both the bucket and the Boulton and Watt "Lap" Engine. As producers, registers, and sensors, plant curatorial agency in the photosynthetic as timbered human–non-human climates register the emergence of fossil-fuelled biochemical lifeforms through the Engine and its originating contexts, the oak trees of the Midlands from which itself and many others were made. For the bucket, plant, curatorial agencies and human–non-human climates are enmeshed with combustion and fire, and etched into its form. Here the material evidence of the emergence of new humanisms becomes visible in which modern populations can no longer be conceptualised as dis-embedded sovereign forcing agents exercising their own agency independently of others but rather decisively put back in their ecological place as dynamic embodied fossil-fuelled biochemical life forms, made tangible through ^{13}C analysis, through measurements of rising CO_2 levels (carbon dioxide molecules per million molecules of air emissions, or ppm) and a trajectory of heating.

Carbon-burning populations and industries are folded into the chemical composition of the atmosphere and declining ^{13}C levels, as well as the changes in the carbon biochemical composition of plant tissue. That is, the many temporalities of the photosynthetic, of molecular bonding of changing weather, and of global heating registered as ^{13}C depletion. The temporalities of the photosynthetic and biochemical change converge not as a representational logic but one of chemical bonding.

Trees, such as oaks and eucalypts, as curatorial agencies, draw modernity's populations in as fossil-fuel biochemical lifeforms burning activities into their histories enmeshed in their tissue integrated into a museum's fabrics in its exhibition objects and building structures. These biochemical fossil-fuelled life forms not only become embedded in technologies and everyday objects, but their emergence also provokes new ways of thinking with the non-human world which opens up heritage collections potentiality to serve as novel cultural carbon mitigation strategies.

Bill Putt's green plastic bucket and the Engine are reflected back on modern populations not only as objects of the unfolding climate crisis and our complicity through our addiction to fossil-fuel intimacy, but also through our bodies which carry a genealogical record of toxic industrial processed hydrocarbons as embodied atmospheric nature cultures within our DNA. As porous bodies, we carry a record of the ecological crisis as another enfolded and enfolding historical record. As media theorist and artist Joanna Zylinska explains, this also includes the enfolding of shards of coal, oil, microplastics, embedded nuclear waste, and so forth.[140] The coordinates and interrelatedness of each of these object's thingness are refigured and folded out to tell embodied stories of partial ecological recuperation and then of and through an understanding of entanglement and climate justice.

Climate change is no longer abstract. Heritage professionals may no longer lament on how to curate cultural collections for climate change because they lack tangible evidence of atmospheric change – these changes are written into the very organic fabric of collection items and the contexts in which they are located. While climate change, as material phenomena, is often presented as charts and figures, in the case of the Engine data it is one of material tracing and sensing. With the bucket, climate change is inscribed into its mutating form and changed composition and its enmeshed histories of human–non-human relations of combustion, extraction, and crafting. The bucket and its contexts registers^{13}C data through the human–non-human photosynthetic atmospheres etched into the trees, buildings, plants, and soils on Bill's property and indeed his own carbon-burning activities that combusted again, and were released into the air that fuelled the wildfire and transformed the bucket. At the time of the Black Saturday bushfires in 2009, the eucalyptus trees on the Putt's property and transformed as wildfire would have likely registered δ^{13}C at –28.80 permille from the simulation and 1 degrees Celsius rise in temperature.[141] The petrochemicals in the fabrics of the bucket and used to fuel its manufacture transform again as wildfire etches its affects into its materials. Fossil-fuelled biochemical lifeforms therefore become enmeshed and inscribed into material culture in multifarious ways, for example, through the bucket's manufacturing process, the extractive carbon-intensive processes through its materials were extracted and to the wildfire that rained down on the bucket and Bill as a register of the biochemical changes inscribed into the flames and its altered, volatile conjoined relations as future climates and impacts.

In provoking new attunements to our atmospheric entanglements, the refiguration of the Engine and bucket opens a new moment in the history of these objects and their potential not just as cultural carbon emitters but also as "cultural carbon mitigators." These collections trigger dispositions of *thinking with* as ecological thoughts, as expressions of radical co-existence, and as new forms of material evidence in storytelling. But they are more than objects that tell stories – they are dense material, semiotic, elemental technical compositions always composing, distributing, and enfolding multiple, entangled temporalities and processes of human and non-human times. The human-ecological relations exposed in their material analysis afford opportunities for engaging broad publics as a lever for invoking what posthuman education theorist Karen Malone calls speculative imaginaries of postcarbon worlds.[142]

Notes

1 Science Museum Collection, accessed February 24, 2023, https://collection.sciencemuseumgroup.org.uk/objects/co50948/rotative-steam-engine-by-boulton-and-watt-1788-beam-engine-steam-engine.

2 British Ecological Society, "Special Feature on Levering Natural History Collections to Understand the Impacts of Climate Change," accessed February 24, 2023, https://www.britishecologicalsociety.org/call-for-proposals-special-feature-on-leveraging-natural-history-collections-to-understand-the-impacts-of-global-change/.

3 Johnson et al., "Climate Change and Biosphere Response: Unlocking the Collections Vault," *BioScience* 61, no. 2 (2011): 147–53. https://doi.org/10.1525/bio.2011.61.2.10.
4 Andrew Urevig, "Seeking Answers on Climate Change, Scientists Venture into the Vaults of the Past," February 24, 2023, https://ensia.com/features/natural-history-museums-climate-change/.
5 The Viking Ship House, Cultural History Museum, Oslo, February 24, 2023, https://www.khm.uio.no/besok-oss/vikingskipshuset/.
6 Fiona R. Cameron, "We Are on Nature's Side? Experimental Work in Rewriting Narratives of Climate Change for Museum Exhibitions," in *Climate Change and Museum Futures*, eds. Fiona R. Cameron and Brett Neilson, Routledge Research in Museum Studies (New York: Routledge, 2014), 62–64.
7 Cameron, "We are on Nature's Side?" 62.
8 Fiona R. Cameron, Ben Dibley and David Ellsworth, "Climate Collections, and Photosynthetic Fossil-Fuelled Atmospheres," *Environmental Humanities* 15, no. 2. (July 2023): 62–84.
9 National Museum of Scotland Collection, accessed February 24, 2023, https://www.nms.ac.uk/explore-our-collections/stories/science-and-technology/boulton-and-watt-engine/.
10 Deutsches Museum Collection, accessed February 24, 2023, https://www.deutsches-museum.de/en/collections/machines/power-engines/steam-engines/early-steam-engines/trestle-steam-engine/.
11 Fiona R. Cameron, "Posthuman Museum Practices," in *Posthuman Glossary*, eds. Rosi Braidotti and Maria Hlavajova (London: Bloomsbury, 2018), 350; Fiona R. Cameron, Ecologizing Experimentations: A Manifesto for a Posthumanist Museum," in *Climate Change and Museum Futures*, eds. Fiona Cameron and Brett Neilson (New York: Routledge, 2014), 18–21.
Cameron, "Ecologizing Experimentations: A Manifesto," 18–21.
12 Cameron et al., "Climate Collections," 62–84.
13 Cameron et al., "Climate Collections," 63.
14 Cameron et al., "Climate Collections," 63.
15 Cameron, "Posthuman Museum Practices," 352; Cameron, "Ecologizing Experimentations: A Manifesto," 16–17.
16 Cameron, "Posthuman Museum Practices," 351–52.; Cameron, "We Are on Nature's Side?" 72–74.
17 Lisa Dale-Hallett, Rebecca Carland and Peg Fraser, "Sites of Trauma: Contemporary Collecting and Natural Disaster," in *Museum Theory*, eds. Andrea Witcomb and Kylie Message (London: Wiley, 2015), 531–52, 533.
18 Cameron, "Posthuman Museum Practices," 349–52.
19 Cameron, "Posthuman Museum Practices," 349–50.
20 Cameron, "Posthuman Museum Practices," 349.
21 Cameron, "Posthuman Museum Practices," 349.
22 Dale-Hallett et al., "Sites of Trauma," 535.
23 Dale-Hallett et al., "Sites of Trauma," 537.
24 Dale-Hallett et al., "Sites of Trauma," 536.
25 Dale-Hallett et al., "Sites of Trauma," 546.
26 Dale-Hallett et al., "Sites of Trauma," 547.
27 Dale-Hallett et al., "Sites of Trauma," 550.
28 Cameron, "Posthuman Museum Practices," 349–52.
29 Cameron, "Posthuman Museum Practices," 349–52.; Cameron, "Ecologizing Experimentations: A Manifesto," 24–28.
30 Cameron, "Posthuman Museum Practices," 350–52; Fiona R. Cameron, *The Future of Digital Data, Heritage and Curation in a More-Than-Human World* (Abington: Routledge, 2021), 143–47.

31 Cameron, "Posthuman Museum Practices," 349–52.
32 Rosi Braidotti, *The Posthuman* (Cambridge: Polity, 2013).
33 Ian Hodder, *Entangled: An Archaeology of the Relationships between Humans and Things* (Malden: Wiley-Blackwell, 2012).
34 Jane Bennett, *Vibrant Matter: A Political Ecology of Things* (Durham, NC: Duke University, 2010).
35 Cameron, *The Future of Digital Data*; Cameron, "Posthuman Museum Practices," 349–52; Cameron, "Ecologizing Experimentations: A Manifesto," 16–33.
36 Cameron, "Posthuman Museum Practices," 351–52.
37 Peter J.T. Morris, *Science for the Nation*, (London: Palgrave Macmillan, 2020), 1–10.
38 Science Museum, London Collection, accessed February 24, 2023, https://collection.sciencemuseumgroup.org.uk/objects/co50900/newcomen-atmospheric-engine-atmospheric-steam-engine.
39 Science Museum, London collection, accessed February 24, 2023, https://collection.sciencemuseumgroup.org.uk/objects/co8084947/stephensons-rocket-steam-locomotive.
40 Science Museum, London collection, February 24, 2023, https://www.sciencemuseum.org.uk/objects-and-stories/everyday-wonders/electric-lighting-home.
41 Cameron et al., "Climate Collections," 64.
42 Cameron et al., "Climate Collections," 64.
43 The Australian Research Council Linkage project, "Curating Museum Collections for Climate Change Mitigation" (2022–2024) is led by Chief Investigator Associate Professor Fiona Cameron with Chief Investigators Professor David Ellsworth, Professor Karen Malone (Swinburne University of Technology), Research Fellow Eloise Florence, and Partner Investigators, Dr Deborah Lawler-Dormer (Powerhouse) and Distinguished Professor Rosi Braidotti (New Institute, Rotterdam). Western Sydney University is the administering organisation and the Powerhouse is the lead partner organisation.
44 Cameron et al., "Climate Collections," 68.
45 Cameron, "Posthuman Museum Practices," 350–52.
46 Cameron, "Ecologizing Experimentations: A Manifesto," 16–33.; Cameron and Neilson, *Climate Change and Museum Futures*; Cameron, "Posthuman Museum Practices," 349–52.
47 Ben Dibley, "Prospects for a Common World: Museums, Climate Change, Cosmopolitics," in *Climate Change and Museum Futures*, eds. Fiona R. Cameron and Brett Neilson (New York: Routledge, 2014), 58–84.
48 Helen Afvidsson and Ann Follin, "Connectedness, Consumption and Climate Change: The Exhibition Human Nature," *Museum Management and Curatorship* 35, no. 6 (2020): 684–96.
49 Cameron et al., "Climate Collections," 68.
50 Claire Colebrook, "Not Symbiosis, Not Now: Why Anthropogenic Change Is Not Really Human," *Oxford Literary Review* 34, no. 2 (2012): 185–209.
51 Marek Tamm and Zoltán Boldizsár Simon, "Historical Thinking and the Human: Introduction," *Journal of the Philosophy of History* 14, no. 3 (2020): 285–309, 285.
52 Cameron et al., "Climate Collections," 68.
53 Cameron et al., "Climate Collections," 76.
54 Stacey Alaimo, "Transcorporeal," in *Posthuman Glossary*, eds. Rosi Braidotti and Maria Hlavajova (London: Bloomsbury, 2018), 435–37, 436; Astrida Neimanis and Rachel Loewen Walker. "*Weathering*: Climate Change and the 'Thick Time' of Transcorporeality." *Hypatia* 29, no. 3 (2014) Special Issue: Climate Change: 558–75.
55 Cameron, "Ecologizing Experimentations: A Manifesto," 16–33.; Isabelle Stengers, *In Catastrophic Times: Resisting the Coming Barbarism*, trans. Andrew Goffey (Open Humanities Press, 2015).

56 Cameron et al., "Climate Collections," 69.
57 Donna Haraway, *Staying with the Trouble: Making Kin in the Chthulucene* (Durham, NC: Duke University Press, 2016), 97.
58 Cameron et al., "Climate Collections," 66.
59 Cameron et al., "Climate Collections," 66.
60 Cameron, "Posthuman Museum Practices," 351–52.; Cameron, "Ecologizing Experimentations: A Manifesto"; Cameron and Neilson, *Climate Change.*
61 Cameron, "Posthuman Museum Practices," 351–52.
62 Cameron, "Ecologizing Experimentations: A Manifesto," 16–33.; Fiona R. Cameron, "Theorising More-Than-Human Collectives for Climate Change Action in Museums," in *L'Internationale Online Climate Change Special Issue*, 2015, accessed February 24, 2023, http://www.internationaleonline.org/research/politics_of_life_and_death/48_theorising_more_than_human_collectives_for_climate_change_action_in_museums; Cameron, "Posthuman Museum Practices," 351–52.
63 Cameron et al., "Climate Collections," 67.
64 Nicolas Shapiro and Eben Kirksey, "Chemo-Ethnography: An Introduction," *Cultural Anthropology* 32, no. 4 (2017): 481–93.
65 Cameron et al., "Climate Collections," 68.
66 Cameron et al., "Climate Collections," 68.
67 Gordon Bonan, "Forests and Climate Change: Forcings, Feedbacks, and the Climate Benefits of Forests," *Science* 320, no. 5882 (2008): 1444–49.
68 Cameron et al., "Climate Collections," 64.
69 National Academies of Sciences, Engineering, and Medicine, "Surface Temperatures for the Last 200 Years," 2006. Washington, DC: The National Academies Press. https://doi.org/10.17226/11676. https://nap.nationalacademies.org/read/11676/chapter/7#46.
70 Cornwell et al., "Climate and Soils Together Regulate Photosynthetic Carbon Isotope Discrimination within C3 Plants Worldwide," *Global Ecology and Biogeography* 27 (2018): 1056–67.
71 Cameron et al., "Climate Collections," 68.
72 Johnsen et al., "Meeting Global Policy Commitments: Carbon Sequestration and Southern Pine Forests," 14–21; Cornwell et al., "Climate and Soils Together."; Cornwell et al., "Climate and Soils Together," 1056–67.; Stephen Leavitt and Austin Long, "An Atmospheric $^{13}C/^{12}C$ Reconstruction Generated through Removal of Climate Effects from Tree-Ring $^{13}C/^{12}C$ Measurements," *Tellus B: Chemical and Physical Meteorology* 35, no. 2 (1983): 92–102, Cernusak et al., "Within Plant Variation in Stable Carbon Isotopes: Why Are Non-photosynthetic Tissues Generally ^{13}C Enriched Compared to Leaves?" *Functional Plant Biology* 36 (2009): 199–213; Cameron et al., "Climate Collections."
73 Marion H. O'Leary, "Carbon Isotopes in Photosynthesis: Fractionation Techniques May Reveal New Aspects of Carbon Dynamics in Plants," *BioScience* 38, no. 5 (1988): 328–36.
74 David S. Ellsworth, "CO_2 Enrichment in a Maturing Pine Forest," *Plant, Cell and Environment* 22 (1999): 461–72.
75 Ellsworth, "CO_2 Enrichment in a Maturing Pine Forest," 461–72.
76 Cameron et al ., "Climate Collections," 62-84.
77 Henry Winram Dickinson, *James Watt Craftsman and Engineer* (Cambridge University Press, 1936), 141.
78 Robin McKie, "James Watt and the Sabbath Stroll That Created the Industrial Revolution," *The Guardian*, May 30, 2015, accessed February 24, 2023, https://www.theguardian.com/technology/2015/may/29/james-watt-sabbath-day-fossil-fuel-revolution-condenser.

79 Paul Crutzen and Eugene Stoermer, "The Anthropocene," *Global Change Newsletter* 41 (May 2000): 17–18.
80 Cameron et al., "Climate Collections," 71.
81 Cameron et al., "Climate Collections," 71.
82 Cameron et al., "Climate Collections," 71.
83 Birmingham Library and Archives, MS 3147 - Boulton and Watt Collection, Letter from James Watt to Henry Goodwyn April 17th, 1784.
84 Birmingham Library and Archives, MS 3147 - Boulton and Watt Collection, Letter from Henry Goodwyn to Boulton and Watt, August 9, 1784.
85 Birmingham Library and Archives, MS 3147 - Boulton and Watt Collection, Letter from Samuel Whitbread to James Watt August 26, 1784.
86 Birmingham Library and Archives, MS 3147 - Boulton and Watt Collection, Letter from James Watt to Samuel Whitbread, Birmingham, December 7, 1784.
87 Cameron et al., "Climate Collections," 72.
88 Eric Robinson, "The International Exchange of Men and Machines, 1750–1800," *Business History* 1 (1958): 3–15.
89 Watt Bicentenary Marked in London: Father of Steam Engine Is Honoured. *New York Times*, January 15, 1936, 17.
90 Science Museum, Energy Hall, https://www.sciencemuseum.org.uk/see-and-do/energy-hall.
91 Science Museum, Energy Hall, https://www.sciencemuseum.org.uk/see-and-do/energy-hall.
92 Gilbert Simondon and Arne De Boever, "Technical Mentality," in *Gilbert Simondon: Being and Technology*, eds. Arne De Boever, A Murry and J Roffe (Edinburgh: Edinburgh University Press, 2012), 1–16, 3.
93 Simondon and De Boever, "Technical Mentality," 1–16, 16.
94 Cameron, *The Future of Digital Data*, 100.
95 Science Museum, London collection, https://collection.sciencemuseumgroup.org.uk/objects/co50948/rotative-steam-engine-by-boulton-and-watt-1788-beam-engine-steam-engine.
96 Rosi Braidotti, *Posthuman Knowledge* (Cambridge: Polity, 2019), 9–10.
97 Vernon Satchell, "Early Use of Steam Power in the Jamaican Sugar Industry, 1768–1810," *Transactions of the Newcomen Society* 67, no. 1 (1995): 221–31.
98 Satchell, "Early Use of Steam Power," 221–31.; Vernon Satchell, "The Diffusion of the Watt Steam Engine in the Jamaican Slave/Sugar Economy 1810–1830," Paper presented at the *29th Annual Conference of the Association of Caribbean Historians*, April 1997, Martinique, Université des Antilles et de la Guyane; Cameron et al., "Climate Collections."
99 Nigel Clark, *Inhuman Nature: Sociable Life on a Dynamic Planet* (California: Sage, 2011).
100 Cameron, "Ecologizing Experimentations: A Manifesto," 16–17; Cameron, "Posthuman Museum Practices," 351. Cameron, *The Future of Digital Data*, 131.
101 Cameron, "Ecologizing Experimentations: A Manifesto," 16–33; Cameron, *The Future of Digital Data*, 131.
102 Cameron, "Posthuman Museum Practices," 351–52; Cameron, *The Future of Digital Data*, 131.
103 Cameron, "Posthuman Museum Practices," 349–52.
104 Cameron, "Posthuman Museum Practices."; Cameron, *The Future of Digital Data*, 137.
105 Cameron, "Posthuman Museum Practices," 349–52.
106 Cameron, "Posthuman Museum Practices," 551–52.; Cameron, *The Future of Digital Data*, 137.
107 Cameron, *The Future of Digital Data*, 137.

108 Cameron, *The Future of Digital Data*, 149.
109 Haraway, *Staying with the Trouble*, 2.
110 Deal was a unit of volume used to measure wood in the late eighteenth and early nineteenth centuries typically referring to a wooden board (usual conifers) between 12 and 14 feet long.
111 Henry Winram Dickinson, *James Watt Craftsman and Engineer* (Cambridge University Press, 1936), 141.
112 Eric Roll. *An Early Experiment in Industrial Organisation. Being a History of the Firm, Boulton and Watt, 1775–1805*. Abington, OX: Frank Cass and Co Ltd, 1930, 161, 167.
113 Colebrook, "Not Symbiosis, Not Now," 36.
114 *Climate Change Science: An Analysis of Key Questions*, Committee on the Science of Climate Change, accessed February 24, 2023. https://nap.nationalacademies.org/catalog/10139/climate-change-science-an-analysis-of-some-key-questions, National Research Council, National Academy Press, Washington DC, 2001, 9.
115 Shapiro and Kirksey, "Chemo-Ethnography," 485.
116 Bob Johnson, *Mineral Rites: An Archeology of the Fossil Economy* (Baltimore, MD: Johns Hopkins University Press, 2019), 18.
117 Johnson, *Mineral Rites*, 18–19.
118 Cameron et al., "Climate Collections," 68.
119 Cameron et al., "Climate Collections," 76.
120 Cameron et al., "Climate Collections," 74.
121 Cameron et al., "Climate Collections," 75.
122 Cameron et al., "Climate Collections," 76.
123 Birmingham Library and Archives, MS 3147/3/377 Boulton and Watt Collection General Correspondence, I, J, K Letter from John Iddins, New Street Birmingham to Matthew Boulton, April 27, 1781.
124 Dickinson, *James Watt* 141, 171.
125 Birmingham Library and Archives, MS 3147/3/85 - Boulton and Watt Collection, Letter from James Watt to Samuel Whitbread, Birmingham, December 7, 1784.
126 Boulton and Watt Order Book, *Muirhead Papers*, 119.
127 Jennifer Tann and Michael J. Breckin, "The International Diffusion of the Watt Engine, 1775–1825," *The Economic History Review* 31, no. 4, 541–64, 545.
128 Tann, "The International Diffusion of the Watt Machine," 545.
129 Dickinson, *James Watt*, 179–80.
130 Birmingham Library and Archives, MS 3147/3/109, - Boulton and Watt Collection, Letter from Boulton and Watt and Co to Whitbread, February 7, 1814.
131 IPCC, "2021: Summary for Policymakers," 6.
132 IPCC, "2021: Summary for Policymakers," 6.
133 National Oceanic and Atmospheric Administration, Global Monitoring Laboratory. Mauna Loa, https://gml.noaa.gov/ccgg/trends/
134 IPCC, "2021: Summary for Policymakers," 28.
135 "Cape Grim, Greenhouse Gas Data, "CSIRO," https://capegrim.csiro.au/ (updated monthly).
136 IPCC, "2021: Summary for Policymakers," 28.
137 "Cape Grim," https://capegrim.csiro.au/.
138 IPCC, "2021: Summary for Policymakers," 28.
139 Rebecca Lindsey, "How Do We Know the Build-Up of Carbon Dioxide in the Atmosphere IS Caused by Humans?" *Climate.gov*, November 5, 2020, accessed February 24, 2023, https://www.climate.gov/news-features/climate-qa/how-do-we-know-build-carbon-dioxide-atmosphere-caused-humans.
140 Joanna Zylinska, *Non-human Photography* (Cambridge, MA: MIT Press, 2017), 125.
141 IPCC, "2021: Summary for Policymakers," 6.

142 Karen Malone, *Children in the Anthropocene* (London: Palgrave Macmillan, 2018), 221–46.

References

Alaimo, Stacey. "Transcorporeal." In *Posthuman Glossary*, edited by Rosi Braidotti and Maria Hlavajova, 435–37. London: Bloomsbury, 2018.

Arfvidsson, Helen and Ann Follin. "Connectedness, Consumption and Climate Change: The Exhibition Human Nature." *Museum Management and Curatorship* 35, no. 6 (2020): 684–96.

Bennett, Jane. *Vibrant Matter: A Political Ecology of Things.* Durham, NC: Duke University, Press 2010.

Birmingham Library and Archives, MS 3147- Boulton and Watt Collection.

Birmingham Library and Archives, MS 3147- Boulton and Watt Collection, Research File, Letter from James Watt to Henry Goodwyn April 7th 1784.

Birmingham Library and Archives, MS 3147- Boulton and Watt Collection, Research File, Letter from Samuel Whitbread to James Watt, August 26, 1784.

Birmingham Library and Archives, MS 3147- Boulton and Watt Collection, Research File, Letter from James Watt to Samuel Whitbread, Birmingham, December 7, 1784.

Birmingham Library and Archives, MS 3147- Boulton and Watt Collection, Research File, Letter from James Watt to Samuel Whitbread, Birmingham, April 11, 1785.

Birmingham Library and Archives, MS 3147- Boulton and Watt Collection, Research File, Letter from BW and Co to Whitbread, February 7, 1814.

Bonan, Gordon B. "Forests and Climate Change: Forcings, Feedbacks, and the Climate Benefits of Forests." *Science* 320, no. 5882 (2008): 1444–49.

Braidotti, Rosi. *The Posthuman.* Cambridge: Polity, 2013.

Braidotti, Rosi. *Posthuman Knowledge.* Cambridge: Polity, 2019.

British Ecological Society. "British Ecological Society: Special Feature on Levering Natural History Collections to Understand the Impacts of Climate Change," February 24, 2023, https://www.britishecologicalsociety.org/call-for-proposals-special-feature-on-leveraging-natural-history-collections-to-understand-the-impacts-of-global-change/.

Cameron, Fiona R. "Ecologizing Experimentations: A Manifesto for a Posthumanist Museum." In *Climate Change and Museum Futures*, edited by Fiona R. Cameron and Brett Neilson, 16–33. New York: Routledge, 2014.

Cameron, Fiona R. "We Are on Nature's Side." In *Climate Change and Museum Futures*, edited by Fiona R. Cameron and Brett Neilson, 51–77. New York: Routledge, 2014.

Cameron, Fiona R. "Posthuman Museum Practices." In *Posthuman Glossary*, edited by Rosi Braidotti and Maria Hlavajova, 379–92. London: Bloomsbury, 2018.

Cameron, Fiona R. *The Future of Digital Data, Heritage and Curation in a More-Than-Human World.* London: Routledge, 2021.

Cameron, Fiona R. "Theorising More-Than-Human Collectives for Climate Change Action in Museums." In *L'Internationale Online Climate Change Special Issue* to coincide with the UNFCCC meeting in Paris, November 30– December 11, 2015, accessed February 24, 2023, http://www.internationaleonline.org/research/politics_of_life_and_death/48_theorising_more_than_human_collectives_for_climate_change_action_in_museums

Cameron, Fiona R. and Brett Neilson. *Climate Change and Museum Futures.* New York: Routledge, 2014.

Cameron, Fiona R., Ben Dibley and David Ellsworth. "Climate Collections, and Photosynthetic Fossil-Fuelled Atmospheres." *Environmental Humanities* 15, no. 2, 2023: 62–84.

"Cape Grim Greenhouse Gas Data." July 2021, accessed February 24, 2023, https://capegrim.csiro.au/.

Cernusak, Lucas A., Guillaume Tcherkez, Claudia Keitel, William K. Cornwell, Louis S. Santiago, Alexander Knohl, Margaret M. Barbour, David G. Williams, Peter B. Reich, David S. Ellsworth, Todd E. Dawson, Howard G. Griffiths, Graham D. Farquhar, Ian J. Wright and Mark Westoby. "Within Plant Variation in Stable Carbon Isotopes: Why Are Non-photosynthetic Tissues Generally ^{13}C Enriched Compared to Leaves?" *Functional Plant Biology* 36 (2009): 199–213.

Clark, Nigel. *Inhuman Nature: Sociable Life on a Dynamic Planet.* California: Sage, 2011.

Colebrook, Claire. "Not Symbiosis, Not Now: Why Anthropogenic Change Is Not Really Human." *Oxford Literary Review* 34, no. 2 (2012): 185–209.

Cornwell, William K., Ian J. Wright, John Turner, Vincent Maire, Margaret M. Barbour, Lucas A. Cernusak, Terry Dawson, David S. Ellsworth, Graham G. Farquhar, Howard Griffiths, Claudia Keitel, Alexander Knohl, Peter Reich, David G. Williams, Radika Bhaskar, J. Hans Cornelissen, Anna Richards, Susanne Schmidt, Fernando Valladares, Christian Körner, Ernst-Detlef Schulze, Nina Buchmann and Louis Santiago. "Climate and Soils Together Regulate Photosynthetic Carbon Isotope Discrimination within C3 Plants Worldwide." *Global Ecology and Biogeography* 27 (2018): 1056–67.

Crutzen, Paul and Eugene F. Stoermer. "The Anthropocene." *Global Change Newsletter* 41 (May 2000): 17–18.

Dale-Hallett, Lisa, Rebecca Carland and Peg Fraser. "Sites of Trauma: Contemporary Collecting and Natural Disaster." In *Museum Theory*, edited by Andrea Witcomb and Kylie Message, 531–552. London: Wiley, 2015.

Deutsches Museum Power Machinery Exhibition, https://www.deutsches-museum.de/en/museum-island/exhibitions/power-machinery#c8502.

Dibley, Ben. "Prospects for a Common World: Museums, Climate Change, Cosmopolitics." In *Climate Change and Museum Futures*, edited by Fiona Cameron and Brett Neilson, 58–84. New York: Routledge, 2014.

Dickinson, Henry Winram. *James Watt Craftsman and Engineer.* Cambridge, UK: Cambridge University Press, 1936.

Ellsworth, David S. "CO_2 Enrichment in a Maturing Pine Forest." *Plant, Cell and Environment* 22 (1999): 461–72.

Hall, Harlan. "Memo to Curator of Conservation, December 17, 1981. Subject: Boulton and Watt Publicity Programme 18432 Steam Engine Boulton and Watt, England 1785 Part 2 of 2." Museum of Applied Arts and Sciences Archive, Sydney, 1981.

Haraway, Donna. *Staying with the Trouble: Making Kin in the Chthulucene.* Durham, NC: Duke University Press, 2016.

Hodder, Ian. *Entangled: An Archaeology of the Relationships between Humans and Things.* Malden: Wiley-Blackwell, 2012.

IPCC. "Summary for Policymakers." In *Climate Change 2021: The Physical Science Basis. Contribution of Working Group I to the Sixth Assessment Report of the Intergovernmental Panel on Climate Change*, edited by Valerie Masson-Delmotte, Panmao

Zhai, Anna Pirani, Sarah Connors, Clotilde Péan, Sina Berger, Nada Caud, Yang Chen, Leah Goldfarb, Melissa Ines Gomis, Mengtian Huang, Katherine Leitzell, Elisabeth Lonnoy, Robin Matthews, Tom K. Maycock, Tignor Waterfield, Ozge Yelekçi, Rong Yu and Baiquan Zhou, Figure SPM 10, 37, Cambridge: Cambridge University Press, 2021, accessed February 24, 2023, https://www.ipcc.ch/report/ar6/wg1/#SPM.

Johnsen, Kurt H., David N. Wear, Ram Oren, Robert O. Teskey, Felipe Sanchez, Rodney Will, John Butnor, Daniel Markewitz, Daniel Richter, Tim Rials, Helen Allen, John Seiler, David Ellsworth, Chistopher Maier, Gabriel Katul and Phillip Dougherty. "Meeting Global Policy Commitments: Carbon Sequestration and Southern Pine Forests." *Journal of Forestry* 99, no. 4 (2001): 14–21.

Johnson, Bob. *Mineral Rites: An Archeology of the Fossil Economy.* Baltimore, MD: Johns Hopkins University Press, 2019.

Johnson, Kenneth, Stephen Brooks, Phillip Fenberg, Adrian Glover, Karen James, Adrian Lister, Ellinor Michel, Mark Spencer, Jonathan Todd, Eugenia Valsami-Jones and Jeremy Young. Climate Change and Biosphere Response: Unlocking the Collections Vault. *BioScience* 61, no. 2 (2011): 147–153. https://doi.org/10.1525/bio.2011.61.2.10.

Leavitt, Stephen and Austin Long. 1983. "An Atmospheric 13C/12C Reconstruction Generated through Removal of Climate Effects from Tree-Ring^{13}C/^{12}C Measurements." *Tellus B: Chemical and Physical Meteorology* 35, no. 2 (1983): 92–102.

Lindsey, Rebecca. "How Do We Know the Build-Up of Carbon Dioxide in the Atmosphere Is Caused by Humans?" *Climate.gov*, November 5, 2020, accessed February 24, 2023, https://www.climate.gov/news-features/climate-qa/how-do-we-know-build-carbon-dioxide-atmosphere-caused-humans.

Malone, Karen. *Children in the Anthropocene.* London: Palgrave Macmillan, 2018.

McKie, Robin, "James Watt and the Sabbath Stroll that Created the Industrial Revolution." *The Guardian*, May 30, 2015, accessed February 24, 2023, https://www.theguardian.com/technology/2015/may/29/james-watt-sabbath-day-fossil-fuel-revolution-condenser.

Morris, Peter J.T. (eds.) "Introduction." In *Science for the Nation.* London: Palgrave Macmillan, 2020, 1–10. "Rotative steam engine by Boulton and Watt, 1788." *Science Museum Group. Collection*, accessed February 24, 2023, https://collection.sciencemuseumgroup.org.uk/objects/co50948/rotative-steam-engine-by-boulton-and-watt-1788-beam-engine-steam-engine.

Museum of Everyday Life collection, accessed February 24, 2023, https://museumofeverydaylife.org/.

National Academies of Sciences, Engineering, and Medicine, "Surface Temperatures for the Last 200 Years." 2006. Washington, DC: The National Academies Press, accessed February 24, 2023, https://doi.org/10.17226/11676. https://nap.nationalacademies.org/read/11676/chapter/7#46.

National Oceanic and Atmospheric Administration, Global Monitoring Laboratory. Mauna Loa, https://gml.noaa.gov/ccgg/trends/.

Neimanis, Astrida and Rachel Loewen Walker. "*Weathering*: Climate Change and the 'Thick Time' of Transcorporeality." *Hypatia* 29, no. 3 (2014) Special Issue: Climate Change: 558–75.

O'Leary, Marion H. "Carbon Isotopes in Photosynthesis: Fractionation Techniques May Reveal New Aspects of Carbon Dynamics in Plants." *BioScience* 38, no. 5 (1988): 328–36.

Penfold, Arthur. *Two Renowned Machines*. Engineering Series 1, Museum of Applied Arts and Sciences, Harris Street, Ultimo, 7 (1–8). Museum of Applied Arts and Sciences Archive, Sydney, n.d.

Satchell, Veront. "Early Use of Steam Power in the Jamaican Sugar Industry, 1768–1810." *Transactions of the Newcomen Society* 67, no. 1 (1995): 221–31.

Satchell, Veront. "The Diffusion of the Watt Steam Engine in the Jamaican Slave/Sugar Economy 1810–1830." Paper presented at the 29th Annual Conference of the Association of Caribbean Historians, April 1997, Martinique, Université des Antilles et de la Guyane.

Science Museum, London. Boulton and Watt 1788 Beam Engine, accessed February 24, 2023, https://collection.sciencemuseumgroup.org.uk/objects/co50948/rotative-steam-engine-by-boulton-and-watt-1788-beam-engine-steam-engine.

Shapiro, Nicolas and Eben Kirksey. "Chemo-Ethnography: An Introduction." *Cultural Anthropology* 32, no. 4 (2017): 481–93.

Simondon, Gilbert and Arne De Boever. "Technical Mentality." In *Gilbert Simondon: Being and Technology*, edited by Arne De Boever, A. Murry and J. Roffe, 1–16. Edinburgh: Edinburgh University Press, 2012.

Stengers, Isabelle. *In Catastrophic Times: Resisting the Coming Barbarism*, translated by Andrew Goffey. London: Open Humanities Press, 2015.

Tann, Jennifer and Michael J. Breckin, "The International Diffusion of the Watt Engine, 1775–1825." *The Economic History Review* 31, no. 4 (1978): 541–64.

United Nations Framework Convention on Climate Change, Subsidiary Body for Implementation, Forty-Eighth Session, Bonn, 30th April to 10th May 2018, Action for Climate Empowerment.

Urevig, Andrew. "Seeking Answers on Climate Change, Scientists Venture into the Vaults of the Past," accessed February 24, 2023, https://ensia.com/features/natural-history-museums-climate-change/ 2017.

The Viking Ship House, Cultural History Museum, Oslo, accessed February 24, 2023, https://www.khm.uio.no/besok-oss/vikingskipshuset/.

Zylinska, Joanna. *Non-human Photography*. Cambridge, MA: MIT Press, 2017.

4

MUSEUMS, CLIMATE POLICY FRAMEWORKS, AND THE PROBLEM OF HUMANIST-DRIVEN SOLUTIONS

Introduction

In an uncertain world facing socio-ecological collapse, museums of all types and forms around the world are committing to climate action and implementing strategies to promote and enact more sustainable futures. These mandates are no longer on the periphery of museum concern, rather have become a driver of museum strategy. This is evident in the fact museums have become named stakeholders on the Paris Agreement (PA),[1] and museum agency is directed to its purposeful implementation in respect to climate change mitigation and its decarbonising processes through pedagogy and international cooperation.

In this chapter, I seek to unpack the dominant thoughts, frames, and systems of value inherent in museums' engagement with climate change mitigation strategies, the PA and its education policy, the Action for Climate Empowerment (ACE) by showcasing a number of implementation strategies. The questions I ask, therefore, are: what are the philosophical premises and value systems underlaying these policy frameworks? And what futures might museum interventions materially enact?

Museums tend to adopt these thoughts and frames and implement them at times without a critical regard for what they mean and their material consequences. The discursive material and policy components of institutions – from infrastructures to legal regimes, to collecting practices and interpretive strategies – materially influence human and non-human communities in profound ways. Modern forms of knowledge, human agency, economy, and culture, which have led to the devastating consequences of climate crisis, are

DOI: 10.4324/9781315212067-4

given authority through museums. These same conative frames continue to be embedded in dominant policy frameworks.

Museums, climate concepts, and envisioned planetary futures

As a rapidly unfolding disaster, global warming reminds us that the planet is no longer solely ours to control and exploit. As globalisation scholar Jessica Schmidt reminds us, "we are no longer the only actors in our history; and our human capacities to drive change in ways that benefit us have been significantly curtailed."[2] What is emerging is a crisis of ecospheric habitability of planetary scale and proportions.

In order to find solutions, in the last 30 years many governments and communities have relied on the directives presented by the climate sciences, technological expertise, and neoliberal economic risk calculations, but these solutions have proven inadequate as warming continues to spiral. Such directives and concerns form the basis of museum exhibition and program offerings.

When climate science defines global warming as simply the accumulation of gases in the atmosphere, solutions are limited to specific greenhouse gas-centric strategies. These include adjustments to carbon-burning practices, the use of neoliberal capitalist market mechanisms and "green energy" transitions, the implementation of pricing policies that put restraints on pollution, better strategies to communicate climate science, and through carbon sink enhancement. Furthermore, technological fixes, such as large-scale geo-engineering projects, currently in development, seek to intervene in the Earth's systems by pulling carbon and methane from the atmosphere. While these approaches might appear radical, and although carbon pollution reduction schemes have had some measure of success in reducing levels of emissions, especially at local levels, the ways of thinking and acting underlying such strategies in western countries are informed by a "business as usual" ideology. These thoughts lack a fundamental appreciation of how western thinking is governed by hubris and exploitative relationships between the human and other-than-human world directed to acquire economic advantage. Alarmingly these practices and ways of thinking created the climate crisis in the first instance.

In climate science research, humans are positioned as sovereign climate-forcing entities as chemical agents generating greenhouse gas emissions changing the chemical composition of the atmosphere thereby forcing heating responses.[3] As a consequence, the atmosphere is monitored and measured for rising greenhouse gas concentrations, and measurements taken are used to model rates of warming and future impact scenarios. This figuration represents a conceptual disconnect – the notion of populations as forcing agents promotes dis-embeddedness and supports the belief in strong human agency and at the same time belies deep enmeshment. Rather the figuration of dynamic biochemical fossil-fuelled life forms as a different conceptualisation introduced in the previous chapter signals humans and their deep material enmeshment in planetary processes as human–non-human climates.

Science calculations instruct governments, communities, and individuals to reduce their carbon footprint, to conserve energy, and to change energy infrastructures. Critically fossil-fuel-dependent populations are relationally configured not only as forcing, and at the same time as stabilising, agents with the capacity to manage and influence Earth systems to create a safe and habitable planet. The exhibition "Atmosphere: Exploring Climate Science" opened in 2010 at the Science Museum, London. Lauded as the leading-edge exhibition on the science of climate change in the world at the time, it framed the dominant narratives of climate action that are still the benchmark for climate governance today, as they are embedded in the PA.[4] "Atmosphere," among other exhibitions, therefore draws audiences into scientific and greenhouse gas-centric forms of thinking and acting.

As a result, climate change thought and action promoted through museums tends to be founded on scientific information and modelling about the capacity of humans, and their technologies, to influence biochemical concentrations in the atmosphere. Green development strategies as presented in and through museum programs provide the impetus to achieve all these ends. The climate crisis is not only an energy or combustion crisis but is borne out of a failure of modern societies to acknowledge and work with the ongoing entanglements, shared vulnerabilities, and consequential futures that emerge as a result in a more-than-human world, a world in which societies co-exist and depend. To accelerate climate action through museum representations, and other forms of engagement, mitigation and adaptation strategies must be amplified and diversified with projects that honour, rather than seek to exploit, these deep entanglements.

It is clear that modern humans, and their activities, are deeply implicated in systemic ecospheric change, and this is brought to light through the concept of the Anthropocene. Museums have been engaged in this space, portraying the systemic damage caused by industrial activity, through exhibitions such as "Welcome to the Anthropocene," held at the Deutsches Museum (2014–2016) and Museo de Amanhã (Museum of Tomorrow), Rio de Janeiro, Brazil (2016–). In addition, museums have engaged in this scholarly space through the Australian Research Council grant, "Understanding Australia in the age of humans: Localising the Anthropocene" led by Professor Iain McCalman.

The concept of the Anthropocene emerged in 2000 and since this time its geological disciplinary focus has shifted to the social sciences and humanities, and most recently into museums.[5] With more than 41 exhibitions on the topic since the development of the concept, mainly in the West and in art, history, and natural history museums, the Anthropocene has become a new keyword in museum space.[6] Most of these exhibitions seek to convince audiences that the Anthropocene is happening through the presentation of complex facts, dramatic visual images, and collections of the technologies of the Anthropocene, rather than through the foregrounding of solution-based interventions.[7] While the concept comes alive in museum space and is central in transgressing the nature-culture divide and in demonstrating the impacts of dis-embeddedness, museum scholars Lotte Isager, Line Vestergaard Knudsen and Ida Theilade argue that these programs do not deal with the controversies of the Anthropocene. As a museum keyword, the concept of the Anthropocene has gone viral, but it remains a distinctly western and geological premise and therefore acts as a colonising concept that erases non-European, or non-Western, ontologies and histories.[8] Rather than solely focusing on the damage inflicted it is important to also promote and enact ways of being that invert the concerns of capital to prioritise ecological and alternative economic practices. Un-minded damage needs to shift to communitarian design and to the implementation of recuperative measures across domains and cultures.

In the same way as the concept of the Anthropocene has been claimed by museums, climate science, the ecological modernisation project, the PA and its associated neoliberal keywords and terms – such as mitigation, carbon pollution, greenhouse gas emissions, cap, and trade – renewable technologies and green capitalism have also migrated into museums. That is not as an "X factor" concept, as seen with the Anthropocene, but as a global methodology for human survival. Earth system scientist Will Steffen and colleagues rightly called for a fundamental change in the role of modern stewardship of the planet.[9] But here Humanity's challenge, which is Humanity with a big H, is to influence the Earth systems to create a habitable, safe, and stabilised Earth. Such an imaginary, again central to museum engagement, casts humans as hubris, as commanders and controllers. In rethinking modern populations and their agency in vital, dynamic, and deeply ecological inhabitations, curatorial practice must reject this kind of entitled thinking, and the idea that modern humans can own and control non-humans or that communal life is resolutely and solely human. Instead curatorial practice must cultivate knowledges and practices that allow modern humans to know, work with and relearn ways to cohabit respectfully with other humans, non-humans, and earthly processes, all of which are stakeholders in planetary life in profound and intimate ways. Such an approach can be achieved through new types of ecologically inflected habitability practices and novel institutional forms that help societies founded on neoliberal and modern projects to think with, live with, and act with an unavoidably damaged world. And as Tim Morton philosopher states, this must be across all sectors, scales, disciplines, and projects.[10]

In addition to a collective will to bring about change through international collaboration such as the PA, modern societies must be reorganised in more communitarian ways. According to Morton, communitarian living can be achieved by reframing the world – and its associated human-centred problems in modern societies – as an ecological project of profound and intense interconnectivity and respect[11] in order to attain the recuperative measures required.[12] This is where museums, as powerful pedagogical institutions, can take a lead role in the communitarian reframing of climate change. To achieve this, curatorial practice needs to encourage the development of what Haraway proposes as new types of earthly politics and a natural-cultural history of a qualitatively different kind.[13] Human geographer Katherine Gibson views such a communitarian earthly politics as a "performative ontological politics for other worlds."[14]

Museums can no longer perpetuate dominant ideas about human agency, or place utilitarian values on "nature" as a resource to serve modern narcissistic interests. Instead, curatorial practice must bring radical co-existence to the fore and put modern populations back in their ecological place – knitted together not just with other biological species, as seen with theories of evolution, but with all manner of animate and inanimate things including non-human beings, materials, and earthly processes. Such a practice would be an acknowledgement of humans' material and relational interdependence with others.[15] The aim must be to build mitigation and adaptation projects that enable us to think as dynamic biochemical lifeforms in movement with others and act according to our relational interdependencies in empathic, ethical, and recuperative ways.[16] Different concepts of sociality and community therefore emerge to include the more-than-human and other-than-human, such as all manner of insect and mammalian life, plants, soils, the Earth's strata, the atmosphere, biosphere, hydrosphere, oceans, and ice – all critical stakeholders in new types of habitability projects to which we must become attuned.[17] The notion that nature is over there, disconnected from us, amendable to being controlled and stabilised must be let go of.

Unpacking the thoughts and frames that currently guide museums and climate futures: The Paris agreement

In analysing the thoughts and frames contained within the PA, and drawing inspiration from policy analyst Matthew Hull, it is prudent to evaluate the discursive, narrative, and material implications of the PA and the political possibilities implicitly or explicitly contained within it. Hull contends that "the producers of government documents, much like scientists, claim to represent, engage with, or constitute realties 'in the world' independent from the processes that produce documents."[18] Herein, it is important to consider the different emerging material realities embedded in each of these policy frameworks and how they are put to work to legitimise a particular set of modern practices, including the multifarious ways museum curatorial practices and networks negotiate those realities. Central to this is a consideration of how the PA endorses particular values and approaches that are at once undisclosed, normalised, and universalised.

The year 2015 was formative for climate policy because of the emergence of three interrelated agendas: the PA seeking to reduce greenhouse gas emissions, Agenda 2030 promoting sustainable growth, and the Addis Abada Action Agenda (of the Third International Conference on Financing for Development) aiming to finance the projects of the previous two. The PA, and its alliance with the Sustainable Development Goals, is presented as a set of devices and instruments to save the planet from inevitable destruction. This alliance of principles, values, and policies drives museum engagement in this space.

The PA, signed in 2015 by 197 countries at the 21st Conference of the Parties, attempts to limit the rise of greenhouse gas emissions to keep temperatures under 2 degrees Celsius and as close to 1.5 degrees Celsius as possible.[19] This collective effort is an extraordinary achievement, however, the agreement is framed by its policy-makers according to human-centred concerns to prevent an irreversible threat to human societies and the planet. Funding sinks, polluter offsets, and technology are recognised as pivotal for mitigation. Modern values are presented as a common concern for all humankind and, in respect to PA obligations, are founded on human rights: the right to health, Indigenous rights, the right to development, and so forth.

The efficacy of "Big climate science" in ecological modernisation strategies

"Big climate science," a large-scale climate science research agenda and program of global extent coordinated by the Intergovernmental Panel on Climate Change, is the authority underpinning the PA. "Big climate science" studies patterns of warming, identified as human emissions and their effects, and considers what are dangerous levels of emissions for the future habitability of the planet. The calculations of "big climate science" seek to predict levels of emissions for each country, as well as likely temperature scenarios and their impacts. It lays out the mitigation measures necessary to limit temperature increases to 2 degrees, and preferably 1.5 degrees, above preindustrial levels. This data, in turn, can be used to inform countries of the monetary and infrastructure contributions and adaptation strategies needed under various scenarios. It suggests where emissions would need to be balanced by technological solutions such as pulling CO_2 out of the air through carbon sequestration in order to keep warming to 1.5 degrees. The calculations of the aggregate effects of national contributions are then directed towards the worldwide goal for emissions reduction.

Furthermore "big climate science" seeks to identify the affects of rising emissions on oceans, land, air, ecosystems, and the ecosphere. As a consequence, the calculation of risks and threats to human populations through "big climate science" informs strategies for climate resilience and low carbon emissions development and its financing.

Museum engagement with climate change overwhelmingly follows the directives of "big climate science" – and its associated economic and technological investments – to seek a solution to the material world impacts of climate change. Drawing on my chapter, "We are on Nature's side," in my 2014 co-edited collection with Brett Neilson, *Climate Change Museum Futures* (Routledge), I revisit my analysis of the "Atmosphere: Exploring Climate Science" exhibition at the Science Museum, London and its engagement with "big climate science."[20] "Big climate science" is presented as the saviour of a now-threatened human race. In this exhibition, the Museum illustrates modern societies and their reliance on the institution of science as the only form of valid knowledge that can reveal how the Earth's system works and as a knowledge practice to make decisions about the future for ourselves and the non-human world.[21]

In the exhibition introductory panel, the Science Museum levers "big climate science" and confirms its commitment to a modernising project:

> Understanding what the science is telling us is crucial to making the right decisions, given the need to balance major costs and risks. Our aim is that "Atmosphere", will provide our visitors with accurate, up-to-date information

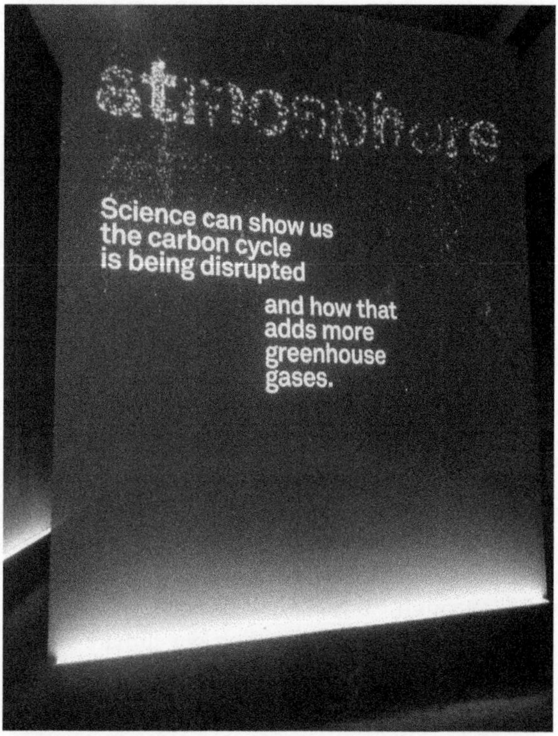

FIGURE 4.1. "Atmosphere: Exploring Climate Science", introductory panel, Science Museum, London, 2010. Photo credit: Fiona Cameron.

> on what is known, what is uncertain, and what is not known… and the ways that science, technology and industry can contribute to a positive future.[22]

In the Museum's articulation of the role of "big climate science" in the governance of the climate problem, complete confidence is given over to scientific reason, as a set of principles and disciplines, that is able to reveal the workings of an already pre-existing climate system that modern, and indeed all populations on Earth, need to be recalibrated with:[23]

> Science can show us that greenhouse gases are increasing and why that makes global temperatures rise… Science can show us how Earth's climate system works and what causes the damage… Science can show us what's already changing… and what might happen next… Science can track what's already changing and help us imagine the future.[24]

"EcoLogic, Creating a Sustainable Future," an exhibition at the Powerhouse (2011–2021), represented "big climate science" data as a carbon dioxide timeline extrapolated from IPCC reports. It included a model of Sydney's CBD that illustrated projected sea level rises based on scientific data, and the display of climate measuring instruments such as Argo floats that monitor ocean currents. The "EcoLogic" exhibition represented "big climate science" in a way similar to that of the IPCC, which articulates activities of biological, earthly, and atmospheric processes in terms of increases in recorded levels of emissions and human-induced climate forcing. Current and potential impacts of climate change on the environment were presented in the exhibition through the computer modelling of these processes.

In the "Atmosphere" exhibition, the underlying scientific rationale informing the means to govern climate change is that "big climate science" climate geographer Mike Hulme argues can generate discoverable objective facts that are socially and politically neutral, can adjudicate between competing claims to truth, and can determine the causes of climate change and climate sensitivity.[25] By revealing the workings of an already re-existing climate system, "big climate science" secures our future by seeking ways to recalibrate our lifestyles with the atmosphere.[26] Science, as a child of Cartesian philosophy, which was an outcome of the separate domains of the Culture/Nature binary, was first authorised to reveal the universal laws of material/physical substances and therefore has the authority to speak for the non-human world.[27]

Finding answers through "big climate science" which "Atmosphere" supports alone is seriously misguided. Presenting big climate science as the solution engenders unrealistic expectations about what science can achieve.[28] According to cultural geographer Mike Hulme, it perpetrates the belief, "that science can calculate and produce definitive statements about what is and what is not dangerous for societies, and accordingly ways to solve the ecological crisis."[29]

In the lead-up to COP 26 in Glasgow in November 2021, the Natural History Museum in London became one of the first museums in the world to align itself directly with the terms of the PA articles. Informed by "big climate science" the Museum now aims to reduce its emissions by 60% from 2015 levels by 2031.[30] In doing so, the Museum seeks to create a sustainable organisation and therefore contribute to a habitable planet through reducing energy consumption and through a consideration of the carbon footprint of the goods and services the museum consumes.[31]

Modern humanisms, ecological modernisation, and climate futures

While museums, through their exhibitions and programs, have been promoting the principles of ecological modernisation for some years now, the enrolment of museums as stakeholders in the PA in 2018 has formalised this strategy for climate governance through policy arrangements.

In the context of the PA, and its ecological modernisation agenda, climate mitigation is advanced through market mechanisms, technology transfer, financing strategies, and the exploitation of plants' capacities to act as sinks to remove greenhouse gases from the atmosphere.[32] The carbon pollution problem is also resolved and offset through the trading of carbon credits between countries. Building emissions reduction capacities is centred around strategies seeking to green citizens and industry as a narrow suite of carbon reduction-centred strategies. The development of green technologies, for example, renewable energy technologies and their financing, is directed to enabling industry to negotiate its way out of the climate crisis and at the same time strengthen the adaptative response. Sustainable development seeks to advance international capital.

The key principles of the PA and its ecological modernisation agenda enshrined in Article 6 are articulated as cooperation between parties in emissions reduction and adaptation and through the implementation of sustainable development measures while maintaining environmental integrity. To achieve this, voluntary and nationally determined emissions reduction plans and contributions are politically encouraged by the parties signed up to the agreement, rather than enforced. These mitigation pledges and emissions pathways seek to regulate greenhouse gases and dramatically reduce dependence on hydrocarbons for energy through the establishment of cooperative carbon markets. Incentives are provided to the private sector to participate in sustainable development, including the sharing of technical knowledge and financing in order to achieve this.

"Atmosphere: Exploring Climate Science" exhibition is divided into five zones. Each is dedicated to different aspects of climate change and its ecological modernisation project. Individual sections are organised according to themes: (1) how the atmosphere works; (2) the history of climate research;

(3) greenhouse gases, the carbon cycle, and adaptation to climate change; (4) effects on earth systems and global communities; and (5) mitigation solutions through carbon markets, scientific and technological innovation, including large-scale geo-engineering projects to extract carbon from the atmosphere, carbon capture and storage, and hydrogen cars.[33] Museum interpretive strategies promote the belief that the enhancement of carbon sinks, technology, and finance are central to solving the climate crisis.

Opened in 2013 The Jockey Club Museum of Climate Change, located on the campus of the Chinese University of Hong Kong (CUHK), also supports ecological modernisation strategies. It seeks to "reach out to and inform more people in Hong Kong and elsewhere about the latest developments in climate change issues." Its mission to inform, engage and enable climate mitigation is driven by a vision to "become a recognised, valued contributor to positive changes in knowledge, attitudes and behaviour in relation to climate change throughout Hong Kong and beyond." The Museum deploys a wide range of projects "to encourage and enable Hong Kong people of all ages and from all walks of life to get involved in carbon-reducing action and living a green lifestyle."[34]

Like the Science Museum, London, the Jockey Club Museum articulates climate change through "big climate science" and promotes mitigation strategies based on an agenda of ecological modernisation directed to greening citizens, reducing greenhouse gases, and commodifying carbon as traces of human carbon-burning activity. In this regard, the Museum promotes strategies of decarbonisation articulated in Article 6 of the PA.[35] Temporary exhibitions highlight current issues and themes such as the Sustainable Development Goals, Zero Waste, and "Eating Greener." An online platform, the Action Monitor, allows visitors to self-review their own green targets and actions and exchange greening information. The exhibition, "Climate Change: Past, Present, Future," includes holograms to explore evidence of climate change over time, as well as exhibits on health, temperature monitoring, and renewable energy technology. [36]

Nature mitigation (the ability of plants to store carbon) is also a key focus of programming. Trees and microalgae are presented as climate regulators because of their capacity to absorb human fossil-fuel emissions but also act as renewable energy sources. These concepts are expressed in a range of ways including through multimedia games "show[ing] nature's power at work in tree planting, wind and solar energy," and through 3D presentations, using augmented reality (AR) technology, presenting the experiences of extreme weather events, as well as the visual history of climate science.[37]

Museum narration continues to routinely focus on green technology, technology transfer, and financing and is therefore imbricated in neoliberal capitalism and its progressive logic. The Museum of Innovation and Science, Schenectady (MISCI), New York, for example, is dedicated to celebrating technological achievements and seeks to inspire people to celebrate and explore science, technology, and its financing.[38]

The MISCI's "Wind and Solar: The Renewable Energy Revolution" (2017) interactive exhibit, sponsored by GE Renewable Energy, sought to encourage visitors' understanding of renewable energy technology and climate change issues in order to inspire advocacy for renewable solutions to society's ever-increasing energy demands.[39] It's focus on renewable technological development resonates with strategies articulated in the PA, Article 6.[40]

The exhibition, "Spark, Australian Innovations Tackling Climate Change," at the Australian Museum (June to October 2021), and currently touring, similarly explores Australian inventions and innovation, including technological solutions for tackling climate change such as affordable, clean energy sourced from wind, hydrogen, and solar power, as well as electric transport.[41] Its exhibits include information on microalgae and seaweed farms which, as explained, act as carbon sinks providing regulatory services by pulling carbon from the atmosphere. Furthermore, the exhibition focuses on practices Australians are using to reduce pollution, and support plant and animal life through regenerative agriculture, sustainable building materials, and First Nations Aboriginal traditional knowledge, including cultural burning as a method for caring and sustaining country.

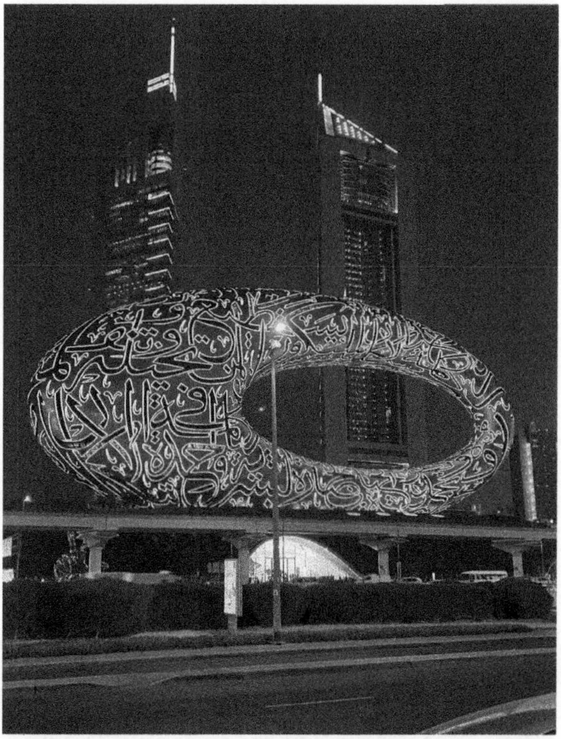

FIGURE 4.2. Museum of the Future, Dubai, September 3, 2022. Photo credit: Shefiq Abdulla.

Opened on February 28th, 2022, the Museum of the Future in Dubai, the United Arab Emirates and founded by the Dubai Future Foundation focuses on projecting preferable futures in 2071. As a professed human-centred institution with a focus on design, science, and technology, the institution seeks to tell stories of a series of preferred futures for global populations and how society could evolve to achieve these visions through technological and scientific innovation.[42] In doing so the Museum makes statements about what are desirable futures and what they look like, and who is involved – a future as a singular technologically mediated and corporatized world. Furthermore, the Museum promotes the idea that a stronger global economy will allow humanity to flourish and enable the development of next-generation technological solutions. In order to attempt to solve problems of existential crises the development team worked closely with corporations and government to produce these technologically mediated future imaginaries.

While the Museum is compelling and impressive in its foretelling of the future, it again promotes the idea of technology as a means of solving the climate crisis through hubris and strong human agency. The solutions the Museum promotes thereby act against ecologically attuned thinking and action. On each of the Museum's floors, the future of healthcare, transportation, aviation, smart cities, government services, and space travel alongside climate change strategies are featured.[43] On Level 2, the Heal Institute is dedicated to climate change and biodiversity. The "Vault of Life," an illuminated immersive installation consisting of a DNA library of 2,400 species as a catalogue of the world's biodiversity as well as a laboratory of experimental species utilises bio-engineering solutions to genetically modify organisms to protect them against extinction in the face of climate change. Climate change and technologically mediated mitigation strategies include geo-engineering projects in space that pull greenhouse gases from the atmosphere permanently. Another solution promotes the idea of Earth powered from space by the capture of solar rays. A digital twin of the Amazon rainforest represents the ambience of complex forest ecologies.

The Arabic script on the building, written by Emirati artist Mattar Bin Lahej based on the writing of poet Sheikh Mohammed, states "The future belongs to those who can imagine it, design it, and execute it. It is not something that you await, but rather create."[44] Inspired by Mohammed, the Museum imagines a future for future generations in which humankind through corporate initiatives can create for themselves. Visitors are asked to think critically about how technology is transforming their lives.

Mitigation, in each of these museum contexts, foregrounds technology, energy transitions, and resource efficiency in which the non-human world provides its raw materials from the plant capacities of seaweed and algae to the chemical affordances of hydrogen and the sun's rays. Pricing carbon rationalises forms of financing as a mitigation measure. The belief is that neoliberal capitalism can be made sustainable through resource efficiency – greening nature, greening citizens and businesses, and switching to renewable energy. Therefore, the programs presented here typical of museum offerings are therefore complicit in articulating the vision of a third green industrial revolution, as proposed by ecological modernisation advocates, Arthur Mol and David Sonnerfeld.[45]

By promoting nature as a regulatory service, carbon trading, and geo-engineering in the case of the Museum of the Future as solutions, the carbon pollution problem is offset and emissions are commodified. Furthermore, many of these museum programs promote the enhancement of carbon sinks, such as soils and plants. Forests are not only sinks but when cleared can become emitters. Promoting carbon offset through tree planting programs or through geo-engineering projects as a means for balancing out emissions is only a temporary measure because greenhouses gases will ultimately be released back into the atmosphere. The capacity of plants to act as effective sinks is also species dependent, thus tree planting as a strategy can be an erroneous commitment to zero emissions. Furthermore, promoting green technology and its transfer as a universal solution, feminist environmental activist Ariel Salleh explains, is homogenising and fails to acknowledge the role of local technologies and Indigenous capacities.[46] By promoting geo-engineering, green technology, and tree planting, museums promulgate the idea of a right to pollute and offset in order to continue on a modern growth trajectory. With these actions, the atmosphere continues to be used as a waste dump and as processes that can be manipulated to serve human ends of enhanced capital growth.

Policy frameworks, museum representations, and programs morally privilege human wellbeing and are indeed in the case of the Museum of the Future remain wholly human-centred, neo-capitalist, and corporatised. Greater attention needs to be paid to the wellbeing of non-humans and to the human and non-human enmeshments that support each's co-survival and flourishing.[47] Non-humans organisms can, however, benefit from climate change. The prospering of microorganisms that consume nitrogen, an outcome of human agriculture, irrigation, and dams, is a case in point.[48] The exploitation of non-human agencies, such as the harnessing of intergalactic solar radiation, must be carefully designed in order to avoid potential consequences. So while the "Spark" exhibition, for example, privileges the concerns of plant life, it is ultimately directed towards supporting human ends.

The PA, and indeed museum strategies such as those outlined in "Spark," and the Museum of the Future encourage capitalist entrepreneurial values and support growth trajectories through solutions directed to retrofitting buildings, audits to enhance energy and water efficiency, sustainable transport infrastructure, and renewables to reduce the carbon footprint and indeed radical plans to modify species. In doing so the exhibitions seek to support the establishment of a new economic sector of green industries for the manufacture of globally competitive products such as green building materials and the right to intervene and modify species DNA.

Such strategies for example are a false green conversion.[49] Salleh warns the availability of land for forests is in competition with that for food production, urban settlement, industrialisation, and biodiversity.[50] The construction of all these technologies, the development of new materials, the retrofitting of buildings, solar urbanisation, the installation of air-conditioning, and the development of high-tech cities all involve the extraction of and modification to non-human resources. These include the mining of metals, the modification of water supplies for component manufacture, impacts on forests and biodiverse ecosystems, the loss of land for food production, and the resettlement of populations, both human and non-human.[51]

Through zero emissions initiatives ecological modernisation strategies view oil, coal, and population growth as bad and destructive, and renewable energy, technology, carbon sequestration, and carbon pricing as good regenerative projects. Energy efficient transport, based on renewables and clean fossil-fuel-generated electricity through carbon capture and storage, comes with its problems and material costs, such as toxicity and environmental degradation. The production of renewable technologies for climate adaptation is erroneous due to the industrial pollution such strategies incur.[52] The scaling up of nature mitigation, by which forests become carbon sinks to create carbon offsets, compromises the livelihoods of forest people.

Like the emergence of dynamic biochemical fossil-fuelled life forms the rise of renewable energy solutions new humanisms emerge in which rare earth elements, cobalt, and lithium tellurium become bonded as emerging dynamic elemental lifeforms. Through this figuration neoliberal capitalist lifestyles become new forms of energy intimacies with the potential to also be destructive through mining and waste.

In promoting these approaches, museums enter alliances that seek to mitigate the collapse of neoliberal capitalism by marketing the idea of green technology, green jobs, green citizens, and green businesses. Dynamic biodiverse populations and earthly processes are treated as commodities, as infrastructures, and as resources subject to human crafting in order to support ecological modernisation. Salleh explains, "The deeply eurocentric and gendered focus on engineering infrastructure and economic growth invert the thermodynamic order of nature, emptying out its metabolic value."[53]

Following earlier initiatives, directed towards the critique of science and technological developments, the Science Museum, London has made nature mitigation and greening technology a matter for debate. The exhibition "Our Future Planet," which opened in 2021 in the lead up to COP 26 in Glasgow showcased the cutting-edge technology and nature-based solutions being developed to trap carbon dioxide released by the burning of fossil fuels. These include preserving ancient woodlands, adding dust to agricultural fields, capturing carbon dioxide from the air, or installing systems that prevent it from leaving power stations and factories.[54] The focus here however is one of critique and debate in respect to these different solutions. Questions include the following. How much difference can new technologies make? What can we do with the carbon dioxide after it has been captured? And why can't we just plant more trees? The approaches outlined above, while commendable remain greenhouse gas-centric.

As an ecological modernisation project, the PA has a substance, a texture, a force, and a modality that enables some and resists other forms of action that facilitate specific socio-material and technological consequences that museums, through their engagement with its articles, resoundingly support. The mainstream discourse of the PA makes no reference to, or indeed has no way of articulating, the material circumstances and metabolic and biological processes that support planetary habitability in which human life – as an ecological organism – is enmeshed and on which our futures in a climate-changed world depend. The United Nations Framework Convention on Climate Change (UNFCCC) and its PA supports the opposite. The PA is economically driven rather than socio-ecologically framed. Mitigation, adaptation, financing and technology transfer, and international cooperation are the key PA and therefore UNFCCC approaches. The UNFCCC also serves the transnational ruling class, enshrined privilege, international financing, and business-as-usual approaches.

Activating a global citizenry, rights-based frameworks, and vulnerable communities

While the "Atmosphere" exhibition represents climate change through the lens of "big climate science," equal weight is given over to solutions based on technological innovation. These include renewable energy systems, transport, industry, and carbon markets also the foci of exhibitions and programs at the Jockey Climate Museum. Furthermore, the exhibition includes images of extreme weather events and personal accounts by people from distant and vulnerable communities articulated within a rights and climate justice framework directed at galvanising action among museum audiences.

The Climate Museum in New York, the first museum in the US dedicated to climate change, is firmly located within an activist rights and climate justice framework. Throughout the museum, climate change is articulated as a common concern of all humanity and as one that requires the building of a collective culture focused on direct action within and through local communities. Driven by visionary Miranda Massie, and informed by her background as a civil rights litigator and activist, the Museum is directed to "bringing people together to learn about solutions and join the fight for a brighter future, providing multiple pathways into civic engagement."[55] An example of a public art exhibition

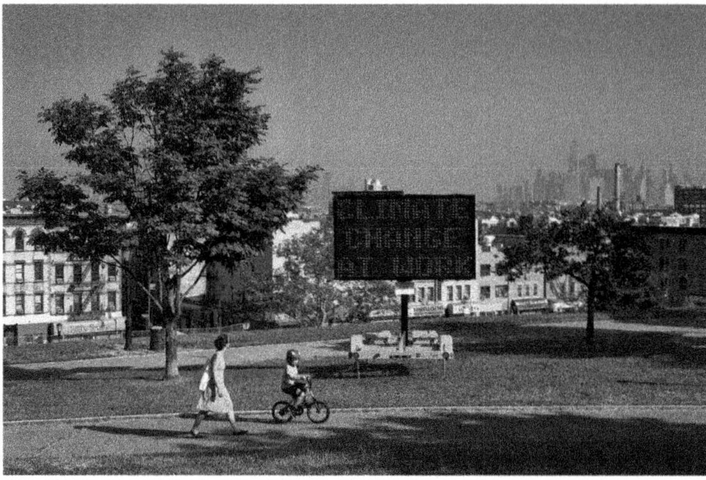

FIGURE 4.3. “Climate Signals”, installation image, Sunset Park, Brooklyn, NY. Image Courtesy of the Artist © 2018 Studio Justin Brice.

developed by the Climate Museum and launched on the streets of Manhattan was “Climate Signals” (2018) by Justin Brice Guariglia. This featured warning signs in five languages, which were viewed by tens of thousands of people. Signs flashed phrases such as “Climate Denial Kills,” “Fossil Fuelling Inequality,” and “Vote Eco Logically” in order to spark reflection and discussion and “draw passers-by into climate conversation” about climate justice.[56]

Attempts to build a global citizenry are enacted through an empathetic engagement with distant others. The first dedicated climate museum in the world, the Klimahaus in Bremerhaven, Germany, established in 2009 seeks to represent the risks and threats to human populations and to build a global citizenry of responsible green individuals by promoting an underlying philosophy of interconnectedness and mutual responsibility to human populations foregrounding personal stories and climate impacts in different parts of the world.[57] To do this, the Museum takes its visitors on a journey beginning and ending in Bremerhaven, following longitude 8° 34’ E. Visitors trek through diverse climate zones encountering people whose daily lives are profoundly affected by the surrounding climate.[58] The Klimahaus creates model “experiences” of countries along this longitude – with appropriate air temperature, lighting, and mocked-up landscapes and buildings. Audiences engage with landscapes from Switzerland, Sardinia, Tunisia, Cameroon, and even an Antarctic camp. They then travel along the matching line of longitude, the 172nd meridian west, via Samoa and Alaska, to complete the circle at Bremerhaven.[59] Dramatic imagery of rising sea levels and environmental destruction is used to illustrate the climate crisis and its human costs to invoke accountability towards distant others and trigger decarbonising behaviour.

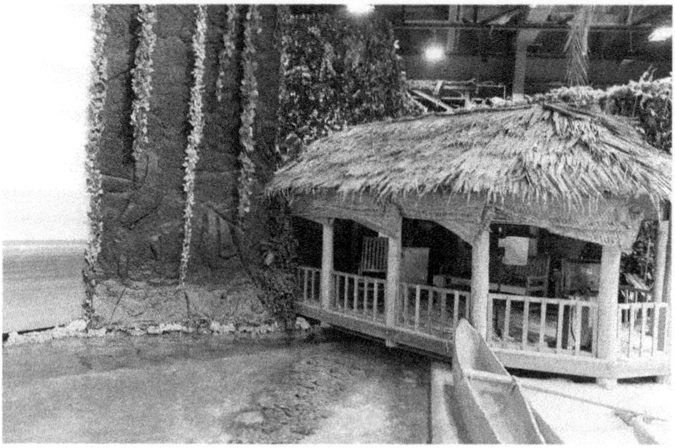

FIGURE 4.4. Klimahaus, Bremerhaven, 2011. Photo credit: Bin im Garten.

Another exhibition, held in honour of the 50th anniversary of Earth Day, opened in October 2019 at the Het Scheepvaartmuseum, the Dutch National Maritime Museum in the Netherlands. Titled "Rising Tide: Visualizing the Human Costs of Climate Change," the exhibition presented works by Dutch documentary photographer Kadir van Lohuizen to illustrate the dramatic consequences of the climate crisis across the world.[60] Through photographs, video, drone images, and sound, visitors experienced the effects of rising sea levels in Greenland, Bangladesh, Papua-New Guinea, Kiribati, Fiji, the Netherlands, the UK, Marshall Islands, and Jakarta in Indonesia.

Museological globalism and museological nationalism

The PA constitutes what sociologist Ulrich Beck calls "methodological nationalism"[61] and what might be called an emerging "methodological globalism." The agreement, and its implementation, rests on a belief in scales of methodological nationalism and methodological globalism in which knowledge and action are divided into domestic, national, and international actors underpinned by the principles and values of political diplomacy. The domain of the nation-state takes territories, borders, and its citizenry as nodes for delineating emissions targets, [62] thereby obviating any other values or the relevancy of any other variable including the marginalising of the non-human world.

The UNFCCC, ACE denotes work under Article 6 of the Convention (1992) and Article 12 of the PA plays a key role in promoting changes in lifestyles, attitudes, and behaviours that are needed to foster low-emissions, climate resilience, and sustainable development. Museums are acknowledged as stakeholders in the PA's list of institutional roles and are tethered to education, training, raising awareness, and international cooperation in respect to mitigation ambitions.[63] As a UN convention, responsibility for the implementation of the ACE initially fell to governments. Over time, however, governments' failure to promote this educational work means that the UNFCCC has now inducted non-state actors, including museums, to take on educational roles and boost ambition by promoting the values and strategies in the PA. The relationship with the ACE and Articles 6 and 12 involves museums promoting climate action, at various scales, in order to reach towards a zero-carbon future. In this way, museums promote the so-called greening of capitalism. The ACE supports both nation-centred responses and methods that are global in their articulation and distribution. Institutions therefore become apparatuses put to work to establish methods of what I call "museological nationalism" and "museological globalization."

The enrolment of museums in the ACE encourages the observation, analysis, and action on climate change by the nation, territory, and state. This promotes the values and strategies of a nation seeking a greenhouse-gas-centric approach at the same time within the aegis of ecological modernisation. For example, the PA articles directed at framing national emissions targets, promulgate nation-centred perspectives and national sovereignty within a global framework and then together as a suite of universal values. These universal values being scientific knowledge, economics, political science, and moral philosophy based on human rights. This, in turn, establishes a global political economy, a global politics, and a global ethics.

Museums, directly or indirectly, therefore promote the PA articles and mitigation goals as they enact, either independently or through international actors such as the UNFCCC, the international mechanisms and non-state actions of a distinct knowledge system for global governance. The promotion of these values and the scaling up of actions by the museum sector in turn form strategies and an orientation as forms of "museological nationalism" and then "museological globalism."

The enrolment of museums in the ACE encourages specific material consequences which endorse particular modern forms of habitability to the exclusion of other thoughts and frames. In this respect, the ACE positions museums as a governing apparatus for greening citizens. Centred on a narrow vision of behavioural change around carbon consumption, this is their main contribution. Museums are not setting the agenda but rather becoming subject to the PA plan.

In becoming subject to an ecological modernisation agenda and the articles of the PA, institutions support the construction of a global knowledge system centred on greenhouse-gas-centric strategies. Apart from the universal concerns of ecological modernisation, such initiatives establish a suite of global climate risk factors. This for example encompasses climate science knowledge, the articulation of global indicators for managing conditions of life through levels of greenhouse gas emissions, storytelling about and with distant others, the documentation of changing conditions in different parts of the world, the future predicaments on which these calculations foreclose, and the adjustments required to national policies to achieve all these things. Museums increasingly frame partnerships for global engagement and at the same time bear witness to the changes that are happening. They also seek to build a community of impassioned and empathetic citizens.

Through these actions, and more recently guided by the ACE, museums in essence provide methodological guidance to the world with the intent of breaking down divisions between parties and encouraging action informed by the principles of the PA. In addressing climate, museums tend to promote one set of values through "big climate science" and at the same time create a single theory, knowledge system and a legal, financial, and moral framework.

As philosopher Claire Colebrook explains neoliberal languages of mitigation, sustainability, cap and trade, and renewable resources all relate to the globe as a milieu for human survival.[64] In endorsing and using such neoliberal terms the museum sector, as emplified through exhibitions such as "Atmosphere" and "EcoLogic," and more recent ones enforce a global view of human survival and showcase the mechanisms to achieve this.

Museums, therefore, support and promote a global governance framework consisting of a global method for mitigation enacted by global governance actors, an established system of global governance made possible through a network of global relations, and global interests established and maintained through global regulation and the establishment of a global rule of law. This network of systems is interdisciplinary through which the disciplines of science, social science, and law legitimise climate actions. But such a framework should involve a broader climate cosmology and comprise many other thoughts and practices within its framework. In doing so, such global frameworks enact the artificial segregation of alternative knowledges and expertise. Museums tend to see these differences in approach to climate action as competing rather than as potentially complementary and capable of working in ways to support each other.

The PA, as an example of methodological globalism, transcends the interests of the nation-state, territory, and interest groups, all of which are locked out of the negotiations in efforts to articulate a global vision for action. Museums support the global and modern vision of transition through energy source change rather than through regeneration. On the other hand, museums increasingly engage with dissenting communities, and specifically youth groups supporting activist initiatives. For example, The Climate Museum, in New York City, and the Museum für Naturkunde in Berlin are working with the Fridays for Future movement by supporting young activists and their teachers.

The Climate Museum in New York designs its exhibitions and programs to address the deficit in understandings of human-induced climate change, the local and global impacts of climate change, and potential solutions. The Climate Museum's New York interactive media literacy toolkit facilitates civic action in response to the lies and influence of the fossil fuel industry and offers a refreshing critique of the interests that limit meaningful action. An example of this was "Taking Action," a five-month interactive exhibition on Governor's Island, which introduced visitors to climate solutions, to the reasons those solutions are not being advanced, and to what individuals can do.[65] Through such exhibitions, the Museum advances a critique of the power relations and vested interests in such constellations. Despite this, however, museums need to critique ecological modernisation as an agenda for climate action, and not just the vested interests of modern capitalist regimes.

Refiguring human agency

Climate change strategies and policy in the west are premised on autonomous modern human-centred thought and design principles.[66] Agency and solution-based thinking tend to be cast in these ways because modern populations promote the idea that they have superior strategies and tools, such as climate science, technology, and economic mechanisms that have the capacity to resolve the climate crisis. Similarly, museum engagement with climate policy adheres to these specific human-centred, modern ecological formations and often endorses them through exhibitions, collecting, and pedagogical activities. But many forms of human agency are not sovereign. They are subject to the profound consequences of environmental injustice. For example Sami in northern Scandinavia, populations of people on Kiribati and the more than 25,000 species that are threatened with extinction are unable to exercise their agency in governing structures.

A reconsideration of the figure of the modern human, and how modern human agency is framed, situated, and enacted within policy frameworks, and in museum engagement with climate change, is critical for thinking and acting in vital attunement with the more-than and other-than-human world. The guiding principles of human agency, embedded in the PA, are outdated. In western legal systems, law academic Ben Mylius explains, "human agency and selfhood exist in an ideal realm where human agency is transcendent, external and independent of ecological context."[67] This figuration of human agency gestures towards the ability to wield power that is infinite, unconstrained, and unilateral and, until recently, that doesn't affect, or undermine, the conditions that sustain human and planetary life.[68] In this sense, western humanism, and its enrolment in the PA and in museums, is an ideology that is profoundly ecologically indifferent to human species embeddedness. The "Atmosphere" exhibition, for example, replicates all other modern mitigation projects in which environmental humanities scholars Lesley Head and Chris Gibson argue separate the western human subject from climate change and the non-human world both in the articulation of the problem and through the suite of solutions that modernity proposes.[69] Instead strong human agency through politics and politicians, science and scientists, the economy and economists, technology and design and technologists, and designers and engineers is shown to be holding supreme power and at the same time promulgating the idea of ecological indifference. Narratives in the "Atmosphere" exhibition explain Earth as a complex system able to be understood through science: "The Earth's climate is very complex, but scientists need to understand how it works. ...Changing one part of our interconnected planet affects the whole system." In one of the interpretive panels, the atmosphere is articulated as a complex biophysical system comprising components of Nature alongside earthly processes:[70]

> Everything on Earth plays a part in creating the climate
> ...the Sun's energy
> ...the movements of the atmosphere
> ...even the land and oceans, vegetation and ice.[71]

While the exhibition through its labels gestures towards the entanglement of human-induced and natural warming through statements such as: "Looking at the pattern of warming suggests it is linked to our carbon emissions" it does not completely reconcile the two systems of "Culture" and "Nature."[72]

As a climate-forcing system, separate from natural systems – human agency in this relational configuration becomes a system of reference and axis for action on which policy is directed. The establishment of these separate worlds, while unable to put the human governance of climate change in its proper relation with the worlds of nature, has led to quite specific ways in which humans formulate knowledge of, and experience of, climate change.[73] It is established as an exclusively human project. In these terms, modern populations are positioned as the sole source of action, but this is founded on hubristic relations of manipulation, domination, and control and orientates the atmosphere, if figured, as an entity that must be objectified, understood, rationalised, and controlled. These divisions fail to adequately acknowledge humans' enmeshments with the non-human world, interacting together as agentic forces.

The "Atmosphere" exhibition speaks of a pure Nature, as a non-individuated Nature where the social is seen as distinct from rather than enmeshed with human designs in a series of complex, dynamic hybrid systems. Narratives, for example, cast the atmosphere solely in terms of the material substance of matter and as particles comprising nitrous oxides, methane, carbon dioxide, water vapour, and as molecules and atoms. Here the atmosphere becomes a scientific nature, separate from humans and indeed not enmeshed in the bodily and cultural practices of modern populations. Science and technology are seen as the instruments that can bridge this divide, as able to attend to, and respond to, an out-of-control nature. One of the labels reads: "Science and technology can help us respond to the challenges...What are our options for tomorrow?"

Understanding the atmosphere as a means for directing strong human agency is reconciled through the scientific and ecological monitoring of Nature by scientists and their tools. The global monitoring of Earth's sea surface and near-ground temperatures is achieved through robotic floats and other devices to extrapolate long-term warming trends. Science discourse acts as a script that directs human action and thought. In this way, the atmosphere becomes an abstraction supported by a variety of formalisms such as objects, models, maps, simulations, and equations that attempt to represent and manage the atmosphere's material qualities as well as the so-called empirical or experiential coherence of them.

Under the banner of science, the modelling of higher levels of greenhouse gases provides the key to future action. In these terms, the future for humanity is predicated on a greater ability to control the atmosphere through more and better science, predictive software, and faster computers. In this scenario, the influence of human agency as outlined in an exhibition label is restricted to anticipated emission scenarios:

> Climate scientists are currently using computer models to help predict what might happen in different parts of the world... they provide useful information about how the planet might respond to higher levels of greenhouse gases in the atmosphere... with increased science knowledge and more powerful computers they're getting better and better... Controlling the atmosphere is possible with more accuracy with better science and technology.

In the "Atmosphere" exhibition, climate modelling and future forecasting are even expressed in a hubristic game that invites the visitor to: "Choose a person and use the climate modelling tools to help them understand what might happen where they live in 2100." This drive towards articulating greater precision, and therefore a higher degree of control of the atmosphere, is used as a script to articulate impacts and therefore adaptation strategies. Other exhibition labels include: "Looking at climate predictions shows the possible impacts on our lives," "Looking at studies around the world shows many things are changing as predicted," and finally, "Looking at models of the future shows how the climate might change."

Strong human agency is also framed exclusively as human agency, where the world becomes a set of objects and economic resources to satisfy human needs and desires. The PA therefore seeks to stabilise human life in the context of an out-of-control nature. In respect to ecological modernisation projects, the way human subjectivity, agency, power, autonomy, intentionality, and moral rights is written into the PA and museum engagements is problematic.

In the PA and the "Atmosphere" exhibition, human agency is framed as transcendent rather than situated and contextual and, in this way, they both represent a policy of global humanism through which all actions are framed. While the PA suggests climate change can be solved and its associated environmental and social problems can be managed through strong and direct human action, including rational decision-making, pollution trading, and the capping of emissions, more is needed, namely, a fundamental societal shift. Essentially climate change is conceptualised as a defined object towards which projects that attempt to stabilise the atmosphere can be directed using tools such as computers and biopolitical compliance. But it is so much more.

Scientific discourse and its technologies therefore come to act as a theology, the religious scripture that directs human action and thought. Hubristic narratives predominate on the basis of the Culture/Nature binary which aligns human subjects with nature as objects whereby humans command and control, measure, and model the atmosphere with the aim of designing interventions that privilege human survival.

The means to govern climate change is presented as focusing on biopolitical policing (the monitoring of sovereign human bodies and their carbon-burning practices) and the reduction of CO_2 emissions. Governance occurs through various apparatus, such as carbon taxes and new technologies, which are operationalised through human will and intent as technical and scientific knowledge. Nature is therefore rendered a subject mediated through human subjectivity.

These narratives do not consider the non-human as agential, as a coordinate of governing, except as a function or instrument put to work to serve human ends. The cost of this approach is that there is only a narrow suite of mitigation solutions to reduce greenhouse gases available through strong human agency and gas-centric ways of thinking. That is, targets and limits, marketing mechanisms such as taxes and trading schemes, and techno-centric investments and interventions destined to command, capture and harness nature's resources such as plant, solar, and wind power. This is demonstrated in more text from the "Atmosphere" exhibition:

> Can science and technology help tackle climate change?
>
> Reducing greenhouse gas emissions in a socially acceptable way requires scientific and technological ingenuity. Cutting-edge solutions will be both global and local in scale with innovation centred on the key sectors of power generation, transport and energy use.

While to some extent this is true, it doesn't engender radical environmental transformation by thinking and acting in attunement. Instead the future of humanity is predicated on technological investments in alternative energy sources – including biofuel, electric, hydrogen, wind and solar power – and measures that ensure improved energy efficiency. In addition futurist technologies are also seen to secure an atmospheric chemical balance such as artificial trees designed to remove CO_2 from the atmosphere.

Other mitigation approaches are framed on the basis of hubris where technologists command, control and manipulate non-human and earthly processes through technology with a big "T," such as carbon capture and storage. Some text from the "Atmosphere" exhibition reads, "Tackling global warming by deliberately manipulating the Earth's system will rely on finding practical, affordable methods for capturing carbon dioxide. This technology could let us keep burning coal and gas but without most of the carbon dioxide emissions".

In the exhibition an example of this at play is interactive games that ask the visitor to limit heating by blocking greenhouse gas particles expelled into the atmsophere to ensure Earth's average surface temperature remains suitable for life. Such a game represents biopolitical compliance and on a world scale is directed at individuals and corporations and mediated through carbon taxes and emissions trading schemes. This is akin to policing not only human and non-human bodies, and their carbon-burning habits, but ecological domains themselves.

Within this practice human agency, climate science, economics and policy-making co-conspire to, geographer Eric Swyngedouw argues, "support a climate stabilization discourse aimed at meeting a pre-defined stabilized temperature and concentration target to which international negotiations are directed."[74] The obsession with a singular "Nature" and stabilization narratives is, in Swyngedouw's words, sustained by "a particular 'quilting' of Nature that forecloses asking political questions about immediately and really possible alternative socio-natural arrangements."[75] These narratives all insist on the existence of Nature's innate stabilising force, that chemical imbalances in the atmosphere must be restored to an equilibrium because they have been disrupted by an external human agency.

In ecological modernisation projects, thus, the rationality that drives these accounts fail to give an adequate explanation of the non-linear ramifications and interpenetration of human activities and the atmosphere, and between technology and nature, except from a hubristic point of view. Modern autonomy, and sanctioned forms of intervention through the PA and museums, therefore, creates a human-centred world.

Framing human agency without embeddedness promotes the idea of structures as fixed, and as entities already constituted, towards which strong interventions can be directed to change the atmospheric composition informed by simple chemical equations based on cause and effect. The key actors present in the "Atmosphere" exhibition's climate change science narratives are humanist formulations of carbon, science, economics, and technology. Many other actors are rendered invisible. By restricting the coordinates of governing to carbon, economics, technology, and science, and accordingly to human apparatus, biopolitical compliance, and technocratic solutions, curators re-iterate these practices as the only viable means to govern. The target of concern, that is the "demon carbon," the chemicals that carbon burning emits becomes the climate question. Visitors to the exhibition are told to cut carbon (Figure 4.5): "Cut the Carbon. Can you reduce the world's greenhouse gas emissions by 2050?"

These scripts often portrayed as radical or innovative, assure "civilization" as we know it can continue, thus calling for a revolution without revolutionary

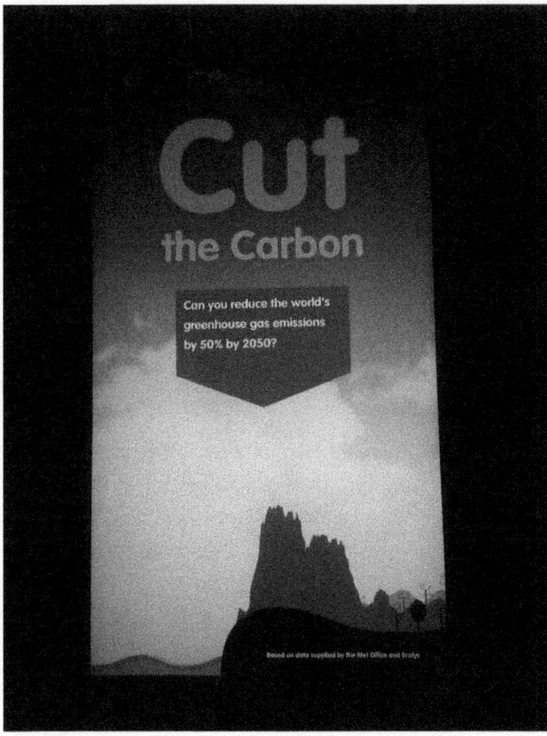

FIGURE 4.5. Interactive game: Cut the carbon in "Atmosphere: Exploring Climate Science". Photo credit: Fiona Cameron.

change."[76] The exhibition's underlying message is to change our energy sources and it is business as usual, thereby consolidating liberal ecological modernisation discourses. In doing so the exhibition acts as a microcosm of global debates on how to best govern climate change. Some more text from the exhibition illustrates this way of thinking:

> How easy would it be to cut carbon dioxide emissions?
>
> The power generated by burning fossil fuels is integral to our way of life. So making rapid emissions cuts would have substantial implications for human infrastructure and society. Instead governments are weighing up the costs of cutting emissions against the costs of adapting to the likely impacts of climate change.

While carbon is acknowledged as a natural element in all things, humans are charged with disrupting the balance of the carbon cycle: "Science can show us carbon's global pathways and how we're causing them to change."

Interestingly, carbon is described as an active agent in the climate system, but when talking about climate change and the human, carbon becomes a passive element, as something to be controlled through human intervention according to a simplistic, linear equation of carbon reduction and the restoration of equilibrium: "We can reduce emissions and restore equilibrium." Energy options are discussed solely in terms of their carbon content. Nature, wind, waves, and sun are harnessed to power humanity:

Narratives are fixated with human targets in mind and timetables are founded on the belief that the climate system can be stabilised to an imagined safe and secure stable level.[77] Another game in the exhibition, for example, asks participants to imagine what their lives will be like in 2025 if emissions are not cut and to start reducing emissions now. This challenges the participant to cut emissions by 50% by 2050.

Unsurprisingly climate stabilisation, as a framing concept, has directed governing at a global level towards the manipulation of emissions and the static quantification of the equilibrium response. This is done while rendering other ideas and approaches relatively invisible.[78] For example, another hubris interactive game asks participants to reduce greenhouse gas emissions in power production, transport, farming and forests, and buildings by 2050. Looking down from above to the earth through the computer interface, the participant commands and controls the atmosphere.

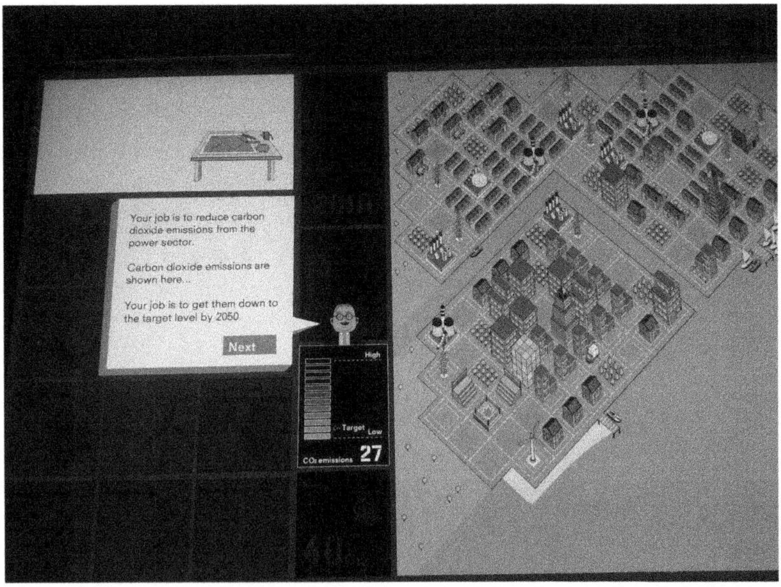

FIGURE 4.6. "Atmosphere: Exploring Climate Science", reducing emissions interactive game, Science Museum, London. Photo credit: Fiona Cameron.

In this narrow, reductionist mitigation paradigm, the coordinates of governing in "Atmosphere," and the technical means to govern are restricted to a closed ecological system comprising carbon, the rules and regulations of a carbon market and its global roll-out, greenhouse gas emissions, targets, international agreements and protocols, behaviour change, cost-benefit analysis, and climate science and market mechanisms such as carbon taxes and emissions trading schemes. This linear and simplistic policy discourse, however, has been consistently undermined by "potentially troubling and unstable biophysical processes that signal indeterminate and irreducible outcomes."[79] Such observations gesture towards the need to engage an expanded range of coordinates, each with different temporalities, durations, agencies, impulses, metabolisms, and effects as enfolded eco-curatorial alliances. The politics of climate change governance is predicated upon the fallacy of a singular, originally harmonious Nature, one that is now out of synch. If the "demon carbon" is properly managed, through a series of technological, managerial, and organisational solutions, then life can be securitised again.[80]

Furthermore, systems thinking, as a form of ecological thinking in the PA articles, comprises interdependent constituent parts involving the incorporation of industrial (energy, transport, technological development, and manufacturing), natural (air, soil, oceans, and biotic), and social systems (human needs, social justice, and equity), as well as the flows of goods, materials, and waste. Museums represent these subsystems. These relations are imagined as a system of systems enacted through aggregations of individual actions, technological innovation, development, and social needs. In this system, the non-human, or natural system, is a passive object to be either cast as a resource or stabilised. Each of these systems, or selected aggregates, is represented in exhibitions such as "Atmosphere," "EcoLogic" or, more recently, "Spark" or as systems linked to social justice and equity at the Climate Museum.

Principles of equity, justice, and fairness are framed within the PA and the "Atmosphere" exhibition through a humanist lens. Communities are only accountable in a moral-rights sense as a human-centred concern. Culturally diverse, vulnerable populations, and their concerns are folded into scientific calculations of risk, equity, and financing. All human populations are presented as sharing a common cause, goal, and destiny through global technocratic and economic measures.

Nature-culture conceptual distinctions are exhausted by the phenomenon of climate change but this position is not reflected in the PA or routinely in museum narratives. They, thus, remain as epistemological blocks for thinking differently about human agency.

Furthermore, while the ACE is framed as an article of empowerment for First Nations peoples, women, and so forth, is it a colonising agenda in which human agency, mitigation projects, and education are defined according to a western agenda and based on relational principles of the sovereign individual, neoliberal capitalism and greenhouse-gas centric thinking. Human needs and rights are placed at the centre, and human capacities in respect to climate change are framed in accordance with individuals' abilities to, through green capitalism, green themselves and their communities. These strategies enshrined in policy are directed to protect the planet so it can "support the needs of the present and future generations."[81]

But most importantly the ACE embraces the individual as autonomous and sovereign within a particular techno-political agenda. In doing so museums set themselves up as a public authority for a particular type of global humanism which focuses on a self-governing agent within an ecological or green capitalism framework. This framing effectively alienates people from their real power, which is the power of their own knowledge practices and the knowledge of the complex interdependencies in which their actions take place. The PA policy document and museum programs, therefore, effectively re-enforce the idea of dis-embedded autonomous agency.

The ACE must therefore be reworked so human agency is not autonomous but enmeshed and embodied across all the articles of the PA. Furthermore, it is not possible to understand how everything in the physical world works and therefore modern societies must let go of the desire to know and control. A profound adjustment must be made, an unlearning of thinking and acting as autonomous and sovereign.

Conclusion: Climate change mitigation and its humanist failings

It is clear that climate change action from a position of a more-than-human and material framing disrupts modern distinctions and notions of human agency but key policy documents and museum programs nevertheless continually seek to reinstate these divisions. The figure of the fossil-fuel-burning modern human, as an external forcing agent, must be rethought in order to adopt more effective and purposeful forms of human agency. When, and if, this occurs, all the entities and interventions in the PA will no longer be states and structures for formulating knowledge, methods, solutions, and their validation, but will constitute a more ontological approach.

Climate is a non-human activity per se that is profoundly different from political, social, and economic concerns that are seemingly controllable.[82] Human agency does not wield the power to intervene in the same way climate itself does and thus strong human agency engenders a sense of urgency, feelings of doom, and unsettling unpredictability predicated on a feeling of a lack of control. In respect to the PA, therefore, agency needs to be reworked. Modern societies need to let go of the erroneous idea that we can comprehend using climate science because, as Mike Hulme explains, the discipline can't do the work we hoped it will.[83]

The problem of strong human agency as a lever for action is indeed highlighted in IPCC reports. The translation of "big climate science," in its full complexity, into policy is viewed as highly problematic. As the IPCC 2018 report states "Heuristic devices and mental models can sometimes inhibit learning by obscuring a problem's full complexity… and complicating policy actions among both experts and… lay people."[84] Therefore, as the IPCC reports explain, such reductionism invites the illusion of strong human agency:

> It is not only the formal decision-making structures themselves that appear as problematic and obstructive; it is also the reductionism inherent in assumptions and mental structures through which we perceive the world and provide answers that fosters our illusion of self-mastery and effective decision-making and inhibits 'interconnectedness learning"[85]

The implementation of the PA through museums has the potential to develop an alternative, ecological conception of human selfhood, agency, and power to which our place in a climate-changed world, and the power we have in it, can be appropriately re-imagined. This would be one that understands humans and their capacities as interconnectedness in more deeply embedded terms.

Building climate change mitigation programming or promoting strategies that seek to bring about behavioural change based on the idea that our power is unilateral are not real options. The axis of relations outlining how the climate constellation is arranged together, and how modern humans might make decisions and frame themselves as agential, requires a different relational understanding of power as dynamic enmeshments, as bio-socio-material emergent processes and entanglements of phenomena. In the PA these coordinates have historically been presented, or thought about, as distinct natural systems and cultural systems.

Notes

1 I would like to acknowledge the work of Henry McGhie, formerly of the Manchester Museum, who was the driving force for enrolling museums as non-party stakeholders to the Paris Agreement in 2018. Paris Agreement, United Nations Framework Convention on Climate Change, 2015, accessed August 2, 2022, https://unfccc.int/process-and-meetings/the-paris-agreement/the-paris-agreement.

2 Jessica Schmidt, "The Empirical Falsity of the Human Subject: New Materialism, Climate Change and the Shared Critique of Artifice," *Resilience, International Politics, Practices and Discourses* 1, no. 3 (2013): 174–92.

3 Will Steffen, Johan Rockström, Katherine Richardson, Timothy M. Lenton, Carl Folke, Diana Liverman, Colin P. Summerhayes et al., "Trajectories of the Earth System in the Anthropocene," in *Proceedings of the National Academy of Sciences of the United States of America* 115, no. 33 (August 6, 2018): 8252–59. https://doi.org/10.1073/pnas.1810141115; IPCC, "Summary for Policymakers," in *Climate Change 2018: The Physical Science Basis.* Contribution of Working Group to the Fifth Assessment Report of the Intergovernmental Panel on Climate Change (2108). Authors: Lisa V. Alexander, Simon K. Allen, Nathaniel L. Bindoff, François-Marie Bréon, John A. Church, Ulrich Cubasch, Seita Emori et al., https://www.ipcc.ch/site/assets/uploads/2018/02/WG1AR5_SPM_FINAL.pdf, 13–14.

4 Fiona R. Cameron, "We Are on Nature's Side? Experimental Work in Rewriting Narratives of Climate Change for Museum Exhibitions," in *Climate Change and Museum Futures*, eds. Fiona R. Cameron and Brett Neilson, Routledge Research in Museum Studies (New York: Routledge, 2014): 56–58.
5 Lotte Isager, Line Vestergaard Knudsen and Ida Theilade, "A New Keyword in the Museum: Exhibiting the Anthropocene," *Museum & Society* 19, no. 1 (March 2021): 88–107, 88.
6 Isager et al., "A New Keyword in the Museum," 88.
7 Isager et al., "A New Keyword in the Museum," 90.
8 Isager et al., "A New Keyword in the Museum," 90.
9 Steffen et al., "Trajectories of the Earth System in the Anthropocene," 8256.
10 Timothy Morton, *The Ecological Thought* (Cambridge, MA and London: Harvard University Press, 2010), 9.
11 Morton, *The Ecological Thought*, 9.
12 Fiona R. Cameron and Brett Neilson, eds., *Climate Change and Museum Futures*, Routledge Research in Museum Studies (New York: Routledge, 2014); Fiona R. Cameron, "Theorizing More-Than-Human Collectives for Climate Change Action in Museums," *L'Internationale Online Climate Change Special Issue* to coincide with the UNFCCC meeting in Paris, November 30–December 11, 2015, accessed August 3, 2022, https://www.internationaleonline.org/research/politics_of_life_and_death/48_theorising_more_than_human_collectives_for_climate_change_action_in_museums; Fiona R. Cameron, "Ecologizing Experimentations: A Method and Manifesto for Composing a Post-Humanist Museum," in *Climate Change and Museum Futures*, eds. Fiona R. Cameron and Brett Neilson, 16–33. Routledge Research in Museum Studies (New York: Routledge, 2014); Cameron, "We Are on Nature's Side?"; Fiona R. Cameron, "The Liquid Museum: New Ontologies for a Climate Changed World," in *Museum Theory: An Expanded Field*, eds. Andrea Witcomb and Kylie Message, 345–62. (Oxford: Blackwell Publishing, 2015); Fiona R. Cameron, "Posthuman Museum Practices," in *Posthuman Glossary*, eds. Rosi Braidotti and Maria Hlavajova, 349–52. (London and New York: Bloomsbury Academic, 2018).
13 Donna Haraway, *Staying with the Trouble: Making Kin in the Chthulucene* (Durham, NC and London: Duke University Press, 2016).
14 Julie Katherine Gibson-Graham, "Diverse Economies: Performative Practices for 'Other Worlds'," *Progress in Human Geography* 32, no. 5 (2008): 614–15.
15 Cameron, "Ecologizing Experimentations," 24–28; Cameron, "Posthuman Museum Practices," 350.
16 Fiona R. Cameron, Ben Dibley and David Ellsworth, "Museum Collections, Photosynthetic Atmospheres," *Environmental Humanities* 15, no. 2, July (2023): 62–84; Marek Tamm and Zoltan Boldizar Simon, "Historical Thinking and the Human: Introduction," *Journal of the Philosophy of History* 14, no. 3 (2020): 285–309.
17 Fiona R. Cameron, "Climate Change, Agencies, and the Museum for a Complex World," *Museum Management and Curatorship* 27, no. 4 (October 2012): 317–39, 332.
18 Matthew S. Hull, *Government of Paper: The Materiality of Bureaucracy in Urban Pakistan* (Berkeley: University of California Press, 2012), 5.
19 Paris Agreement 2015, https://unfccc.int/sites/default/files/english_paris_agreement.pdf.
20 Cameron, "We Are on Nature's Side?"
21 Cameron, "We Are on Nature's Side?"
22 Cameron, "We Are on Nature's Side?" 58.
23 Cameron, "We Are on Nature's Side?" 58.
24 Cameron, "We Are on Nature's Side?" 58.

25 Mike Hulme, *Why We Disagree About Climate Change: Understanding Controversy, Inaction and Opportunity* (Cambridge: Cambridge University Press, 2009), 73.
26 Cameron, "We Are on Nature's Side?" 58–59.
27 Cameron, "We Are on Nature's Side?" 58–59.
28 Fiona R. Cameron, "Climate Change, Agencies and the Museum and Science Centre Sector," *Museum Management and Curatorship* 27, no. 4 (2012): 325–27.
29 Hulme, *Why We Disagree About Climate Change*, 73; Cameron, "Climate Change, Agencies and the Museum and Science Centre Sector," 325–27; Cameron, "We Are on Nature's Side?" 59.
30 Bea Mitchell, "Natural History Museum Sets Science-based Carbon Reduction Targets," *Blooloop*, November 5, 2021, accessed August 2, 2022, https://blooloop.com/sustainability/news/natural-history-museum-science-based-reduction-target/.
31 The Natural History Museum announces science-based carbon reduction target, https://www.nhm.ac.uk/press-office/press-releases/the-natural-history-museum-announces-science-based--carbon-reduc.html.
32 Ariel Salleh, "Climate Strategy: Making the Choice between Ecological Modernisation or Living Well," *Journal of Australian Political Economy* 66 (January 2010), 118–43, 124.
33 Cameron, "We Are on Nature's Side?" 56.
34 Jenny Newell, "Climate Museums: Powering Action," *Museum Management and Curatorship* 35, no. 6 (2020), 599–617, 604–5.
35 Paris Agreement, Article 6, https://unfccc.int/process-and-meetings/the-paris-agreement/the-paris-agreement, 24.
36 Newell, "Climate Museums: Powering Action," 604–5.
37 Jockey Club Museum of Climate Change, https://www.mocc.cuhk.edu.hk/en-gb/.
38 Museum of Innovation and Science, Schenectady (MISCI), New York, "Wind and Solar: The Renewable Energy Revolution," *Kids Out and About*, July 12, 2017, https://albany.kidsoutandabout.com/content/wind-and-solar-renewable-energy-revolution.
39 Museum of Innovation and Science, Schenectady (MISCI).
40 Paris Agreement, Article 6, https://unfccc.int/process-and-meetings/the-paris-agreement/the-paris-agreement, 26.
41 Australian Museum, "Spark: Australian Innovation Tackling Climate Change," *Exhibition*, accessed August 3, 2022, https://australian.museum/exhibition/spark/.
42 Visit Dubai, https://www.visitdubai.com/en/places-to-visit/museum-of-the-future.
43 Nicola Chilton, "Defying Gravity: How Dubai's Museum of the Future Was Built," *CNN Travel*, June 20, 2022, https://edition.cnn.com/travel/article/museum-of-the-future-dubai/index.html.
44 Dubai Future Foundation, https://www.dubaifuture.ae/about/.
45 Arthur P. J. Mol and David A. Sonnenfeld, eds., *Ecological Modernisation Theory around the World* (London: Routledge, 2000).
46 Salleh, "Climate Strategy," 136.
47 Toby Svoboda and Jacob Haqq-Misra, "Is Climate Change Morally Good for Non-Anthropocentric Perspectives?" *Ethics, Policy & Environment* 21, no. 2 (2018).
48 Svoboda and Haqq-Misra, "Is Climate Change Morally Good for Non-anthropocentric Perspectives?" 215–16.
49 Salleh, "Climate Strategy," 120–21.
50 Salleh, "Climate Strategy," 120–21.
51 Salleh, "Climate Strategy," 120–21.
52 Salleh, "Climate Strategy," 120–21.
53 Salleh, "Climate Strategy," 122.
54 Science Museum, "Our Future Planet: Can Carbon Capture Help Us Fight Climate Change?", accessed August 3, 2022, https://www.sciencemuseum.org.uk/see-and-do/our-future-planet.

55 The Climate Museum, "Culture for Action: A Note on Our Vision," accessed August 3, 2022, https://climatemuseum.org/mission.
56 The Climate Museum, "Climate Signals," Exhibition by Justin Brice Guariglia, September 1–November 6, 2018, https://climatemuseum.org/climate-signals.
57 Newell, "Climate Museums: Powering Action," 602–3.
58 Newell, "Climate Museums: Powering Action," 602–3.
59 Newell, "Climate Museums: Powering Action," 602–3.
60 Museum of the City of New York, "Rising Tide: Visualizing the Human Costs of the Climate Crisis," *Exhibition*, April 15, 2021, https://www.mcny.org/risingtide.
61 Zhenye Liu, "The Critique of Methodology: Nationalism and Discipline Construction of Global Governance Research," *Journal of Global Policy and Governance* 1 (2012): 4. https://doi.org/10.1007/s40320-012-0001-4.
62 Liu, "The Critique of Methodology," 4.
63 UNFCCC, "Glasgow Work Programme on Action for Climate Empowerment," November 13, 2021, https://unfccc.int/documents/310896.
64 Claire Colebrook, *Deleuze and the Meaning of Life* (London: Continuum, 2010), 61.
65 The Climate Museum, "Taking Action," *Exhibition*, June 1–October 27, 2019, https://climatemuseum.org/takingaction.
66 The use of human-centred design for climate change solutions is widespread: an illustration of my point can be found in Kate Maher, Carissa Carter and Melissa Miranda, "Human-Centred Design in the Age of Climate Change," *Earth and Space Science Open Archive (ESSOAr)*, February 17, 2019, https://www.essoar.org/doi/pdf/10.1002/essoar.10500828.1.
67 Ben Mylius, "Ecological Indifference: Thinking about Agency in the Face of Ecological Crisis," *Minding Nature* 8, no. 2 (Spring 2015): 27, https://www.humansandnature.org/ecological-indifference-thinking-about-agency-in-the-face-of-ecological-crisis.
68 Mylius, "Ecological Indifference," 27–28.
69 Lesley Head and Chris Gibson, "Becoming Differently Modern: Geographic Contributions to a Generative Climate Politics," *Progress in Human Geography* 36, no. 6 (December 2012): 700.
70 Cameron, "We Are on Nature's Side?" 59.
71 Cameron, "We Are on Nature's Side?" 59.
72 Cameron, "We Are on Nature's Side?" 60.
73 Cameron, "We Are on Nature's Side?" 60.
74 Erik Swyngedouw, "Trouble with Nature – Ecology as the New Opium for the People," in *Conceptual Challenges for Planning Theory*, eds. Jean Hillier and Patsy Healey (Farnham, Surrey: Ashgate, 2010), 308.
75 Swyngedouw, "Trouble with Nature," 310.
76 Swyngedouw, "Trouble with Nature," 319.
77 Max Boykoff, Dave Frame and Samuel Randalls, "Discursive Stability Meets Climate Instability: A Critical Exploration of the Concept of 'Climate Stabilization' in Contemporary Climate Policy," *Global Environmental Change* 20 (2010): 58.
78 Boykoff, Frame and Randalls, "Discursive Stability Meets Climate Instability," 61.
79 Cameron et al., "Climate Collections," 62–84.
80 Swyngedouw, "Trouble with Nature," 309.
81 United Nations, "Transforming Our World: The 2030 Agenda for Sustainable Development," A/Res/70/1, accessed August 3, 2022, https://sustainabledevelopment.un.org/content/documents/21252030%20Agenda%20for%20Sustainable%20Development%20web.pdf.
82 Jessica Reilly, "The Substance of Climate: Material Approaches to Nature under Environmental Change," *Wires* 9, no. 6 (November/December 2018), https://wires.onlinelibrary.wiley.com/doi/10.1002/wcc.550.

83 Hulme, *Why We Disagree About Climate Change*, 178–310.
84 IPCC, *Managing the Risks of Extreme Events and Disasters to Advance Climate Change Adaptation,* Special Report of the Intergovernmental Panel on Climate Change (2018), eds. Christopher B. Field, Vicente Barros, Thomas F. Stocker, Qin Dahe, David Jon Dokken, Kristie L. Ebi, Michael D. Mastrandrea et al., https://www.ipcc.ch/site/assets/uploads/2018/03/SREX_Full_Report-1.pdf, 53.
85 IPCC, *Managing the Risks*, 53.

References

Australian Museum. "Spark: Australian Innovation Tackling Climate Change." *Exhibition*, accessed August 3, 2022, https://australian.museum/exhibition/spark/.

Boykoff, Max, Dave Frame and Samuel Randalls. "Discursive Stability Meets Climate Instability: A Critical Exploration of the Concept of 'Climate Stabilization' in Contemporary Climate Policy." *Global Environmental Change* 20 (2010): 53–64.

Braidotti, Rosi. *The Posthuman.* Cambridge: Polity Press, 2013.

Cameron, Fiona R. "Climate Change, Agencies and the Museum and Science Centre Sector." *Museum Management and Curatorship* 27, no. 4 (2012): 317–39.

Cameron, Fiona R. "Climate Change, Agencies, and the Museum for a Complex World." *Museum Management and Curatorship* 27, no. 4 (October 2012): 317–39.

Cameron, Fiona R. "Ecologizing Experimentations: A Method and Manifesto for Composing a Post-Humanist Museum." In *Climate Change and Museum Futures*, edited by Fiona R. Cameron and Brett Neilson, 16–33. Routledge Research in Museum Studies. New York: Routledge, 2014.

Cameron, Fiona R. "We Are on Nature's Side? Experimental Work in Rewriting Narratives of Climate Change for Museum Exhibitions." In *Climate Change and Museum Futures*, edited by Fiona R. Cameron and Brett Neilson, 51–77. Routledge Research in Museum Studies. New York: Routledge, 2014.

Cameron, Fiona R. "The Liquid Museum: New Ontologies for a Climate Changed World." In *Museum Theory: An Expanded Field*, edited by Andrea Witcomb and Kylie Message, 345–62. Oxford: Blackwell Publishing, 2015.

Cameron, Fiona R. "Posthuman Museum Practices." In *Posthuman Glossary*, edited by Rosi Braidotti and Maria Hlavajova, 349–52. London and New York: Bloomsbury Academic, 2018.Cameron, Fiona R. "Theorizing More-Than-Human Collectives for Climate Change Action in Museums." *L'Internationale Online Climate Change Special Issue* to coincide with the UNFCCC meeting in Paris, November, 30–December, 11 2015, accessed August 3, 2022, https://www.internationaleonline.org/research/politics_of_life_and_death/48_theorising_more_than_human_collectives_for_climate_change_action_in_museums.

Cameron, Fiona R., Ben Dibley and David Ellsworth. "Climate Collections and Photosynthetic, Fossil-Fueled Atmospheres." *Environmental Humanities* 15:2 / July 2023

Cameron, Fiona R. and Brett Neilson, eds. *Climate Change and Museum Futures.* Routledge Research in Museum Studies. New York: Routledge, 2014.

Chilton, Nicola, "Defying Gravity: How Dubai's Museum of the Future Was Built." *CNN Travel*, June 20, 2022, https://edition.cnn.com/travel/article/museum-of-the-future-dubai/index.html.

The Climate Museum. "Climate Signals." Exhibition by Justin Brice Guariglia, September 1–November 6, 2018, https://climatemuseum.org/climate-signals.

The Climate Museum. "Taking Action." *Exhibition*, June 1–October 27, 2019, https://climatemuseum.org/takingaction.

The Climate Museum. "Culture for Action: A Note on Our Vision," accessed August 3, 2022, https://climatemuseum.org/mission.

Colebrook, Claire. *Deleuze and the Meaning of Life*. London: Continuum, 2010.

Dubai Future Foundation. https://www.dubaifuture.ae/about/.

Gibson-Graham, Julie Katherine. "Diverse Economies: Performative Practices for 'Other Worlds'." *Progress in Human Geography* 32, no. 5 (2008): 613–32.

Haraway, Donna. *Staying with the Trouble: Making Kin in the Chthulucene*. Durham, NC and London: Duke University Press, 2016.

Head, Lesley and Chris Gibson. "Becoming Differently Modern: Geographic Contributions to a Generative Climate Politics." *Progress in Human Geography* 36, no. 6 (December 2012): 699–714.

Hull, Matthew S. *Government of Paper: The Materiality of Bureaucracy in Urban Pakistan*. Berkeley: University of California Press, 2012.

Hulme, Mike. *Why We Disagree About Climate Change: Understanding Controversy, Inaction and Opportunity*. Cambridge: Cambridge University Press, 2009.

IPCC. *Managing the Risks of Extreme Events and Disasters to Advance Climate Change Adaptation*. Special Report of the Intergovernmental Panel on Climate Change (2018). Editors: Christopher B. Field, Vicente Barros, Thomas F. Stocker, Qin Dahe, David Jon Dokken, Kristie L. Ebi, Michael D. Mastrandrea et al., https://www.ipcc.ch/site/assets/uploads/2018/03/SREX_Full_Report-1.pdf.

IPCC. "Summary for Policymakers." In *Climate Change 2018: The Physical Science Basis*. Contribution of Working Group to the Fifth Assessment Report of the Intergovernmental Panel on Climate Change (2108). Authors: Lisa V. Alexander, Simon K. Allen, Nathaniel L. Bindoff, François-Marie Bréon, John A. Church, Ulrich Cubasch, Seita Emori et al. https://www.ipcc.ch/site/assets/uploads/2018/02/WG1AR5_SPM_FINAL.pdf.

Isager, Lotte, Line Vestergaard Knudsen and Ida Theilade. "A New Keyword in the Museum: Exhibiting the Anthropocene." *Museum & Society* 19, no. 1 (March 2021): 88–107.

Jockey Club Museum of Climate Change, https://www.mocc.cuhk.edu.hk/en-gb/.

Liu, Zhenye. "The Critique of Methodology: Nationalism and Discipline Construction of Global Governance Research." *Journal of Global Policy and Governance* 1 (2012): 3–15, https://doi.org/10.1007/s40320-012-0001-4.

Maher, Kate, Carissa Carter and Melissa Miranda. "Human-Centred Design in the Age of Climate Change." *Earth and Space Science Open Archive (ESSOAr)*, February 17, 2019, https://www.essoar.org/doi/pdf/10.1002/essoar.10500828.1.

Mitchell, Bea. "Natural History Museum Sets Science-based Carbon Reduction Targets." *Blooloop*, November 5, 2021, accessed August 2, 2022, https://blooloop.com/sustainability/news/natural-history-museum-science-based-reduction-target/.

Mol, Arthur P. J. and David A. Sonnenfeld, eds. *Ecological Modernisation Theory around the World*. London: Routledge, 2000.

Morton, Timothy. *The Ecological Thought*. Cambridge, MA and London: Harvard University Press, 2010.

Museum of Innovation and Science, Schenectady (MISCI), New York. "Wind and Solar: The Renewable Energy Revolution." *Kids Out and About*, July 12, 2017, https://albany.kidsoutandabout.com/content/wind-and-solar-renewable-energy-revolution.

Museum of the City of New York. "Rising Tide: Visualizing the Human Costs of the Climate Crisis." *Exhibition*, April 15, 2021, https://www.mcny.org/risingtide.

Mylius, Ben. "Ecological Indifference: Thinking about Agency in the Face of Ecological Crisis." *Minding Nature* 8, no. 2 (Spring 2015): 27–30, https://www.humansandnature.org/ecological-indifference-thinking-about-agency-in-the-face-of-ecological-crisis.

The Natural History Museum Announces Science-Based Carbon Reduction Target. https://www.nhm.ac.uk/press-office/press-releases/the-natural-history-museum-announces-science-based--carbon-reduc.html.

Newell, Jenny. "Climate Museums: Powering Action." *Museum Management and Curatorship* 35, no. 6 (2020): 599–617.

Paris Agreement, United Nations Framework Convention on Climate Change, 2015, accessed August 2, 2022, https://unfccc.int/process-and-meetings/the-paris-agreement/the-paris-agreement.

Reilly, Jessica. "The Substance of Climate: Material Approaches to Nature under Environmental Change." *Wires* 9, no. 6 (November/December 2018), https://wires.onlinelibrary.wiley.com/doi/10.1002/wcc.550.

Salleh, Ariel. "Climate Strategy: Making the Choice between Ecological Modernisation or Living Well." *Journal of Australian Political Economy* 66 (January 2010): 118–43.

Schmidt, Jessica. "The Empirical Falsity of the Human Subject: New Materialism, Climate Change and the Shared Critique of Artifice." *Resilience, International Politics, Practices and Discourses* 1, no. 3 (2013): 174–92.

Science Museum, "Our Future Planet: Can Carbon Capture Help Us Fight Climate Change?" accessed August 3, 2022, https://www.sciencemuseum.org.uk/see-and-do/our-future-planet.

Steffen, Will, Johan Rockström, Katherine Richardson, Timothy M. Lenton, Carl Folke, Diana Liverman, Colin P. Summerhayes, Anthony D. Barnosky, Sarah E. Cornell, Michel Crucifix et al. "Trajectories of the Earth System in the Anthropocene." *Proceedings of the National Academy of Sciences of the United States of America* 115, no. 33 (August 6, 2018): 8252–59, https://doi.org/10.1073/pnas.1810141115.

Stengers, Isabelle. *In Catastrophic Times: Resisting the Coming Barbarianism.* Open Humanities Press, 2015, http://openhumanitiespress.org/books/download/Stengers_2015_In-Catastrophic-Times.pdf.

Svoboda, Toby and Jacob Haqq-Misra. "Is Climate Change Morally Good for Non-anthropocentric Perspectives?" *Ethics, Policy & Environment* 21, no. 2 (2018): 215–28.

Swyngedouw, Erik. "Trouble with Nature – Ecology as the New Opium for the People." In *Conceptual Challenges for Planning Theory,* edited by Jean Hillier and Patsy Healey, 299–320. Farnham, Surrey: Ashgate, 2010.

Tamm, Marek and Zoltan Boldizar Simon. "Historical Thinking and the Human: Introduction." *Journal of the Philosophy of History* 14, no. 3 (2020): 285–309.

UNFCCC. "Glasgow Work Programme on Action for Climate Empowerment." November 13, 2021, https://unfccc.int/documents/310896.

United Nations. "Transforming Our World: The 2030 Agenda for Sustainable Development." A/Res/70/1, accessed August 3, 2022, https://sustainabledevelopment.un.org/content/documents/21252030%20Agenda%20for%20Sustainable%20Development%20web.pdf.

Visit Dubai. https://www.visitdubai.com/en/places-to-visit/museum-of-the-future.

5

COMMUNITARIAN DESIGN

Eco-curating climate change in attunement

Introduction

Chapter 4 was an occasion to critically investigate two policy frameworks: the Paris Agreement (PA), and its education policy, the Action for Climate Empowerment (ACE) and how the principles, values and notions of human agency embedded in these policies and climate mitigation strategies are mobilised through museum curation. In this chapter, I explore the PA and the ACE and how these policy frameworks might be reworked and expanded beyond their modern frameworks and at the same time highlight other practices that might be put to work to curate climate for planetary habitability. Furthermore, I promote new humanisms, ecological thinking, and living practices that take account of the metabolic processes of the non-human world in relationality through curatorial practice on which our future prospects come to depend.

Our lives and histories are inextricably ecological because all things shape other things but many societies in the West have lost their way. Modern societies are not truly thinking, feeling, or ecologically minded in an embedded sense because they see humans as the dominant governing system. Nor do western societies, cultural infrastructures, and industrial and political systems that formally govern their interactions with the non-human world, view others as cohabitants in a shared world that promotes ethical forms of conjoint action that, according to Haraway, has the potential to render non-humans capable.[1]

The current crisis is an outcome of fundamental errors in philosophy, economy, and representation, and it is a crisis that museums are complicit in. Museums have promulgated moral and ontological mastery over earth others through disciplines and established systems on which social and ecological relationships come to rest. Even institutions themselves are presented

DOI: 10.4324/9781315212067-5

as detached structures that are lofty and hierarchical, hovering above society and the non-human world. Critically, environmental humanities scholar Val Plumwood argued that dis-embeddedness,[2] also a key feature of museum engagement, prevents humans' very survival because of the destructive ways of thinking and acting such concepts promote and the obsessively anthropocentric and narcissistic orientations they support.

The divide between nature and culture widened in the nineteenth and twentieth centuries with the development of enlightenment sciences.[3] Cultural institutions promulgated ideas using disciplinary humanist orientations. Boundaries between humans, animals, cultures, and technologies were, and still are, described in a way that maintain the privileging of humanist thinking rather than promoting a shared sense of radical copresence, of shared events and processes. Furthermore, through representational strategies, the modern human as dominantly western and male became a figure of accumulation and planetary destruction underpinned by advanced capitalism. Taxonomic and classificatory practices of museum research tended to disregard, as Haraway suggests, the fact that organisms, including human populations, come into being through "complex webs of constitutive relatings."[4] Museums helped to make these fictions go viral and at the same time tended to obfuscate the idea of human agency as embedded and situated, and the non-human world as vital.

Museums in some cases represent the consequences of modern humanism. The representation of the Anthropocene in museums tends to showcase the immoral and unethical destructive practices wrought by narcissistic individuals and societies. Institutions, however, do not remedy the humanisms that have created the problem and the injustices and exclusions that result from these limited modes of being, thinking, and acting.

Not all societies are enrolled in this figuration of human agency and accumulation. Some are more ecologically minded because humans are not viewed as an isolated and hubristic species but as being part of a series of intimate relationships deeply shared with other things and beings.[5] The more-than-human, as a figuration to think differently and as an empirical reality, are activated through First Nations knowledge but these knowledge practices have historically been excluded from museum programming or sidelined as anthropological or pre-modern forms of life. This, however, is fast changing. Museum humanisms have become another historical error of totality and most significantly a theoretical and empirical imposition that seeks to overlay other types of ecological thought, the latter that now must be nurtured. First Nations' knowledges are seen as containing critical insights into how to negotiate the emissions crisis, relate to and work with the non-human world in recuperative, respectful and reciprocal ways, to refine practices of observation, and to invoke memories of environmental change and knowledge renewal in alignment with emerging conditions of planetary life.

Ecological thinking and acting as cultural and material realities

It is abundantly clear that material evidence of ecological thought and action surrounds us and indeed constitutes us in emergence. The energy sources that underpin modern capitalism and western ways of life are deeply ecological. We make our electricity, heat our homes, cook our meals, heat our water, drive industrial plants, and search the Internet from the energy that plants, long dead, absorbed from the sun many millions of years ago.[6] Our cars are fuelled by billions of dead non-human zooplankton and algae that were trapped in sand and mud for millions of years and then converted into oil and gas under immense heat and pressure.[7] We are deeply and profoundly ecological beings. These symbiotic relations signal our living together in intimate interrelatedness as opposed to the figuration of dis-embedded humanisms portrayed in museums.

Museums have historically relied on the authority of the scientific disciplines and promulgated the validity of certain knowledge practices. All the sciences museums rely on have their own ecological thoughts and practices but they are too restrictive because they focus on particular subjects and forms of dominant human agency. "Big climate science" promotes ecological thinking about the interaction between Earth and human systems, however, modern populations and their representation in museums as discussed in the previous chapter are viewed and presented as external forcing agents driving changes in the Earth system through, for example, greenhouse gas emissions that lead to higher surface temperatures.[8] Physics research focuses on the atomic, and principles of motion and these concepts are often directed towards making science and technology profitable. An example is the Deutsches Museum's exhibition, titled "Physics," which illustrates the physical principles of earthly processes and how they are harnessed and applied to mechanical devices to serve human ends. The principles of hydrodynamics and the properties of solid bodies, liquids, and gases are showcased to illustrate their application in aircraft and ship design.[9] In regard to the natural sciences, biological evolutionary thought has promoted the idea of the human as an animal through the sharing of DNA with other species. This has enabled us to think, feel, and act, to a degree, as beings that are interconnected with other life forms, but unfortunately has been put to work to portray humans as superior hominids. Accordingly, specieist hierarchies are routinely represented in museum exhibitions on evolution, such as the Hall of Human Origins at the American Museum of Natural History[10] and the Natural History Museum in London, to illustrate the emergence of the human species from earlier hominids.[11] The very same principles of evolution have been applied to demonstrate the progress and social development of societies and have supported profound social injustices and inequalities such as slavery, genocide, and Indigenous policies of assimilation. Biodiversity thinking examines the interconnectedness of biological systems through a scientific lens but this is often for the purpose of directing conservation efforts towards systems management in patriarchal terms. For example, the goal of the permanent "Biodiversity" exhibition at the National Museum of Natural Sciences in Madrid[12] draws on the museum's collections of diverse species to illuminate the theory of evolution and to bring attention to wildlife conservation and its protection. The exhibition poses two questions, "What specimens and items shall we curate, and where?," and "How and why should we curate them?" The Anthropocene, as a philosophical field of enquiry, thinks through humans' relations with non-humans and earthly processes with the aim of alerting us to the damaging effects we have had on the planet. These effects are akin to geological forces rather than a refiguration of human–non-human agencies. Much of this "humanist," ecological thinking, however, still privileges human thought, strong human agency, and the promotion of anthropocentric capabilities and action as a lever for change.

Interestingly, as philosopher Tim Morton observes Darwinian evolutionary thought is an example of ecological thought gone viral because such thinking has permeated deep into the modern imaginary and into concrete practices, such as medical science.[13] Even 160 or so years ago biological evolution stood against what most people thought of as the philosophy of science. Theorist Ernst Mayr explains that it took an ideological revolution in biology to enable the widespread acceptance of the ideas promoted by Darwin himself through taxonomic observation, practices of classification, and a genetic program of development.[14]

Many natural science museums spearheaded this ideological revolution, notably the Natural History Museum in London, through the development of collections for the purposes of research, however, in some circles these evolutionary thoughts still remain controversial. A revolution in thinking and refiguration of human–non-human relations, as occurred with Darwin's work but expanded to all domains of life, process, and phenomena, can happen again, and indeed is already underway, and museums can again play a pivotal role.

The profusion of ecological thoughts

Ecological thoughts of various kinds are proliferating in the creative arts, humanities, social sciences, and economic practice and these have the potential to provide alternative and valuable knowledge practices that can be inducted into museums. We are witnessing the rise in Science and Technology Studies (STS),[15] new materialisms,[16] feminist and geomaterialism,[17] environmental humanities,[18] critical posthumanities,[19] affect theory,[20] bio-art (collaborations between artists and living organisms),[21] the ontological turn in anthropology,[22] and eco-criticism[23] in all its multifarious manifestations. This is just to name a few theoretical approaches. For more than 15 years, I myself have pioneered and intensified my research in critical posthumanities and new materialisms, museum studies, and post-anthropocentric (more-than-human) museologies.[24] Post-capitalist and post-socialist economies are being rethought and envisioned in multiple forms and as sites for ethical action.[25] This is being done through the commoning of resources to ensure the survival of both humans and earth's others who are at risk due to a capitalist mindset as one that promotes the commodification and exploitation of the non-human world.

Potawatomi scholar and activist Kyle Whyte argues that climate change is an intensification of environmental change imposed on Indigenous people by colonialism and as the result of disruptions to the ecological conditions that support life by the expansion of capitalism, industrialisation, and carbon-intensive economies.[26] Developed over millennia, Indigenous knowledge comprises systems of monitoring, recording, communicating, and learning about relationships between humans, non-human animals, plants, and ecosystems that enable a community to survive and flourish in situated ecosystems.[27] The revision of Indigenous' and First Nations' ecological knowledge and practices seeks to strengthen self-determined adaptation planning for climate change and altered conditions of life.

Indigenous ecological perspectives, as philosopher Simone Bignall explains, embody different ecological understandings of the interconnected forces of human and non-human agencies operating within a complex system. They promote a materially situated and vitalist ethics of human responsibility, attunement, and care.[28]

Indigenous knowledge practices typically view the non-human world as kin in social communities and as displaying values in spiritual communion with humans. Some institutions, such as the National Museum of the American Indian and, more recently, the Australian Museum, showcase different ecological styles of thought in respect to Indigenous knowledge practices. In regard to Australian First Nations people, this includes an understanding of relations with Country and the non-human world.

Transition movements, such as Buen Vivir in South America, combine a range of values and seek to promote an economically, socially, and environmentally sustainable future.[29] Drawn from ancestral conceptions of "sumak kawsay" in the Quechua and "suma qaman" in the Aymara language, translated as Buen Vivir in Spanish, and "good living" in English, these transition movements are concerned with the collective well-being of human and all earth others and encouraging an alternative to development and consumption. The aim is on developing a sense of the collective. The concept of Buen Vivir also connects with anti-capitalist philosophies in the west that reject economic growth, political philosophy based on feminist thought and environmentalism, and Indigenous cosmovisions. Within this philosophy, a system of knowledge and living is encouraged that is based on the communion of humans and the non-human world in an interconnected community concerned with the spatial-temporal-harmonious totality of existence.

Buen Vivir therefore combines different cosmovisions, philosophical, spiritual, and philosophical premises. This movement inspired the 2008 revision of the Ecuadorian constitution in which nature was attributed rights. It reads: "We… hereby decide to build a new form of public coexistence, in diversity and in harmony with nature, to achieve the good way of living."[30]

Another example is the project titled Museums for Climate Action which held a competition called "Reimaging Museums and Climate Action." One of the winning proposals, titled "Existances" imagined a network of small, temporary structures spreading ecological knowledge developed by African and Amerindian communities across the Brazilian state of Minas Gerais. The proposal outlined a series of distributed museum structures being put to work to help prevent the destruction of ecosystems by large-scale agricultural businesses.[31] This was presented as being achieved by combining ecological knowledge for renewal and recuperation through narratives and pedagogical programs.

All of these approaches promote values, thoughts, and practices fully aware of how humans are connected with other beings and things, geological, animate, and inanimate. These profuse ecological thoughts have the potential to spark curatorial inspiration and are indeed doing so most recently seen in artworks and installations in art museums.[32] An approach based on embeddedness without hubris has the potential to reinstate not just the museum institution and its material form but all its inhabitations on a broad communal scale. The liquid museum concept I developed is both a mode and method that can be used to rethink institutional forms for climate action folded into earth processes.[33]

Through all these frameworks, western disciplinary work, institutional forms, and curatorial practices become what I call ecologising experimentations in museums directed to reframing modern humanisms into an ecological more-than-human framework.[34] This follows Bruno Latour's question as to whether we should modernise or ecologise worlds.[35] Accordingly, human populations, modern humanisms, and lofty institutions can become decentred in material and social relations and thereby create a new sense of community, embeddedness, associated relational responsibilities, accountabilities, possibilities, and ethical practices.

Under this model, a multitude of recuperative and mutually supportive values and principles in western disciplinary and First Nations contexts in respect to the non-human world exist or are emerging. New curatorial concepts and disciplinary practices can provide important insights into how to survive and flourish in a climate-changed world. These practices are linked to living ecologies and communities in mutual reciprocity and/or are informed by Indigenous cosmovisions.

Most broadly all these philosophies and practices register an intimate connection between humans and the ecological health of the environment and are directed to sustaining planetary processes and life forms. They offer multifarious options to move beyond human exceptionalism, strong human agency, and museum representations that promote interventions solely based on a command and control approach to mitigation and adaptation. These knowledge practices, scholarship, and pedagogies also represent a continuum of ecological practice. The critical posthumanities for example seek to return human bodies, knowledge, actions, and powers to their ecological place; knitting together with all manner of things, animate and inanimate, including non-human beings, materials, and earthly processes.[36] Critical here is Indigenous First Nations environmental stewardship and its emphasis on renewing knowledge in emerging conditions, instilling new sensibilities, and attuning thoughts and observations. Critical posthumanism, as a form of ecological practice, can be put to work in the interface of technological apparatus, empirical applications, and experimental work, and be applied across a range of museum practices in order to create respectful and sustaining solutions.[37]

Many of these "ecologizing" thoughts and acts are not solely directed to utilitarian concerns and service human needs of growth and extraction. Rather ongoing ecospheric habitability is viewed to be first and foremost a question of local and situated ecological thought and practice. These particular worldviews, and the human–non-human inhabitations that these socio-ecological systems and novel forms of scholarship invoke, and the rich ontologies and forms of interconnectivity, they promote have the potential to creatively diversify and refigure alternative, more equitable, situated mitigation and adaptation strategies. Rather than strategies framed according to conventional modern relations and human-centred policy, the engagement with climate change, through diverse and alternative knowledge practices and humanisms, enlarges, intensifies, and makes visible the diversity of human-climate natures and conceptual ideas involving human–non-human agency and recuperative and adaptative practices. It does not matter what form of attuned ecological thought and design practice are used in global, regional, or local policy and museum responses to climate change as long as they follow similar principles. There are many examples of Indigenous adaptive projects and emerging knowledge practices for example that can do similar work in different contexts or in combination as long as they decentre human needs, lessen environmental impacts, and invoke aligned sensibilities of care, respect, mindfulness, attunement, and responsibility. Radical interconnectivity and communitarian sociality within human–non-human collaborations are key.

Human agency, in mitigation and adaptation strategies, is rethought in the context of these more-than and other-than-human social collectives as distributed across a heterogeneous field of things and entities. Rather than just factoring in human labour, mitigation initiatives must take account of the conative force fields and effects of all things animate and inanimate, that is, human, more-than-human and other-than-human beings such as the atmosphere, oceans, and terrestrial forms.[38]

When human agency takes place in more-than-human collectives, feminist theorist Rosi Braidotti argues rationality becomes collective, and subjectivity multiple. Therefore, climate action becomes a collective effort that takes place transversally in flows that dislocate binaries such as nature/culture, subject/object, and so forth.[39] Many forms of agency exist and not all of them can be categorised as human–non-human transversal subjects. Rather what emerges is various forms of humanness nestled transversally. The agencies of the situated human are entangled with other agencies comprising different forms of cognition and intelligence as enmeshed processes that also include sensing, calculative processes, and various forms of human and non-human crafting. Together these collectives enact eco-curatorial distributed processes that include the agencies of the non-human in many forms.[40] Through such a framework more-than-human framework institutions become liquid, nimble, mobile, and ecological compositions, promoting new forms of compositional design.[41] The articles of the PA then must be reframed as involving a diverse array of distributed coordinates, agencies, and eco-curating processes that are never completed but that beget other coordinates and processes.

Relational human agency as described imposes new forms of ethical constraints that require the exercise of power and intent in mindful and careful ways. The exercise of agency takes the form of, what I call, eco-curatorial thinking and practice that involves the multiple coordinates in which humans are enmeshed and dependent.[42]

What all these practices share, and what feminist philosopher Donna Haraway observes, is that, "all things beget other things as processes of 'becoming with' a world" in which "natures, cultures, subjects and objects do not pre-exist their intertwined worldings."[43] Thinking and acting in different ecological forms in which values are shared represents, what Haraway suggests, a radical and ethical commitment to thinking with, rather than for, one's own ends.

Ecologically and materially inflected experimental design practices must permeate all domains of modern life, including philosophical thought, everyday practices, finance, consumption, global trade, economic governance, and investment. This is work museums are well positioned to contribute to. As James Williams explains,

> A simple human ethical responsibility is thereby replaced by a joint ethical task in a shared ecosphere; we move from questions of what is right for us humans to questions of how to care for an ecology as a complex series of interdependencies.[44]

Attuning modern climate narratives and projects

In my 2014 co-edited collection, *Climate Change, Museum Futures,* with colleague Brett Neilson and specifically my chapter "We are on Nature's side:

Experimental work in rewriting narratives of climate change for museum exhibitions," I progressed a new type of politics attuning modern narratives of climate change and mitigation in which fossil-fuelled populations were folded into climate processes as embodied and enmeshed entities. A central feature of my ontological refashioning of these climate narratives was the reworking of the humanist idea of man, of human agency, into what constitutes the significant relations in which we are bound and of community in which we depend, acknowledging the vitality of matter and the agential role of the non-human worlds.[45]

Climate change governing coordinates and strong human agency directed towards the control of the atmosphere are no longer restricted to, nor founded on, a modern constitution of climate science, technological innovation, and economic and political practices operating under the aegis of ecological modernisation directed towards the abatement of human CO_2 emissions. Rather, governing practice is now more intimate, attentive to a comprehensive range of coordinates that are not just the human, but include the more-than and other-than-human world and the radically asymmetrical interrelated processes in which they are enfolded, and that were all previously categorised and collapsed into the signifier Nature.[46] Human agency and subjectivity are now rethought as unmediated contingencies that incorporate all manner of worldly coordinates and earthly processes. These have always been there and significant but until now western societies have failed to adequately accept this ecological reality. Decision-making can no longer be presented as a solely human pursuit. Events and phenomena, previously traced through linear accounts of human agency, can no longer be sustained. The atmosphere as an entity "over there and outside of us" in which curating with carbon released through anthropogenic burning activities is articulated as one type of external forcing agency driving the climate to change but such interventions must be one of communion as part of us in a way it always was and will be as dynamic biochemical fossil-fuelled life forms in motion discussed in Chapter 3. A consideration of these complex and entangled performativities as eco-curatorial processes out in the world in curatorial practice become, what I call, eco-curatorial methods and narrative forms.

Through an engagement with climate change as an empirical reality, and its narration through museum representations as eco-systemic performances, it is possible to challenge claims of dominant human agency and refigure such perceptions as agential dynamism. Strategies and narratives made explicit in the "Atmosphere" exhibition are then perceivable as relational material enactments. The atmosphere and its coordinates, comprising the concerns of modern forms of governing and distinctions between human agency and the non-human world, no longer exist. It is not just a question of a relational refiguration between human and non-human forces in climate governance and its narration, all strategies must also involve material accountability. Governing arrangements thus comprise an array of vital actants of interrelated agencies and forces. This includes, for example, earthly, non-human, and chemical processes, such as water, clouds, nitrogen, oxygen, carbon dioxide, coal, malaria, cyclones, oceans, terrestrial systems, trees, ice, radiation, heat, electrons, and gravity. With this in mind, human agency becomes humanness in its various forms. These forms are multifarious. They are embedded and embodied practices of crafting and composing, they interact with other material agencies and forces as stakeholders. Such considerations must be central to lifestyle decisions. This includes the formulation of legislation and economic strategies, the multifarious ways electricity is generated, the concerns of climate science, how computer modelling is conducted, and who is involved. Also how carbon taxes and emissions trading are composed, the materialising affects of coal-fired power stations, carbon capture and storage, nature mitigation, and tree ethics. A refiguration of the ideologies, beliefs, and rationalities that underpin such decisions is vital and this is where museums have a powerful role to play.

Each of these coordinates – from social actors, such as scientists, rationalities, legislation, economic theory, and climate science, to human and non-human machines such as coal-fired power stations, to earthly and chemical processes such as gravity – operates as multi-scaled, extensive, radically interoperable webs of heterogeneous coordinates, forces, and agencies.[47] These ecological processes are rarely, if ever, fully Nature or fully human. Eco-curating, narration, and curatorial practice must take into account the multiple durations of these distributed climate coordinates within these agential tangles.

Empowerment moves from strong human agency to interventions based on the distribution of humanness across climatic ecological compositions as multiple entangled composing forces – unfolding and enfolding and begetting other coordinates in emergent processes. New modes of governance become processes of composing in intimate and vital attunement, guided by values of care and respect. Representational practices develop into material and materialising enactments. The practice of museum curating has always been material enactments. The relations explicating the interaction between human–non-human worlds are composed through exhibitions but also narrative, pedagogies, and documentation practices.

Reworking climate change governance, and its narration in museums through an eco-systemic framework, considers how the atmosphere and climate mitigation is composed, conjoined, or transformed by human, more-than-human, and non-human agencies, and the interrelatedness involved in their emergence. Drawing inspiration from theorist Jane Bennett's work on the vitality of matter,[48] all these coordinates operate together as constituted and constituting forms within the extended, distributed webs of intelligibility that comprise them.[49] The atmosphere, as a singular entity, as an organising logic for climate change governance, and according to a simplistic vision of cause and effect to achieve stability, is therefore no longer productive as a sole source to direct action.[50]

The modern precepts that frame the PA and the ACE thus become another form of colonial imposition in which universal science, ecological modernisation, and nature capital usurp all other knowledge practices. The ideologies and practices of ecological modernisation that inform the PA and ACE are complicit in racial, gender and economic exclusions, and environmental injustices.

These policy frameworks must be refigured to accommodate different genres of being human and of human agency, collective responsibility, and action. The ACE must, for example, refigure its stakeholders – Indigenous communities, youth, women, museums, and so forth – as contributors to climate governance. Not as passive receptors of ecological modernisation strategies and PA principles around decarbonisation, but as agents in considering the multiple, practical, socio-cultural, material practices, and expertise that comprise its constituencies. There are many ways diverse communities might contribute to new methods and progress fundamental values of the social and collective by situating the non-human world alongside methods of attunement, respect, and care.

Climate change, therefore, emerges as a vast sprawling ecological composition comprising disrupting and transforming forces in dynamic processes of infinite reach penetrating deep into the geological substrates of the earth, through mining and industrial processes, and facilitating chemical interactions in the upper atmosphere.[51] Human populations are just one of many agential coordinates operating in these eco-curatorial processes.[52]

This refiguration of climate change and governance directs planetary solutions away from global policy abstractions and economic growth towards considering the multiple ways of being in the world as part of more-than-human climate collectives and the methods that might be respectfully enacted as culturally diverse solutions. Such a refiguration works against perpetuating the universal and modern on a global scale. The problem of climate change, and indeed museum engagement, encourages a reconsideration of how new subjects, new concepts of the human, and new planetary relations are conceptualised as irrupting agents, even within and through the conventions of the IPCC.

New humanisms, transcorporeal subjects, and museum engagement

As part of the Australian Research Council Linkage project, *Hot Science, Global Citizens: The Agencies of Museums in Climate Change Interventions* (2008–2013), the first international research project on museums and climate action, surveys were conducted in Australia and the USA and showed how, confirmed alongside more recent studies, that the perceived risk of climate change, or changes in public opinion, is not associated with measures of general public knowledge of climate change. Nor is it linked to their emotional response in terms of how people view and evaluate the issue.[53] Engagement with the topic through museums, therefore, is dependent on the framing of worldviews based on one's own locus and expressed as a personal responsibility and priority.[54] This rests on an environmental focus, or an altruistic concern for the environment and non-human others. It is distinct from the "hip pocket" focus on individual and economic self-enhancement.[55] Based on our Australian Research Council findings of an economic versus environmental prioritisation (economic focus – environmental focus), and on a sense of empowerment, we identified six population segments.[56] The US sample (dark grey) showed a greater economic focus and feelings of disempowerment (27% of the population) with a smaller cohort of environmentally focused individuals. The largest group in Australia (light grey) was economically focused, disempowered comprising 21% of the population. The most engaged were those in the segment environmental focus – disempowered, 19.7% of the Australian cohort and 18.8 of the US cohort.

We identified these six segments based on a two-dimensional framework. One dimension was about the relative importance of climate change compared to economic exigencies. The second reflects the degree to which people felt disempowered or helpless in light of climate change.[57] They question whether it is something they feel is beyond their control.

The public debate about climate change consists, on the one hand of an engaged subset of the population who express a "we must do something"

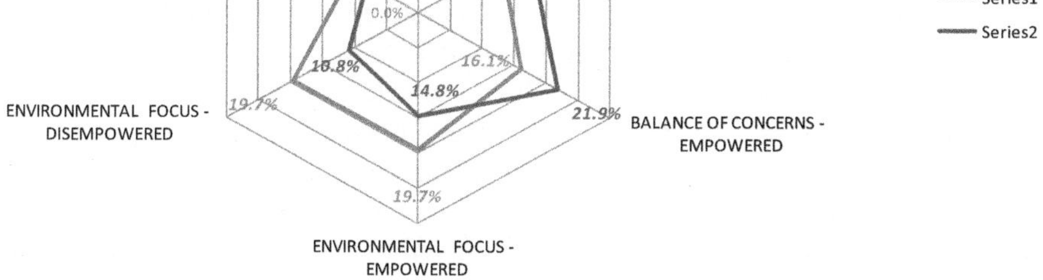

FIGURE 5.1 Climate change population segmentation, Australia and New York City, New York State and New Jersey, Hot Science Segmentation Analysis Report, Fiona Cameron, Institute for Culture and Society, Western Sydney University, 2016.

sentiment, and on the other hand, a group who tend to be less vocal and less interested. For a portion of the public, the dialogue is unlikely to be understood and processed unless points are expressed in terms of the climate change issue relative to economic issues. If publics are to get behind mitigation and adaptation initiatives, then more than half have to be convinced that the issue is at least as important as their economic priorities. Communicating the co-benefits of addressing climate change, in economic terms, would motivate action to a similar degree as believing anthropogenic climate change is real.[58] Here ecological compositional design practices have an important role to play in the development of mitigation technologies and adaptive strategies that can express climate change in terms of its negative economic impacts or adaptive strategies as positive economic benefits.

Environmentally focused individuals generally hold egalitarian, communitarian worldviews. This is in comparison to those that reveal that they devote greater attention to individual needs. As revealed in Figure 5.1, 39% of those surveyed in Australia and 33.6% in the US fell into these categories. Our findings thus suggest the possibility of the emergence of new humanisms as embedded and embodied human subjects. Changing behaviour without what Plumwood calls material or terrestrial intimacies and thoughts,[59] the case for the majority of climate programs lacks substance and material accountability, but in combination with earthly thoughts and the refiguration of modern populations as dynamic biochemical lifeforms bonded with the non-human, they would have the potential to alter politics, disciplines, and ethical practices. The environmentally focused segment of the population is akin to Haraway's "sym or sympoietic people,"[60] individuals who are likely to hold worldviews open to a new type of earthly politics.as opposed to non-sympoietic people who hold hierarchical, individualistic, and economically focused worldviews.

In folding ecological thinking into strategies for climate action in light of these findings, environmental humanities scholar Stacy Alaimo's transcorporeal subject emerges as a new figuration of the human after humanism, a figure conscious of materialism and transcorporeal responsibility, in which thoughts, acts, and bodies are intermeshed with the dynamic material world that crosses them, transforms them, and is transformed by them.[61] Aligned with Aliamo's porous transcorporeal subject is Rosi Braidotti's view of the posthuman subject in which agency is transfigured as embedded, embodied, transversal, and affective. Therefore thinking, reasoning, and acting become situated and emerging within diverse ecologies.[62] This is not as a singular capacity and agency but rather as a more-than-human subject produced by and within broader collectives and inter-relational dependences.[63] This collective of agencies and temporal durations inflect the direction of events and make a difference that cannot always be known in advance.[64]

Climate change is a profound failure of human agency. Rather than greening citizens, modern populations become ecologising lifeforms or at least those that are already deeply connected as sym people, as our findings show they already are. Museum's roles then include opening up worldviews to communitarian entanglement and facilitating eco-behavioural change.

Through the new figuration of humanism, in which modern populations become dynamic fossil-fuelled biochemical lifeforms, thinking and acting on climate change is enacted materially as a transcorporeal, attuned, affective, material lifeform, and subject passionate about doing environmentalism differently. The collective of agencies and temporal durations in which the transversal, transcorporeal subject is enmeshed and embodied therefore emerges in social movements as materialist, as deeply materialising, materially accountable, and responsible.

In curatorial practice reframing, human agency as ecological research, which has integrity and is bio-materially diverse, represents new conditions of possibility because of the enaction of a dispersed sense of agency. The physical sciences for example can promote new forms of relational research in collaboration with the embodied and situated human and more-than-human and other-than-human agencies. Research work across disciplines, both inside museums and out in the world, involves the observation and study of the relations and material affects of what agencies, including our own, do in deep copresence. Indeed this is what climate science seeks to achieve. But this disciplinary focus is too concerned with the tendencies of greenhouse gases and the mechanisms of heating. Such studies must be extended, at least partially, to include the entangling of infrastructures, cultural values, and political decisions about how all these things are folded into carbon-burning practices and their material affects in emergence.

Linear time and human agency are rethought in duration, in dynamic multiplicity, and in emergence, meaning human agencies are considered with climate change times that are at once multiple and are differential. A consideration of the wider distribution of climate-related agencies as material processes has the potential to reconfigure our relationships with the world and form new habitability practices. By thinking, acting, and crafting in a mesh of trans-corporeality with lively forces, new ethical frameworks and governing practices can be fostered in which non-humans become kin in shared futures. This confederation of agencies as thingness and in eco-curating processes through which multifarious coordinates in collaboration distribute, usurp, transform, become frictious, destablise, vitalise simultaneously and at the same time exhibit different rhythms, temporalities, momentum, and intensities. In order to curate consciously, with an expanded repertoire of agencies, and as a distributed form of agency, strategies must acknowledge and accommodate a multitude of non-humans and their enmeshments and must consider the porosity between these interactions and their affects in complex durations. Habitability projects, therefore, become interventions or cuts into the complex webs and multiple durations of these more-than and other-than-human processes.

The worldviews and human–non-human inhabitations that socio-ecological systems, novel forms of scholarship, and eco-curating invoke have the potential to creatively diversify and refigure alternative, more equitable, and situated mitigation and adaptive strategies in museums and beyond. This can be achieved by rethinking human agency as radical atmospheric inhabitations and curatorial crafting as eco-curating with all manner of disciplines, including the biological sciences, economics, technology, and so forth. By proposing a different conceptual framework an expanded set of coordinates for climate change governance can be developed in which human and non-human destinies are understood as entangled political systems and revisions are made to such governing strategies in respect to this array of agencies and who is negotiated with. Rethinking humanness as transcorporeal and climate action as complex eco-systemic processes requires deep mindset changes and museums can play a leading role.

Eco-curating for planetary habitability

Acts of curating have always been worlding practices even if the relation between thought, action, and material consequences has not always been acknowledged. Therefore, while museums have been complicit in modern forms of thinking and acting through curatorial work that has been destructive, institutions have an important role to play as places for reframing knowledge, modifying disciplinary, research regimes, and changing the very concept of curating so that it becomes attuned eco-curating crafting practices "in-worlds."

Importantly rather than judiciously following the principles of the PA, and other global policy documents as humanist practices, institutions can become active agents in formulating policy and context-specific and attuned climate research by engaging a range of knowledge practices. Of course, none of these should necessarily be presented as exclusive or as being authoritative over other practices. Through such an approach, museums can promote thinking and acting beyond the anthropocentrism of modern humanism, that currently underpins governance initiatives and programming, and move towards working with climate change as an entangled socio-technical, material, biological, and compositional process.

Conceptualising museum research, narrative, documentation, curating practices, and museologies with, and as part of, more-than-human social collectives and diverse knowledge practices can transform time-honoured interventions that centre around human will and intent. The challenge, however, is to build new narratives and to meld climate science with thoughts that situate humans in ecological arrangements. In doing such arrangements foreground our radical interconnectedness and co-dependence with others, shared predicaments as emergent processes, the social and community as more-than-human and other-than-human and thinking-feeling communities that engender respect and act responsibly towards others and that render others capable through their interactions.[65]

In reframing museum representations as dynamic ecological thoughts and actions in more-than-human social collectives, institutions can work with alternative concepts of social inclusiveness that acknowledge the many non-human coordinates that now form part of civic life, and indeed always have done.[66] By engendering respect for various forms of life and inanimate things a new position emerges from which transcorporeal curatorial transactions can operate that are based on composing respectful and ethical enmeshments. From this baseline better decisions about how best to deal with the ecological crisis can be made.

Ecological communitarian design and eco-curating go beyond the sharing of DNA and genetic variation in populations to the communal sharing of all manner of thoughts, affects, and actions. Through an eco-systemic framework, for example, curatorial practice can create an awareness of different social, cultural, material, political, and technical agencies, registers and circumstances in which climate governance resides and through which it can be enacted. Climate change governance, research, and eco-curatorial practices therefore become acts of composing by and with entangled bundles of coordinates. Through this process, the registers of sensing and calculating, the discursive, atomic, and materialising are born out of an infinite variety of conjoined material, mineral, chemical, discursive, energetic, social, human, non-human, natural, and cultural forces as domains of influencing.[67]

In Haraway's Chthulucene (precarious times in which a reconsideration of human and multispecies relations must be reworked as practices of becoming-with)[68] ecological communitarian design becomes a question of how modern populations are bound up, and who with, and in what ways. But most importantly it is about how we might bind ourselves with others, the types of worlding practices we might engage with and what relations matter in specific contexts. A consideration of the vital relations that arise in specific circumstances can be achieved through expanded forms of communitarian design. Central to the concept and methods of eco-curating is what Haraway calls a process in which "things beget other things."[69] For example, in the context of the documentation and interpretive work Haraway's "begetting" as eco-curatorial becomes cuts into the temporalities and durations of thingness, and the tracing of connections in past and emergent processes. Eco-curating is also directed to thinking about new ways we can be bound up in relationships of begetting in respect to both human agencies and those of non-human worlds. Thinking and curating as dynamic compositional lifeforms, as transcorporeal or transversal subjects, requires a consideration of what is involved, and who is involved – and their agencies, entanglements, requirements and dynamic exchanges. This, thinking and acting with others is no longer just about what is best for modern populations and the economy.

The physical and biological sciences are materialist in their aims and seek to understand the non-human world, and its agencies in relation to human populations. Scientific modelling as a worldling practice, for example, can be complemented with experimental practices informed by the critical and environmental posthumanities, and Indigenous cosmologies. The combining of knowledges and practices that share principles of attunement has the potential to make visible, at least partially, what consequences may emerge from certain entanglements and the risks and dangers of composing with forces we don't fully understand. Experimental forms of eco-curatorial design can be a method for considering what goes with what, what has its own agencies, how they are expressed, and how individuals and communities can work in collaboration with more-than-human processes and non-human agencies. Therefore, eco-curating and communitarian design as a method involves the study of relationalities, interdependencies, and enmeshments and then the consideration of the different types and forms of negotiation required with others. The expanded sense of community that emerges beyond the human-social comprises multiple ecologies of inhabitation, comprising elements, phenomena, processes, multispecies, and life forms in deep copresence.

Eco-curating, as a method for governance and representation in climate contexts, is, as suggested in my previous work on collections documentation, for example, on digital heritage and more-than-digital compositions in Chapter 2 and the green plastic bucket and the Boulton and Watt Engine in Chapter 3 is primarily onto-ethnographical. By tracking the processes – both the ontology (the affecting capacities and emergent consequences of the various coordinates within an expanded and dynamic field) and the ethnographical specificities (the specific elements and agencies, including the meanings and values attached to them) – it is possible to investigate the many different ways to gain an understanding of what climate practices might emerge and how they might be articulated and given new meanings.[70] Eco-curating can also become ethnographical form of sampling that comprises, inhabits, mobilises, or is mobilised by as an extended climate composition in the field, and a consideration of its multiple agencies, sequences, interrelations, interoperabilities, and spatial and temporal dimensions and durations.[71] The onto-cartographic mapping of these embedded, materialist constellations – as different kinds and forms of relations, and patterns of distribution that arise alongside their intended and unintended consequences – become their emergent histories and effects.[72] Eco-curatorial experimentation, in respect to this process, can be conducted through cutting, crafting, mixing, knitting, weaving, and mending in recombination and attunement with lively agencies and forces. Having said this, capturing the full range of coordinates, and their enmeshments, can never be complete because the influence of others in emergence cannot be fully known or visible. In eco-curating practice, the expanded field of coordinates that emerges, or comes into view, might become what scholars Woolgar and Lezaun contend in STS practice, for example, as way of representing the world as an approximation,[73] and will therefore retain a certain distance from what Swyngedouw calls "the Real of natures."[74] "Real natures," Swyngedouw explains, are complex, chaotic, often unpredictable, radically contingent, historically and geographically variable, risky, patterned in endlessly complex ways and ordered along "strange" attractors.[75]

Conclusion

Creative projects have always been practices of composing and crafting, of knowing the agencies of the non-human world and harnessing them for human purposes. All knowledge is situated in terms of relations with other kinds of knowledges which are necessarily entwining together and have the potential to create new ways of doing politics, economy, making-culturing, and curating. Composing through research, and for attunement, draws on knowledge from the sciences, and Indigenous practices of observation and interpretation all of which are mobilised through various socio-ecological knowledges. Eco-curating therefore becomes new curatorial experimentations that work with all manner of knowledges, thingness, forces, and their various temporal durations. Curating ecologically can be enacted in different forms as long as the principles that facilitate observation, a consideration of enmeshment, and a willingness to act in attunement, and most importantly a disposition of thinking with others and their agencies as ethical relational possibilities, are at the fore.

Museums, therefore, can emerge as agencies for the promotion of new ecological design and empathetic sensibilities. They can be perceived as eco-curating experimental laboratories for pedagogies, exhibition design, empirical experimentation, content development, distributed institutional forms and the combining of knowledge practices. And most importantly they can act as a means of thinking as different lifeforms in worlding work.

In climate change, governance scientific and biodiversity research can be put to work in mitigation projects to gain partial knowledge of how all the coordinates and agencies, and their entanglements can support the design of projects for attunement rather than for extraction and commodification. Such work can be complemented by adaptation projects that consider the tangle of radical coexistence in which humans are embedded, how these have triggered events and situations in local settings and how this knowledge can be levered differently to render partial recuperative measures.

Detailing the properties, and the form, of coordinates and their composing forces must be taken into account in eco-curating climate change occurs in the context of particular, unique ecological circumstances. Each ecological composition in emergence has its own unique features and elements. Each comprises unique forms of thingness such as technologies and ecologies, and their own deep and long histories of relatedness and enmeshment many of which are difficult to categorise – as either distinctly human, conceptual or ideational, technological, biological or mineral – because they are radically interoperable. None of these coordinates and agencies have an independent existence. Neither are they capable of being delineated into discrete things or objects.

As new forms of curating emerge as eco-curatorial processes, curators are decentred, embedded, and embodied in these processes.[76] Thinking, sensing, acting, composing and crafting, for example, can be done with soils, carbon burning, trees, diverse communities, distant others, disciplinary knowledge, technology, the economy, and so forth.

The Science Museum Group (SMG) has pledged to reach its net zero target across all of its five museums by 2033. Emissions from SMG buildings and operations account for 10% of the group's overall carbon footprint; the rest comes from goods and services. The group aims to reduce this figure through a combination of cutting emissions and carbon offsetting schemes, the latter through the planting of 1,000 trees a year.[77] But what the Museum fails to do is to think beyond a greenhouse gas mindset and work in attunement – that is consider the SMG as material climates. Such forms of attunement can be activated in exhibition programming. "Atmosphere," and indeed all the Science Museum exhibitions are planetary in distribution and impact. The museum's very fabric, its exhibitions and display furniture, the stone, concrete, plastic, bricks, wood and metals that comprise it are material indices of climate change. They are cultured more-than-human climatic compositions. In this regard, they are legacies of carbon-burning practices in material spaces, as well as capitalist ideologies, consumerism, mineral extraction, industrial processes, and carbon emissions.

As a consequence, the Museum's deep co-presense – from its collections, to its fabrics, earthly elements, species, plants, organic material, and chemicals – is resoundingly non-human. The Museum is thus an exemplar of multiple processes of human–non-human crafting and convergence through which it is made and remade. Despite this, it isn't always in respectful attunement. Wood, for example, is curated for human ends and serves the need of building construction. And significantly museum time is the duration of "real" natures of material, human-discursive, political, policy, and social temporalities, and their entanglements with non-humans.

The SMG does not only green itself by reducing its carbon footprint through energy transitions. Sustaining practice can be achieved in more radical ways by interpreting institutional forms in respect to their many facets and activities as coordinates in metabolic processes and as examples of more-than-human and non-human crafting and compositions. This relates to all processes, that being institutional social, political, biological, technological, and economic contingencies. Attunement is enacted through a consderation of the affordances and agencies of the non-and more-than-human in collaboration.

Returning humans to their ecologically embedded position in situated contexts is critical for enacting conceptual and material change and this can be achieved by thinking with and through museum fabrics and activities as transcorporeal subjects. The transcorporeal subject for example museum staff, audiences, and so forth become emergent in social movements, in material circumstances, and through routine living practices attuned to thinking, acting, and composing with a range of physical materials including fossil fuels, plastics, transport modes, CO_2 emissions, multispecies, plants, distant others and others. Thus, the transcorporeal subject is an example of radical interdependency.

Empirical data from the Hot Science research reveals that 69.3% of people surveyed in Australia and 63.3% in the USA would be open to a new habitability politics as outlined in this chapter. The Six America's study (2018) on Global Warming showed that 69% of people now believe global warming is happening. The economically focused profile is also now shrinking.[78]

In support of the PA, and in light of these findings, institutions can now promote a new transcorporeal, transversal earthly politics that is different from that based on the sovereign individual that supports living well on a damaged planet.[79] Museum activism is not only about social humanism. Activism can become more-than-humanist as activities are directed to more-than-human flourishing. This registers a shift in value and orientation that inverts hubris. Haraway's idea of non-human others as kin, and of a deep connectedness with others, in addition to the sharing of DNA, and the application of these ideas to governing projects and museum representations, also have the capacity to incite careful and responsible worlding. Building a new environmental politics will involve thinking, acting, living and working with material attunement. For example, alleviating poverty might be achieved through understanding the entangled and recuperative relations between food resources, plants, soils, multispecies, and economies.

Such strategies, critical to planetary survival, have the potential to counter-narratives of competition, control, exploitation, patriarchal benevolence and progress, as well as representations of nature that showcase non-humans and earthly processes as being over there and outside of us. Museums can also promote different types of ecological thinking and practices demonstrated by diverse cultural groups in which different natures and cultures as socio-ecological practices and projects come to the fore. In their activist capacities, museums are also in the position to promote ecological compositional design practices as interdisciplinary collaborations with others in local, practical contexts and to raise questions about what type of communitarian politics do we want to nurture and inhabit.[80] This can all be achieved through storytelling, visualisations representing interconnectedness, and the use of collections as material indices of climate change processes – not just in respect to the Anthropocene and its damaging effects but also as a lever for action as new types of cultural carbon mitigation strategies. Immersive experiences, including posthuman education, ecological play, bio-art interventions, and forums on the design of more-than-human mitigation strategies, can be put to work to foreground human populations as ecological beings. New liquid institutional forms,[81] and embedded community museums of different types, and forms could promote alternative ecological knowledge practices, including those of Indigenous First Nations, the posthuman practices of specific cosmovisions relevant to particular climate contexts, and those in which the transcorporeal and transversal subject is at the centre.

Museums can revisit their collections and present them not as old or obsolete technologies within an anthropological frame, but rather as examples of alternative economies and low-carbon strategies. These forms of more-than-human heritages represent a heterogeneity of low-carbon economies that can be linked to pedagogies to form an inventory of more-than-human alternatives.

As human geographers Katherine Gibson and Julie Graham explain, despite the social inequities and degraded ecologies capitalism produces, it appears to many as a system that is here to stay and to which there is no alternative.[82] Museums are historically and resoundingly deeply capitalist. Revisioning museum representations of economy is thus vital here.

As discussed in the previous chapters and early in this one, monolithic capitalism and ecological modernisation are the frames and strategies museums use to present climate change mitigation and adaptation. Taking inspiration from the Community Economies project, it is clear to me that reframing museum involves two main steps. First is a critique of museums' capitalist framing, which is embedded in institutional processes and representations and which stands in the way of instituting attuned habitability practices. Second, developing new visions of economy in multiple forms that can enable different kinds of action and incorporate the multifarious ways economic futures might be shaped.[83] Drawing on the wisdom of community economies these two steps represent a strategy for "doing economy" which starts with an understanding of the economy as something that is diverse and heterogeneous, rather than monolithic, and that is mutable rather than fixed in form.[84]

Museums, through a community economies framework, could play a role in empowering and supporting economic diversity and, in contrast to capitalism, promote ethical, and solidaristic modes of interdependence. They could help mitigate environmental destruction and increasing inequality and become sites of ethical action. The representation and practice of diverse economies in museums could allow for certain activities to be highlighted and thus valued and encourage audiences to reflect on the multiple dimensions of economic life that make up our shared world.

Eco-curating, as a method, can enable and represent the economy differently by illuminating multiple economic and non-extractive trajectories in a multiplicity of situated contexts. So the aim would not be to return to an ideal pre-modern way of life for all but to demonstrate the potential for capital degrowth and speculation in alternative worlds, systems of value and economies. Healing damaged ecological and social relations would centre the transcorporeal subject rather than the humanist and economic one. It would be a system in which the economy would become more equitable and environmentally sustainable.

Notes

1 Donna Haraway, *Staying with the Trouble: Making Kin in the Chthulucene* (Durham, NC and London: Duke University Press, 2016).
2 Val Plumwood, *Environmental Culture: The Ecological Crisis of Reason* (London: Routledge, 2002).
3 Fiona R. Cameron, "We Are on Nature's Side? Experimental Work in Rewriting Narratives of Climate Change for Museum Exhibitions," in *Climate Change and Museum Futures*, eds. Fiona R. Cameron and Brett Neilson, Routledge Research in Museum Studies (New York: Routledge, 2014), 51–77, 53.
4 Haraway, *Staying with the Trouble*, 6.
5 Phillipe Descola, *Beyond Nature and Culture* (Chicago, IL: University of Chicago Press, 2013); Isabelle Stengers, *In Catastrophic Times: Resisting the Coming Barbarianism* (Open Humanities Press, 2015), accessed February 25, 2023, http://openhumanitiespress.org/books/download/Stengers_2015_In-Catastrophic-Times.pdf.
6 Timothy Morton, *The Ecological Thought* (Cambridge, MA and London: Harvard University Press, 2010), 29–30.
7 Morton, *The Ecological Thought*, 29–30.
8 Cameron, "We Are on Nature's Side?" 60.
9 Deutsches Museum, Physics, accessed August 3, 2022, https://www.deutsches-museum.de/museumsinsel/ausstellung/physik.
10 American Museum of Natural History, "The Anne and Bernard Spitzer Hall of Human Origins," *Exhibition*, accessed August 3, 2022, https://www.amnh.org/exhibitions/permanent/human-origins#.
11 Natural History Museum, The Museum of South Kensington, "Human Evolution," *Exhibition*, accessed August 3, 2022, https://www.nhm.ac.uk/visit/galleries-and-museum-map/human-evolution.html.
12 National Museum of Natural Sciences, Madrid, "Biodiversity," *Exhibition*, March 27, 2012, accessed August 3, 2022, https://www.mncn.csic.es/en/visit-us/exhibitions/biodiversity.
13 Morton, *The Ecological Thought*, 13.
14 Ernst Mayr, "Darwin's Influence on Modern Thought," *Scientific American*, November 24, 2009, https://www.scientificamerican.com/article/darwins-influence-on-modern-thought1/, 1–2.
15 Bruno Latour, *Reassembling the Social: An Introduction to Actor-Network-Theory* (Oxford and New York: Oxford University Press, 2005).
16 Jane Bennett, *Vibrant Matter: A Political Ecology of Things* (Durham, NC: Duke University Press, 2010); Diana Coole and Samantha Frost, *New Materialisms: Ontology, Agency and Politics* (Durham, NC and London: Duke University Press, 2010).
17 Jussi Parrika, *A Geology of Media* (Minneapolis: University of Minnesota Press, 2014).
18 Deborah Bird Rose, Thom van Dooren, Matthew Chrulew, Stuart Cooke, Matthew Kearnes and Emily O'Gorman, "Thinking Through the Environment, Unsettling the Humanities," *Environmental Humanities* 1, no. 1 (2012), accessed August 22, 2023, https://doi.org/10.1215/22011919-3609940.

19 Rosi Braidotti, *Posthuman Knowledge* (Cambridge: Polity Press, 2019); Rosi Braidotti and Simone Bignall, *Posthuman Ecologies: Complexity and Process after Deleuze* (London and Lanham, MD: Rowman & Littlefield, 2019).
20 Melissa Gregg and Gregory Seigworth, *The Affect Theory Reader* (Durham, NC and London: Duke University Press, 2010).
21 Eduardo Kac, ed., *Signs of Life: Bio Art and Beyond*, Leonardo Book (Cambridge, MA: MIT Press, 2007).
22 Martin Holbraad and Morten Axel Pedersen, *The Ontological Turn* (Cambridge: Cambridge University Press, 2017).
23 Lawrence Buell, *Writing for an Endangered World: Literature, Culture, and Environment in the U.S. and Beyond* (Cambridge, MA and London, England: The Belknap Press of Harvard University Press, 2001).
24 Fiona R. Cameron, "Ecologizing Experimentations: A Method and Manifesto for Composing a Post-Humanist Museum," in *Climate Change and Museum Futures*, eds. Fiona R. Cameron and Brett Neilson, Routledge Research in Museum Studies (New York: Routledge, 2014): 16–33.; Fiona R. Cameron, "Posthuman Museum Practices," in *Posthuman Glossary*, eds. Rosi Braidotti and Maria Hlavajova (London and New York: Bloomsbury Academic, 2018), 349–51; Fiona R. Cameron, "Climate Change, Agencies and the Museum and Science Centre Sector," *Museum Management and Curatorship* 27, no. 4 (October 2012): 317–39; Fiona R. Cameron, "Liquid Governmentalities, Liquid Museums and the Climate Crisis," in *Hot Topics, Public Culture, Museums*, eds. Fiona Cameron and Lynda Kelly (Newcastle on Tyne: Cambridge Scholars, 2010), 112–28; Fiona R. Cameron, "From Mitigation to Complex Reflexivity and Creative Imaginaries – Museums and Science Centres in Climate Governance," *Museum and Society Special Issue: Hot Science, Global Citizens: The Agency of the Museum Sector in Climate Change Interventions* 9, no. 2 (2011), accessed June 3, 2018, http://www.le.ac.uk/ms/museumsociety.html; Fiona R. Cameron, *The Future of Digital Data, Heritage and Curation in a More-Than-Human World* (London: Routledge, 2021), 130.
25 Julie Katherine Gibson-Graham, *The End of Capitalism (As We Knew It): A Feminist Critique of Political Economy* (Oxford: Blackwell Publishers, 1996).
26 Kyle Whyte, "Indigenous Climate Change Studies: Indigenizing Futures, Decolonizing the Anthropocene," *English Language Notes* 55, no. 1–2 (Fall 2017): 153–62, 154.
27 Whyte, "Indigenous Climate Change Studies," 157.
28 Simone Bignall and Daryle Rigney, "Indigeneity, Posthumanism and Nomad Thought: Transforming Colonial Ecologies," in *Posthuman Ecologies: Complexity and Process after Deleuze*, eds. Rosi Braidotti and Simone Bignall (London and Lanham, MD: Rowman & Littlefield, 2019): 159–82.
29 Juan Francisco Salazar, "Buen Vivir: South America's Rethinking of the Future We Want," *The Conversation*, 2016, accessed February 23, 2023, https://theconversation.com/buen-vivir-south-americas-rethinking-of-the-future-we-want-44507.
30 Eduardo Gudynas, "Value, Growth, Development: South American Lessons for a New Ecopolitics," *Capitalism, Nature, Socialism* 30, no. 2 (2019): 234–43, 236.
31 Reimaging Museums for Climate Action, *Existances Exhibition*, accessed August 2, 2022, https://www.museumsforclimateaction.org/reimagine/exhibits/existances.
32 Fiona R. Cameron, "Theorizing More-Than-Human Collectives for Climate Change Action in Museums," *L'Internationale Online Climate Change Special Issue* to coincide with the UNFCCC meeting in Paris, November 30–December 11, 2015, accessed August 3, 2022, http://www.internationaleonline.org/research/politics_of_life_and_death/48_theorising_more_than_human_collectives_for_climate_change_action_in_museums.
33 Cameron, "Liquid Governmentalities, Liquid Museums and the Climate Crisis."

34 Fiona R. Cameron, "Ecologizing Experimentations: A Method and Manifesto for Composing a Post-Humanist Museum," in *Climate Change and Museum Futures*, eds. Fiona R. Cameron and Brett Neilson, Routledge Research in Museum Studies (New York: Routledge, 2014), 16–33.
35 Bruno Latour, "To Modernize or to Ecologize? That's the Question," trans. Charis Cussins, in *Remaking Reality: Nature at the Millenium*, eds. Noel Castree and Bruce Braun (London and New York: Routledge, 1998), 221–42, 221.
36 Cameron, "Ecologizing Experimentations: A Method and Manifesto for Composing a Post-Humanist Museum," 16–33.
37 Cameron, "Ecological Experimentations."
38 Fiona R. Cameron, "Climate Change, Agencies and the Museum and Science Centre Sector," *Museum Management and Curatorship* 27, no. 4 (2012), 317–39; Cameron, "Theorizing More-Than-Human Collectives for Climate Change Action in Museums;" Fiona R. Cameron, "The Liquid Museum: New Ontologies for a Climate Changed World," in *Museum Theory: An Expanded Field*, eds. Andrea Witcomb and Kylie Message (Oxford: Blackwell Publishing, 2015), 112–28.
39 Rosi Braidotti, "A Theoretical Framework for the Critical Posthumanities," *Theory, Culture & Society* 36, no. 6 (2019): 31–61, 33.
40 Cameron, *The Future of Digital Data.*
41 Cameron, "Liquid Governmentalities, Liquid Museums and the Climate Crisis."
42 Cameron, *The Future of Digital Data.*
43 Haraway, *Staying with the Trouble*, 13.
44 James Williams, "Time and the Posthuman," in *Posthuman Ecologies: Complexity and Process after Deleuze*, eds. Rosi Braidotti and Simone Bignall (London and Lanham, MD: Rowman & Littlefield International, 2019), 105–22, 107.
45 Cameron, "We Are on Nature's Side?" 51–77.
46 Cameron, "We Are on Nature's Side?" 71.
47 Cameron, "We Are on Nature's Side?" 72–74.
48 Bennett, *Vibrant Matter.*
49 Cameron, *The Future of Digital Data, Heritage*, 130.
50 Cameron, "We Are on Nature's Side?" 71–74.
51 Cameron, *The Future of Digital Data*, 145.
52 Cameron, *The Future of Digital Data*, 149.
53 Cameron, *Hot Science Segmentation Analysis Report*, 3–5.
54 Cameron, *Hot Science Segmentation Analysis Report*, 4–5.
55 Cameron, *Hot Science Segmentation Analysis Report*, 8.
56 Cameron, *Hot Science Segmentation Analysis Report*, 4–5.
57 Cameron, *Hot Science Segmentation Analysis Report*, 5–8.
58 "Politics of Climate Change Belief," *Nature Climate Change* 7, no. 1 (January 4, 2017), accessed August 4, 2022, https://www.nature.com/articles/nclimate3198.
59 Plumwood, *Environmental Culture*, 121.
60 Haraway, *Staying with the Trouble*, 140.
61 Stacy Alaimo, "Trans-corporeality," in *Posthuman Glossary*, eds. Rosi Braidotti and Maria Hlavajova (London and New York: Bloomsbury Academic, 2018), 435–37, 436.
62 Braidotti, *Posthuman Knowledge*, 40.
63 Cameron, *The Future of Digital Data*, 233.
64 Cameron, *The Future of Digital Data*, 6.
65 Cameron, "Ecologizing Experimentations;" Cameron, "We Are on Nature's Side?"; Cameron, "Liquid Governmentalities, Liquid Museums and the Climate Crisis;" Cameron, "Climate Change, Agencies,"317–39.; Fiona R. Cameron, "Posthuman Museum Practices." In *Posthuman Glossary*, eds. Rosi Braidotti and Maria Hlavajova (London and New York: Bloomsbury Academic, 2018), 349–52.

66 Cameron, "Climate Change, Agencies."
67 Cameron, "Posthuman Museum Practices."
68 Haraway, *Staying with the Trouble.*
69 Haraway, *Staying with the Trouble*, 5.
70 Cameron, *The Future of Digital Data*, 130.
71 Cameron, *The Future of Digital Data*, 214.
72 Cameron, "Posthuman Museum Practices," 352.
73 Steve Woolgar and Javier Lezaun, "The Wrong Bin Bag: A Turn to Ontology," *Social Studies of Science June* 43, no. 3 (2013), 321–40.
74 Erik Swyngedouw, "Trouble with Nature – Ecology as the New Opium for the People," in *Conceptual Challenges for Planning Theory*, eds. Jean Hillier and Patsy Healey (Farnham, Surrey: Ashgate, 2010), 299–320, 316.
75 Ilya Prigogine and Isabelle Stengers, "Order Out of Chaos," *British Journal for the Philosophy of Science* 36, no. 3 (1985): 352–54.
76 Cameron, *The Future of Digital Data*, 13.
77 Science Museum Group, "Towards Net Zero," accessed August 3, 2022, https://www.sciencemuseumgroup.org.uk/our-work/sustainability-approach/towards-net-zero/.
78 Livia Albeck-Ripka, "How Six Americans Changed Their Minds about Global Warming," *New York Times*, February 21, 2018, accessed August 3, 2022, https://www.nytimes.com/interactive/2018/02/21/climate/changed-minds-americans.html.
79 Cameron, "Theorizing More-Than-Human Collectives for Climate Change Action in Museums." Cameron, "Posthuman Museum Practices."; Cameron, "Theorizing More-Than-Human Collectives for Climate Change Action in Museums."; Cameron, "Posthuman Museum Practices."
80 Cameron, "Climate Change, Agencies," 317–39.
81 Cameron, "The Liquid Museum."
82 Julie Katherine Gibson-Graham, "Diverse Economies: Performative Practices for 'Other Worlds'," *Progress in Human Geography* 32, no. 5 (2008): 614–18.
83 Gibson-Graham, "Diverse Economies," 614–18.
84 Gibson-Graham, "Diverse Economies," 616.

References

Alaimo, Stacy. "Trans-corporeality." In *Posthuman Glossary*, edited by Rosi Braidotti and Maria Hlavajova, 435–37. London and New York: Bloomsbury Academic, 2018.

Albeck-Ripka, Livia. "How Six Americans Changed Their Minds about Global Warming." *New York Times*, February 21, 2018, https://www.nytimes.com/interactive/2018/02/21/climate/changed-minds-americans.html.

American Museum of Natural History. "The Anne and Bernard Spitzer Hall of Human Origins." *Exhibition*, accessed August 3, 2022, https://www.amnh.org/exhibitions/permanent/human-origins#.

Bennett, Jane. *Vibrant Matter: A Political Ecology of Things*. Durham, NC: Duke University Press, 2010.

Bignall, Simone and Daryle Rigney. "Indigeneity, Posthumanism and Nomad Thought: Transforming Colonial Ecologies." In *Posthuman Ecologies: Complexity and Process after Deleuze*, edited by Rosi Braidotti and Simone Bignall, 159–82. London and Lanham, MD: Rowman & Littlefield, 2019.

Bird Rose, Deborah, Thom van Dooren, Matthew Chrulew, Stuart Cooke, Matthew Kearnes and Emily O'Gorman. "Thinking Through the Environment, Unsettling the Humanities." *Environmental Humanities* 1, no. 1 (2012): 1–5. https://doi.org/10.1215/22011919-3609940.

Braidotti, Rosi. "A Theoretical Framework for the Critical Posthumanities." *Theory, Culture & Society* 36, no. 6 (2019): 31–61.

Braidotti, Rosi. *Posthuman Knowledge*. Cambridge: Polity Press, 2019.

Braidotti, Rosi and Simone Bignall. *Posthuman Ecologies: Complexity and Process after Deleuze*. London and Lanham, MD: Rowman & Littlefield, 2019.

Buell, Lawrence. *Writing for an Endangered World: Literature, Culture, and Environment in the U.S. and Beyond*. Cambridge, MA and London, England: The Belknap Press of Harvard University Press, 2001.

Cameron, Fiona R. "Liquid Governmentalities, Liquid Museums and the Climate Crisis." In *Hot Topics, Public Culture, Museums*, edited by Fiona Cameron and Lynda Kelly, 112–28. Newcastle on Tyne: Cambridge Scholars, 2010.

Cameron, Fiona R. "Climate Change, Agencies and the Museum and Science Centre Sector," *Museum Management and Curatorship* 27, no. 4 (2012): 317–39.

Cameron, Fiona R. "Ecologizing Experimentations: A Method and Manifesto for Composing a Post-Humanist Museum." In *Climate Change and Museum Futures*, edited by Fiona R. Cameron and Brett Neilson, 16–33. Routledge Research in Museum Studies. New York: Routledge, 2014.

Cameron, Fiona R. *Hot Science Segmentation Analysis Report*. Sydney: Institute for Culture and Society, Western Sydney University, 2014.

Cameron, Fiona R. "We Are on Nature's Side? Experimental Work in Rewriting Narratives of Climate Change for Museum Exhibitions." In *Climate Change and Museum Futures*, edited by Fiona R. Cameron and Brett Neilson, 51–77. Routledge Research in Museum Studies. New York: Routledge, 2014.

Cameron, Fiona R. "The Liquid Museum: New Ontologies for a Climate Changed World." In *Museum Theory: An Expanded Field*, edited by Andrea Witcomb and Kylie Message, 345–62. Oxford: Blackwell Publishing, 2015.

Cameron, Fiona R. "From Mitigation to Complex Reflexivity and Creative Imaginaries – Museums and Science Centres in Climate Governance." *Museum and Society Special Issue: Hot Science, Global Citizens: The Agency of the Museum Sector in Climate Change Interventions* 9, no. 2 (2011), accessed June 3, 2018, http://www.le.ac.uk/ms/museumsociety.html.

Cameron, Fiona R. *The Future of Digital Data, Heritage and Curation in a More-Than-Human World*. London: Routledge, 2021.

Cameron, Fiona R. "Posthuman Museum Practices." In *Posthuman Glossary*, edited by Rosi Braidotti and Maria Hlavajova, 349–52. London and New York: Bloomsbury Academic, 2018.

Cameron, Fiona R. "Theorizing More-Than-Human Collectives for Climate Change Action in Museums." *L'Internationale Online Climate Change Special Issue* to coincide with the UNFCCC meeting in Paris, November 30–December 11, 2015, accessed August 3, 2022, http://www.internationaleonline.org/research/politics_of_life_and_death/48_theorising_more_than_human_collectives_for_climate_change_action_in_museums.

Coole, Diana and Samantha Frost. *New Materialisms: Ontology, Agency and Politics*. Durham, NC and London: Duke University Press, 2010.

Descola, Phillipe. *Beyond Nature and Culture*. Chicago, IL: University of Chicago Press, 2013.

Deutsches Museum, Physics, accessed August 3, 2022, https://www.deutsches-museum.de/museumsinsel/ausstellung/physik.

Gibson-Graham, Julie Katherine. *The End of Capitalism (As We Knew It): A Feminist Critique of Political Economy*. Oxford: Blackwell Publishers, 1996.

Gibson-Graham, Julie Katherine. "Diverse Economies: Performative Practices for 'Other Worlds'." *Progress in Human Geography* 32, no. 5 (2008): 613–32.

Gregg, Melissa and Gregory Seigworth. *The Affect Theory Reader*. Durham, NC and London: Duke University Press, 2010.

Gudynas, Eduardo. "Value, Growth, Development: South American Lessons for a New Ecopolitics." *Capitalism, Nature, Socialism* 30, no. 2 (2019): 234–43.

Haraway, Donna. *Staying with the Trouble: Making Kin in the Chthulucene*. Durham, NC and London: Duke University Press, 2016.

Harrison, Rodney and Colin Sterling. *Deterritorializing the Future: Heritage in, of and after the Anthropocene*. London: Open Humanities Press, 2020.

Holbraad, Martin and Morten Axel Pedersen. *The Ontological Turn*. Cambridge: Cambridge University Press, 2017.

Kac, Eduardo, ed. *Signs of Life: Bio Art and Beyond*. Leonardo Book. Cambridge, MA: MIT Press, 2007.

Latour, Bruno. "To Modernize or to Ecologize? That's the Question." Translated by Charis Cussins. In *Remaking Reality: Nature at the Millenium*, edited by Noel Castree and Bruce Braun, 221–42. London and New York: Routledge, 1998.

Latour, Bruno. *Reassembling the Social: An Introduction to Actor-Network-Theory*. Oxford and New York: Oxford University Press, 2005.

Mayr, Ernst. "Darwin's Influence on Modern Thought." *Scientific American*, November 24, 2009, https://www.scientificamerican.com/article/darwins-influence-on-modern-thought1/.

Morton, Timothy. *The Ecological Thought*. Cambridge, MA and London: Harvard University Press, 2010.

National Museum of Natural Sciences, Madrid. "Biodiversity." *Exhibition*, March 27, 2012, accessed August 3, 2022, https://www.mncn.csic.es/en/visit-us/exhibitions/biodiversity.

Natural History Museum, The Museum of South Kensington. "Human Evolution." *Exhibition*, accessed August 3, 2022. https://www.nhm.ac.uk/visit/galleries-and-museum-map/human-evolution.html.

Parrika, Jussi. *A Geology of Media*. Minneapolis: University of Minnesota Press, 2014.

Plumwood, Val. *Environmental Culture: The Ecological Crisis of Reason*. London: Routledge, 2002.

"Politics of Climate Change Belief." *Nature Climate Change* 7, no. 1 (January 4, 2017), accessed August 3, 2022, https://www.nature.com/articles/nclimate3198.

Prigogine, Ilya and Isabelle Stengers, "Order Out of Chaos." *British Journal for the Philosophy of Science* 36, no.3 (1985): 352–54.

Reimaging Museums for Climate Action. *Existances Exhibition*, accessed August 2, 2022, https://www.museumsforclimateaction.org/reimagine/exhibits/existances.

Salazar, Juan Francisco. "Buen Vivir: South America's Rethinking of the Future We Want." *The Conversation*, 2016, accessed August 3, 2022, https://theconversation.com/buen-vivir-south-americas-rethinking-of-the-future-we-want-44507.

Science Museum Group. "Towards Net Zero," accessed August 3, 2022, https://www.sciencemuseumgroup.org.uk/our-work/sustainability-approach/towards-net-zero/.

Stengers, Isabelle. *In Catastrophic Times: Resisting the Coming Barbarianism*. Open Humanities Press, 2015, accessed June 3, 2018. http://openhumanitiespress.org/books/download/Stengers_2015_In-Catastrophic-Times.pdf.

Swyngedouw, Erik. "Trouble with Nature – Ecology as the New Opium for the People." In *Conceptual Challenges for Planning Theory,* edited by Jean Hillier and Patsy Healey, 299–320. Farnham, Surrey: Ashgate, 2010.

Whyte, Powys, Kyle. "Indigenous Climate Change Studies: Indigenizing Futures, Decolonizing the Anthropocene." *English Language Notes* 55, no. 1–2 (Fall 2017): 153–62.

Williams, James. "Time and the Posthuman." In *Posthuman Ecologies: Complexity and Process after Deleuze*, edited by Rosi Braidotti and Simone Bignall, 105–22. London and Lanham, MD: Rowman & Littlefield International, 2019.

Woolgar, Steve and Javier Lezaun. "The Wrong Bin Bag: A Turn to Ontology." *Social Studies of Science June* 43, no. 3 (2013): 321–40.

6

VIRAL MUSEOLOGIES

Curating human-species-viral worlds in sympoiesis

Introduction

Covid-19 is a significant curatorial agent, co-making a new milieu within us borne out of previous forms. Coronaviruses exist in the background as silent lifeforms, and in our bodies are made significant when their contagion spills over and threatens human populations, their health, the social and political order, or when they violently kill.[1]

Despite predictions of emerging potentially catastrophic viral agents and a strong history of such occurrences, human populations were unprepared. In discussing the H5N1 avian influenza strain in Indonesia, viral ethnographer Celia Lowe observes, "microbes are made significant in given contexts, and the material properties play an iterative role in shaping the milieu in which they come to exist."[2]

The virus, its intent to replicate and its ability to suddenly disassemble and re-assemble, to take control and take life, operates in more-than-human sympoietic relations. It inserts itself and infects not just our bodies but the complex ecologies of life itself, including social and political systems, economies, and the material circumstances of life as a radical, bio-cultural rhizomic curatorial agent. What is emerging are dense more-than viral eco-curating processes and domains of influencing in becoming.[3]

Covid-19, and its eco-curating processes, shook world political systems and global capitalism to the core, leading to the collapse of whole industries, the loss of jobs, the destabilisation of health systems, the killing of millions, and the situation whereby many more were left starving. This viral agent, its intent to replicate through its collaborations with human and more-than-human systems, has highlighted socio-economic disparities, brought underlying racist

DOI: 10.4324/9781315212067-6

ideologies to the fore and driven an upsurge in domestic violence. During lockdowns, aircraft sat on tarmacs, and cars remained stationary. The virus, and its drive to replicate, reorganised museum spaces leading to closures, pushing some into administration, changing attendance patterns through physical distancing, and replaced face to face and tactile activities with digital engagement. Such interventions by a viral agent, alongside regimes of contagion control that sought to contain its spread, instructed museum staff to consider how to curate programs and exhibitions differently in light of changing patterns of visitation and interaction.[4]

Museums portrayed themselves as vulnerable victims of the pandemic. However, I contend museums are complicit in the creation and spread of this new viral kin because they have, in many instances, promulgated thoughts and acts of dis-embeddedness, hubris, and un-mindedness in respect to the non-human world in which we inhabit and embody and upon which we depend. Here the porosity, vulnerability, and symbiotic nature of our own bodies are highlighted, and it becomes clear that the sovereign individual is no longer tenable. Museums therefore became radical viral inhabitations and potential spreading venues and events.[5]

In this chapter, I conduct an ecologising experimentation of planetary worlding[6] in the context of Covid-19, the Delta variant and most recently Omicron and in doing so draw on the arguments in my chapter contribution, "Viral agencies and curating worldly life differently in museum spaces," published in the 2021 collection, *The Posthuman Pandemic* edited by Saul Newman and Tihomir Topuzovski (London: Bloomsbury Academic).[7] Drawing inspiration from multispecies research, the work of Eben Kirksey, Stefan, Helmreich, Celia Lowe, Donna Haraway, and my work on ecological more-than-human thinking, new types of profoundly intimate communities, and eco-curating processes are made visible in museum spaces. Curatorial practice, museum space, exhibitions, collections, audiences, curators, and all manner of non-human stakeholders emerge as transcorporeal entities in multiple material, biological, energetic, cultured, and discursive co-constitutive enmeshments. These are knitted together at a molecular level, genetically, chemically, elementally, energetically but also virally.[8] Further to this, the Anthropocene during the pandemic became the Virosphere,[9] an unimaginatively vast and potent agent shaping political, economic, and social forces outside human capacities of control, and with the potential to become a new topic for museum engagement, and social, political, and agential reform.

Viral inhabitations and digital transformations

Covid-19 as a curatorial agent in its most potent Delta form – in respect to enhancing or impinging on human prospects of living and dying – radically reorganised museums. It led to closures, either temporary or permanent, shifted traditional attendance patterns through physical distancing, and social and tactile hands-on activities were replaced with digital engagement. In this context, museum structures became dispersed, more liquid, and rhizomic[10] as institutions focused on expanding and accelerating digital forms of interaction.

The 2020 survey of European museums by NEMO, the Network of European Museum Organisations, illustrated that the majority of museums worldwide were closed from March 2020 and these closures led to a dramatic drop in income particularly for larger institutions in tourist destinations. The findings revealed that more than 60% of museums increased their online presence through social media, and exhibited a renewed focus on digital cultural heritage. Digital representations of collections were put to work to produce virtual tours, online exhibitions, podcasts and talks on zoom, Google handouts, live content, and game creation which together increased online visitation.[11] Greater attention was directed to crisis awareness, including the formulation of pandemic emergency plans for both the public and internal work processes, and more flexible work methods in museums in general.

Similarly, the findings of UNESCO's global survey revealed that out of an estimated 95,000 museums around the world around 90% of institutions were affected by temporary closures.[12] In order to continue their work and activities many institutions bolstered their existing strategies through online engagement, drawing on previous digitised content, digitising current exhibitions and events, and exploring the potential of Information and Communication Technology (ICT) tools and social media in new ways to communicate with their audiences.[13] Hastings Contemporary, in the United Kingdom, for example, took advantage of its closed spaces to promote museum tours with telepresence robots to overcome barriers of isolation. These tours specifically targeted people in lockdown and those with disabilities.[14] The Getty Museum has made its online collection available for use in Nintendo's Animal Crossing, New Horizons video game, which allowed users to transform their homes into art galleries, fill their islands with art, or create their own custom patterns.[15]

The UNESCO report illustrated how many museum institutions devised new ways of maintaining contact with their audiences and generating alternative financial income using the digital environment through virtual visits and the use of social networks. Like other areas of the world, Russian museum professional Anna Guboglo explained, museums adopted multifarious digital strategies from virtual museum tours, social media activities, artistic video and photo installations, online lectures, children's interactive and educational sessions, and posts relating to the quarantine experience. Other institutions expanded their social service activities to support community mental health and wellbeing in combination with digital methods. The Garage Museum of Contemporary Art in Moscow, for example, launched the "Everything will be good" campaign, delivering lunches for elderly people, medical personnel, volunteers, and immigrants. Other museums addressed the needs of their communities by donating masks and gloves to hospitals and increased digital services to engage and comfort people staying at home.[16] The Catalyst Science Centre, and its MindLab project, provided telephone support and mentored vulnerable young people during lock down.[17] Fundamentally institutions rethought their operations and social roles in terms of accelerating digital transformation, leading to a new normal for museum practice and operation in order to increase their accessibility and maintain relevance.

Digital transformation, however, highlighted inequities, especially between genders, in regard to access to digital tools and technologies. The UNESCO report made several recommendations as a response to these inequities. These recommendations included the implementation of a more balanced digital policy, focused on the digitisation of collections including an up-to-date inventory, a minimum level of IT infrastructure, sufficiently stable internet access, and adequate staff skills.

The digital solution in museums, like many other sectors, was a direct response to social distancing, in which communal and creative programming sought to create proximity and intimacy with their collections through high-quality digitisations in order to enable engagement by those who are not geographically proximate. This represented an effort to create museum experiences of a different kind.

The 2020 *Museum and Society* journal special issue showcased museum scholarship directed to exploring the pandemic as a social, psychological, economic, and cultural phenomenon.[18] This scholarship sought to promote a discussion about, and a critique of, the digital solution and, in doing so complicate the relations between virtual and physical experiences in the museum realm. Media scholars Areti Galani and Jenny Kidd[19] explained how the pandemic had "forced a de-prioritization of touch and physicality" as individuals were forced to isolate or socially distance themselves as institutions shifted to digital formats. These scholars focused on the integration of digital and material collections to create hybrid materialities, and at the same time highlighted the inadequacy of digital engagement in such situations. Similarly, museum scholar Lindsay Balfour[20] addressed material and auric experiences in her discussion of the pandemic in the context of the 9/11 Museum and Memorial in New York. Balfour asked: how does one convey the "enormity and gravity" of absence and death through digital media? While praising Russian museums' efforts to serve "as models of public service and collaboration," Anna Guboglo recognised the limits of virtual museum tours or exhibitions and argued for the importance of virtual offerings that educate, reduce isolation, and decrease boredom.[21]

Structural inequality revealed during the pandemic made explicit the deep systemic oppressions, inequalities, and privileges most societies are built upon.[22] Some museums responded to the identity, gender politics, and racial ideologies that were made manifest as a result of the pandemic and sought to rectify the structural inequalities that resurfaced, many of which are historically their own. They sought to do this through a mandate to highlight prejudice, combat racism, and engage diverse communities in pandemic issues and often deployed digital solutions to achieve these ends. For example, The Museum of Chinese in America (MoCA), New York sought to combat anti-Chinese prejudice towards Asians and Asian Americans that arose in light of the Covid pandemic which was cast as a Chinese flu by collecting stories of Asian Americans through their OneWorld project. These resources were then made available in schools.[23]

The brutal killing of George Floyd by Minneapolis Police Department officer Derek Chauvin further highlighted structural barriers, institutional racism, racial injustice, isolation, frustration, and high unemployment among black communities. The Black Lives Matter movement comprised communities most at risk of Covid-19. The consequences of the pandemic for these communities included high infection and death rates provided an opportunity to air these grievances. The Orange County Regional History Center staff in Orlando, Florida collected a face mask with anti-Trump messaging (one, for example, said "Trump Resign"), a photo of a protester holding up a sign with an image of July Perry who was lynched 100 years ago in the Ocoee massacre in Florida, and a postcard with a list of the victims of police brutality so people could chant their names together at a Black Lives Matter protest in downtown Orlando.

The Smithsonian National Museum of American History in Washington D.C. extended its collection of protest art to document and to bear witness of these events for future generations. Furthermore, the New-York Historical Society in Manhattan collected protest items related to the killing of Floyd, the subsequent protests that his death ignited, and how people responded to systematic racism across the world. The collection included T-shirts, political leaflets, and protective eyewear worn by protesters to protect themselves from pepper spray and tear gas.[24]

Museums across the world radically reorganised their operations, engagement practices, and prioritised digital transformation as a way to remain relevant and in order to survive the pandemic. At the same time, institutions were co-opted into global biosecurity agendas, thereby implementing a new normal based on risk management, containment, physical distance, and the protection of the sovereign individual, while at the same time documenting structural inequalities.

In doing so, museums narrated and shaped new realities, by instituting novel physical formations and spatial arrangements, producing new knowledge, and using new media platforms for storytelling and at the same time through collections gathering emotional experiences of the "new normal." All these strategies highlighted uncertainty within an institution that has traded on offering certainty and at the same time modelled new forms of community organisation.

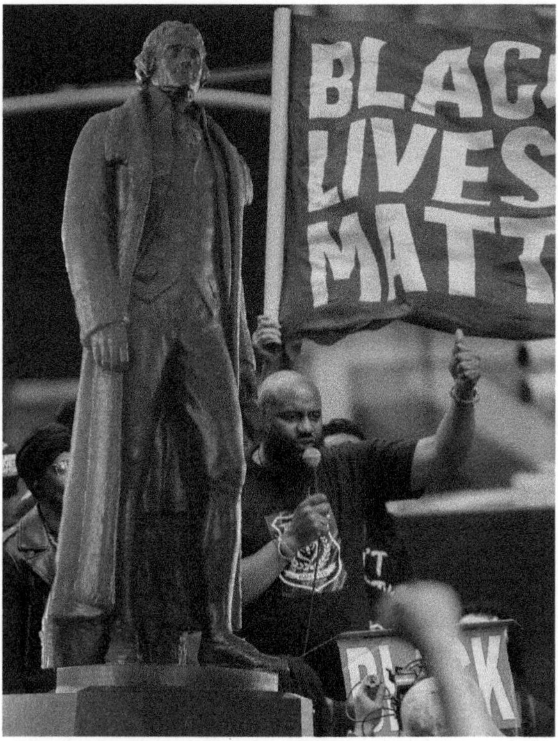

FIGURE 6.1 Black Lives Matter protest, Times Square New York City, June 7, 2020. Photo credit: Anthony Quintano.

As cultural theorist Brett Neilson reflects, the virus did not respect boundaries. In formal settings, the virus was presented as an entity outside of us and that sovereign power could control and protect us from it.[25] As a result, we witnessed a hardening of borders in an attempt to externalise the virus and to contain it within urban spaces through the identification of hotspots and through the use of tracing technology. Attempts to stop the spread of the virus focused on new sanitisation regimes with an emphasis on hand washing, disinfectant, and masks. What emerged were contagion regimes that sought to protect the porous body. Accordingly, a new vision of porosity emerged that acted against the principles of taxonomy that many natural history museums have historically promulgated. Museum institutions alongside shopping malls, supermarkets, libraries, and theatres became sites for sovereign power through hand cleaning, the use of tape, and stencils, and all manner of markings on the floors to designate social distancing through tracing regimes such as QR registration and protection such as mask wearing. All of these measures were directed to the governance and control of populations.

All these practices, from digital transformation to regimes of biosecurity, sovereign power, and control within museum spaces, are examples of viral curation born out of scientific, biochemical, and medical research on the mechanisms of viral spread. These mechanisms, in which viral curation enfolded, rest fundamentally on the genetic material of the virus, its RNA, and its radical inhabitations with human hosts. This was measured through the presence of RNA detected using RT-PCR (the reverse-transcriptase polymerase chain reaction, or serology tests for antibodies), the identification of clusters and hot spots, genomic relations between strains and contact tracing, the routes and mechanisms of transmission and efficacy of vaccines. Tracing the eco-curatorial movements of the virus was contingent on knowing more about it, how it behaved, its mutating strains, and its host. The dominant forms of viral curation rested on physical distancing and were most starkly illustrated through the radical attention given over to digital technologies in museum engagement.

Viral curation centred around safety mechanisms and was directed to the organisation of space, the monitoring of movement, the surveillance of surfaces and of touch, of bodily intimacy, and aerial dispersal of viral droplets in an effort to disrupt the virus in its never-ending search for a new host. Museums became sites for biomedical bodies and polities in which the permeability of self was highlighted and technological surveillance, in respect to physical distancing, was implemented.

All these more-than-human curatorial practices are humanist responses to viral contagion. The biological redesign of museum spaces and the rhythms of museum practices and operations in the interim was informed by research conducted by scientists and clinicians in recognition that aerosols, in addition to larger droplets, could transmit the novel coronavirus. Identifying viral agency, and the routes and sequences of transmission resulted in significant changes in how communities engaged in museum spaces, and their activities and indeed the future of museum work and individual job descriptions (in which a focus on wellbeing was increasingly present). While social distancing and mask wearing in the short term had some success in controlling transmission and enabling many museums to open their doors, the redesign of museum spaces and activities was indeed significant. This included online ticketing, queue lines, a greater use of outdoor spaces for activities, a shift from free-flow exhibition spaces to directed routes, the reduction of wall text, a greater use of phone apps, and the conversion of touch interactives to touchless.[26]

Collecting social contagion

Many museums also put in place Covid-19-related collection policies and forged new acquisitions procedures and practices to collect contagion. These collections, and their preservation, acted as forms of documentary evidence and included photographic archives, artefacts, and oral histories, so that we might recall, and future generations might understand, that moment in history.

The National Museums of Scotland Covid-19 collecting activities were directed to representing the impact of Covid-19 across Scotland as it unfolded, through tangible, 3-dimensional objects, supported by digital and print material, all of which were acquired and gathered up in a highly sensitive manner. Examples included a face shield made by a teacher at Lochaber High School in the Highlands using a 3D printer, then finished with plastic panels and delivered to local Belford hospital by the Lochaber Mountain Rescue Team. This object, curator Sarah Laurenson explained, "crosses several areas outlined in our plan: the impact on the National Health Service (NHS), personal protective equipment (PPE) shortages, the community response and finally, if tangentially, school closures."[27] A pair of knitted hearts from NHS Greater Glasgow and Clyde, for example, metaphorically connected patients with the loved ones from whom they were separated.[28] This collection focused on the vulnerable health worker, the carrier, the patient in quarantine, community mutuality, populations in lockdown, emotional responses to social isolation, the death of loved ones, the porous body, and mechanisms of containment such as the mask and shield.

The Museum of the Home, formerly the Geffrye, documented the lives of those living under lockdown through a series of personal accounts, focused on everything from improvised workspaces to haircuts.[29] The Museum of London issued an open call for both objects and first-hand experiences of Londoners in lockdown as part of an initiative entitled "Collecting Covid" and sought to acquire anything "from clothing to hair clippers, from diaries to memes."[30]

The Powerhouse in Sydney Covid-19 collection,

> captures the urgency of the time; the collaborative nature of the responses; medical treatments and research; systems of communication and technological innovation, and cultural and artistic expressions including data interpretation, and expressions of the care, fear and resilience demonstrated by people and communities.

This collection's temporal framing reflected the sudden onset of the pandemic in Australia in March 2020 and the first lockdown through to the present day and items that illustrated everyday living with the virus. Examples included "bush billboards," created by women artists Anne Thompson, Vivian Thompson, Marissa Thompson, Nicole Rupert, and Lynette Lewis working at Ernabella Arts in the Pukatja Community, APY Lands, South Australia. In this series of works, the women collaborated to make signs in the Pitjantjara language of the Anangu community to communicate important health and wellbeing messages in the early stages of the pandemic in Australia. Hand-painted on repurposed car bonnets, turned upright and positioned in prominent places, these "billboards" in Pukatja, read "Mamanya mukulya kura tjuta kulira wantima (Look after one another, love one another)," from the Pika nyanga kura wiyaring kunyyjaku (Follow these rules to help keep you safe from that virus.

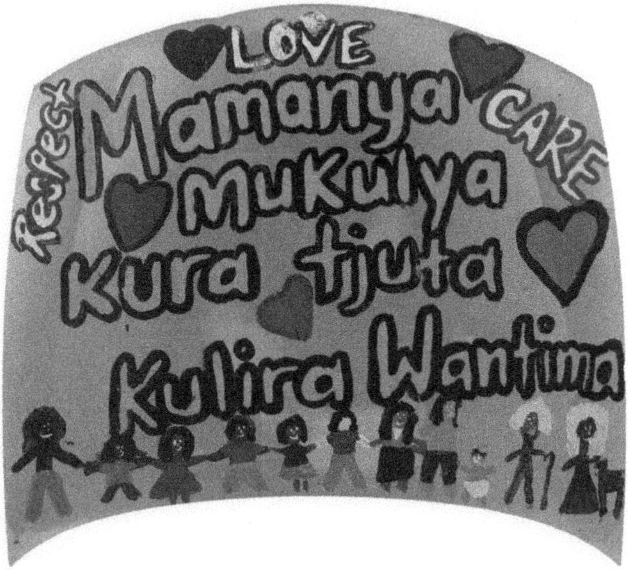

FIGURE 6.2 Painted sign, "Mamanya mukulya kura tjuta kulira wantima (Look after one another, love one another)", from the Pika nyanga kura wiyaring kunyyjaku (Follow these rules to help keep you safe from that virus). COVID-19 series, repurposed car bonnet, hand painted in acrylic paint, Anne Thompson, Powerhouse collection, 2020/124/6. Gift of the artists and Ernabella Arts Inc, 2020. Photo credit: Michael Myers.

The impact on First Nations and diverse communities was also represented through multi-lingual public health posters made by the Northern Land Council and the Aboriginal Medical Services Alliance NT, home-based teaching materials in English, Arabic, Dari, and Turkish, and video of online performances and social events.

Other acquisitions included masks, scrubs, and other PPE made by Australian fashion labels including Cue. Paste-ups and signs, which appeared around cities and suburbs with messages of care and togetherness, were gathered up alongside stories on the effect on retail, such as the exponential rise in online shopping including sales in lingerie and adult toys. Artistic works reflected on periods of lockdown and quarantine; design objects and photographs documented the streets of Sydney during lockdown, alongside items that captured the changing nature of visual communication and the built environment in this era. Working with the Westmead Institute for Medical Research (WIMR), the Powerhouse is building a biomedical collection, including medical and scientific material related to the development of a Covid-19 vaccine, contact tracing research and technology, and stories from lead researchers and patients with whom they are working.[31]

Like the National Museums of Scotland collection, the Powerhouse collecting initiative reflected the human-viral interface as a social response to the material properties and behaviour of the Covid-19 virus. They are collections of contagion control and design, focused on physical distancing, containment and safety, the responsible citizen, the reliable diagnosis, at-risk people, carriers, infected people, infection prevention and vaccination, and regimes of care. What is evident is how the perceived agency of Covid-19, and the notion of contagion control through distancing, barriers, vaccines, and so forth (including through a cultural lens), changed the status and image of people and society: from expressions of fear of one another because of contagion, to the stigmatisation and shaming of those with the disease. These collections were categorised by behaviours of the individual in lockdown, individuals as risk-takers, as carers and as companions, or as Covid safe individuals. For example, fashion masks represented fashion-conscious individuals and home-schooling materials conscientious young people and their carers who practiced social distancing as a cultural form. The economic impact of Covid on retail, the rise of online shopping and retail behaviour to the effects of social distancing in lockdown, and intimate relations were illustrated through collections of lingerie. Behaviour, safety, and the reconfiguration of space and mobility were represented by the photographs that documented the streets of Sydney during lockdown. Some were motivated by, or sought to capture, the psychological effects and a whole range of emotions experienced by people such as fear, despair, isolation, and empathy. The latter were documented in paste-ups and signs that appeared around cities and suburbs with messages of care and togetherness. Other collections were classified according to an individuals' Covid status and the intimate relations they shared with the virus. The biomedical collection in development at the Powerhouse, for example, was configured around radical or potential inhabitations. Some individuals, therefore, became Covid positive, carriers, super spreaders, or at risk. Others were returned travellers in hotel quarantine, front-line workers, survivors, elderly people in ICU or daily reported mortality figures. Other collections represented medical workers who sought to prevent infection, through the display of objects such as PPE gear, safety protocols, masks, shields, gloves, and gowns.

Alongside these Covid-19 collections, a new vocabulary of visual imagery, narration and description emerged, or re-emerged. This visual imagery is historically embedded in and borne out of previous pandemics or co-opted from political, medical, governmental, and cultural contexts and embedded in museum spaces and activities and in collections documentation. This vocabulary is not just restricted to the term coronavirus, later Covid-19, the "Delta" and "Omicron" variants, but also to words such as "pandemic," "contagion," "case fatality rate," "immunity," "mutations," "host," "Covid safe," "infection," "mobility," "lockdown," and "vaccination status." A distinctive discursive and also spatial turn emerged within museum spaces, one that was made and remade in response to shifting viral temporalities and biomedical advice. This is in contrast to the concerns museums have been occupied with such as temporal markers and linear concepts of history and ways of thinking based on a progressive logic of advancement.

These Covid-19 collections produced "the new normal," relationally configured as spatially or structurally discrete. While an expression of viral enmeshment, or indeed potential contagion, they, and the way they are interpreted, held us apart from the non-human world. They are expressions of control, that is, controlling the rate of infection and collectively representative of the war against the virus.

These collections and forms of thinking are enshrined as cultural heritage, which is what museums as societal, officiating institutions see as important to pass on to future generations. They could however be reinterpreted otherwise, in which viral inhabitations are and become a condition of life. Here we witness the emergence of a new type of Viroscene heritage and waste formed around a human agency of containment and control. Virospheric heritage are the masks and shields we choose to discard, the obsolete ICU machines, decommissioned aircraft, and the ruins of industrial-scale farming facilities becoming viro-fossil, in which the hard parts of all these things appear in the geological record recovered from the archaeological sites in the future.[32]

Museums and complicity with the pandemic

The meeting of Covid-19 and people in museum space highlighted modern populations and their dysfunctional relationship with the non-human world as a series of profound failures. This is because museums sanctify modernity and its ambitions, including its destructive practices and forces. Museums are no longer safe spaces, that is if they ever were. Covid-19 fundamentally changed the perception of modern populations and their relationship to the non-human world in which non-hierarchical and sympoietic relations are made visible as a condition of biological, cultural, and political life. Modern populations and individuals are not sovereign individuals or singular organisms.

Museums as pedagogical institutions have played a fundamental role in organising modern life and defining what worldviews, narratives, and ways of life are sanctioned and given importance. Typically, in the West, this includes global capitalism, human hubris, nature as resource, industrialisation, and technological progress. These unequal structures, thoughts, and acts are complicit in supporting and normalising the nature and forms of livelihood based on destructive practices that led to the emergence of viruses.

It is well known that modernity and its organising logics alienated humans from the natural world in such a way that has allowed destructive practices to be sanctioned. Structures, thoughts, and acts of un-mindedness and hubris enable viruses to emerge and spread. Furthermore, many museums have been complicit in forming multiple othering categorisations, not only in terms of humans based on "racial" differences, in which many are seen as closer to nature than others, but also a whole host of animate and inanimate things including plants, animals, fungi, viruses, viruses, and bacteria, all of which are set apart from dominant forms of humanism, that is the western male species.

Collections and museum exhibitions on history, natural history, and technology often represent, and therefore endorse, the political, economic, social forces, and destructive practices that led to the increased risk of new viruses emerging, and thus to the recent pandemic. Collecting is an activity of discernment directed at deciding what is important to save for future generations. In this regard, museums decide what objects best showcase contemporary life and therefore perhaps unintentionally normalise such thoughts, frames, and acts.

While many biological exhibitions and collections showcase biodiverse ecosystems, museums at the same time have been complicit in enrolling their audiences in ecological thoughts and relationships that are human-centric and dis-embedded. Through these forms of relating audiences come to see themselves as commanders in relationships with non-human others and earthly processes.

The museum form is founded on Western metaphysics. The entire museum project is dependent on objects, defined entities, and the classification of life and inanimate things into non-interacting categories. They project a view of life as composed of distinct entities, objects, and substances, with well-defined and reasonably stable boundaries, rather than as objects and entities as embedded, embodied dynamic ecological processes.[33]

The way many museums present biological individuality is sometimes a problem. Human bodies are often presented as structurally discrete and distinct from other entities. Taxonomies, analytical categories, and narratives create spatial and temporal distinctions between forms of life, and hold animals and microbes, for example, in hubristic, relational categories with humans, rather than presenting them as kin in embodied, enmeshed, and interdependent relationships. We are not singular organisms; indeed, all human populations host multiple entities including billions of viruses and bacteria as part of integrated biological systems and processes.

The problem of identifying and naming taxonomies in museums is often based on the relatedness between, and sharing of, DNA as a key property. Taxonomy holds open homo sapiens as the special ape separate from other animals, as a more intelligent product, and as a sovereign entity. This inhibits the ability to think of humans as existing in what Donna Haraway calls a "thick presence"[34] with others less closely genetically related.

All things are classified as living or non-living. This approach has had major implications for our understanding of biological entities such as viruses. As biological scientists John Dupre and Stephan Guttinger explain, "viruses are often portrayed as stable distinct individuals that do not fit into a more integrated, collaborative picture of symbiosis as individuated living entities." Viruses instead should be understood as processes, rather than as individual entities, and offer up an opportunity to understand the collaborative and integrated nature of the world.[35]

Most importantly, taxonomy is based on phylogeny and, is a form of categorisation that does not take into account the fact that organisms come into being through complex webs of constitutive relatings. While distinctions do indeed exist between species, they are not as clear cut as taxonomic categorisation would suggest, nor do they exist outside their ecological situations. Furthermore, the study and presentation of animal evolution in museums generally take prominence over other domains of life because they are visible organisms and higher up in the so-called evolutionary chain.

The basis of museum biological research is founded on the study of vertical genetic evolution over hologenomic association. Natural history collections have often been presented in taxonomic series in showcases that seek to highlight their structural differences and similarities as a hierarchy of evolution rather than their co-constitutive relations. Curators at the Natural History Museum, London, for example, use specimens and taxonomic categories to help identify new species and understand how life on Earth evolved. Many of these collections originate from Charles Darwin's expeditions, and include his finches, collected from the Galápagos Islands during the second voyage of the Beagle, that informed his theory of genetic evolution.[36]

Through pedagogical activities, museums promote the notion of the individual as sovereign. Importance is placed on humans, the genealogies of human thoughts, actions, and their moral reform, rather than as species-being as co-constitutive, as having feelings of attunement to enact ***worlding with*** practices.

Multispecies ethnographers Eban Kirksey and Stephan Helmreich remind us of the host of organisms whose lives and deaths are dependent on humans and how organisms and their livelihoods are shaped by political, economic, and social forces. Modern society, Kirksey and Helmreich explain, has stigmatised certain types of non-humans and animals such as bats, birds, and even pigs as pathogens living on the margins. This has promoted fear, disrespect, and has led to these animals' exploitation or the casting of them as dangerous multispecies relations that threaten human life. It has not led to respect and conviviality. Species have been cast as dangerous hosts of zoonotic viruses and as demonstrating an increased risk of transmission due to an upscaling of industrial-scale farming of such animals for food or pelts. Many of these animals live on the margins but as part of the natural environment, as food for humans, and as resources to use. Viruses are seen as barely alive, even as non-living, simply because humans can't see them and, while they appear alongside humans in the bios, are seen as killable rather than as symbiotic. Although sometimes they are perceived as beneficial collaborators in life.[37] Museums have historically been complicit in promoting these categories of non-human others. Speciesism is rife in museum collections and exhibitions in which discrimination manifests as cognate and physical violence against "inferior" non-humans in the way they are excluded, collected, and interpreted. In a similar manner as colonialism racism, sexism, or discrimination based on ability, this mode of thought renders certain lives inferior and therefore subject to discrimination and subjugation, and even death – museums normalise such thoughts and acts of killing, torture and exploitation to benefit some categories of human. The invocation of speciesist logics and hierarchisation of species, bodies, needs, and fundamental rights is often made manifest through taxonomies driven by speciesist logics. Historically natural history museums killed native fauna and collected them as specimens displaying them taxidermed, in cases pinned to boards, or in the case of birds on their backs in order to illuminate their structure and physical characteristics. In doing so museums promoted, and continue to promote, speciesist beliefs in which birds, bugs, and butterflies are lesser and by displaying injustice and inequality in death in this manner they justify and drive ecocides. Covid-19, and most dramatically the Delta variant, has dispelled the myths museums promulgate of independence and of autonomy and supplanted this with an awareness of inevitable symbiotic relations. Countering these narratives of hierarchisation, fear, and superstition, including the idea that bats are dirty disease carriers, the Smithsonian National Museum of Natural History has reinterpreted its 140,000 bat specimen collection as a resource to understand and protect bats whose numbers are declining.[38] A thorough interrogation must be conducted into the ways museums value different forms of life and for example so-called animate and inanimate things, this includes for example geological specimens and the values placed on certain types of minerals as exploitable resources.

A report published by the Intergovernmental Science-Policy Platform on Biodiversity and Ecosystem Services (IPBES) says that:

> the same human activities that drive climate change and biodiversity loss also drive pandemic risk through their impacts on our environment. Changes in the way we use land; the expansion and intensification of agriculture; and unsustainable trade, production and consumption disrupt nature and increase contact between wildlife, livestock, pathogens and people is the path to pandemics.[39]

The collections that museums acquire oftentimes are complicit in creating these new ecological circumstances. The destruction of ecosystems and large-scale industrial farming, which bring humans and non-humans together, involve the exchange of microbes, exposure to new ones, and the arrival of new forms through spontaneous mutations in humans. When ecosystems are destroyed, non-humans are exposed to humans at the edges of their habitat. Or, when ecosystems are destroyed non-humans expand their territories into urban areas, increasing the likelihood of zoonotic transmissions.

Through their collections and exhibitions, museums have traditionally followed the routes and sequences of neoliberal capitalism, often supporting a global capitalist ethos through displays of wealth and privilege. All these collections and activities represent the acceleration of capitalism enabled through and within museum spaces. Through collecting discarded or obsolete technology, agricultural equipment, and industrial and steam machines – documented and presented as exemplars of modern progress, economic growth, industrialisation, and colonialism – museums become complicit in the endorsement of structural inequity, environmental destruction, climate change, and the pathogen, neoliberal capitalism.

The Science Museum, London collection comprises steam machines, transport, technology, scientific and industrial processes, and consumer products. The Museum therefore is an artefact of the industrial revolution, a celebration of industrial progress, and of the supreme ability of humans to harness the non-human world as resources for economic growth. The Science Museum in short, is a celebration of modernity and its British accomplishment. In the context of the climate crisis, the Science Museum collections now have other resonances: colonial expansion and dispossession, extraction, and ecological collapse. These collections are embedded in the physical substrates of the biosphere and therefore in ecospheric problems that have profound ecological consequences. For example, the Boulton and Watt "Lap" rotative steam engine, discussed in Chapter 3, is a climate machine or climatic ecological composition in the sense that it is composed of the relations between: the metals belonging to the geological substrates of the planet and their extraction; the combustion integral to its mechanic function; the acceleration of capitalist industrial production embedded in its design; the exploitation of the land and colonial dispossession; air pollution; and rising CO_2 levels.[40] Built in 1788, in the Boulton and Watt factory in Birmingham, this engine has long been celebrated as an emblem of the "age of steam," but also recently as a device deeply embedded in an emerging global carbon economy. It is now as much a harbinger of climate change as it is an icon of the Industrial Revolution. Indeed 1784, as the year of the invention of the rotative steam engine, has been proposed as the date the Anthropocene began as a new geological epoch and the beginning of the Great Acceleration.[41]

The Powerhouse, Sydney wool collection includes wool samples and technology spanning the period from 1804 to 2003, documents the history of Australia's wool industry, industrial-scale wool breeding programs, and the emergence of the Australian merino in the context of the new environmental conditions and its historical importance in the development of the national economy.[42] Likewise the National Wool Museum in Geelong, Victoria is dedicated to the history of the Australian wool industry and how the sheep's fleece became central to Australia's economic growth.[43] However, this collection of wool, textiles and machinery is also burdened with a destructive environmental history, including the demise of habitats and biodiversity loss, the rise of agri-capitalism destroying local ecologies and creating new animal genomes embedded in the composition of the wool. The acceleration of global climate change as emissions from agriculture accounts for a significant proportion of historical emissions.The Museum of the Goldfields in the mining town of Kalgoorlie and its collection of mining equipment, gold bars and nuggets and other geological samples, represent the opening up of the Australian continent to the extractive industries which have been historically pivotal in the development of the Australian economy but have been major contributors to the country's global emissions and the destruction of ecosystems.

Machines of various types and forms are not only human inventions but rather are founded on engineering and culturing the non-human world through the appropriation and incorporation of animate and inanimate materials. The making of early steam machines, like the Boulton and Watt engines, household items, and cultural infrastructures such as water pipes, involves the incorporation and culturing of timber complicit in deforestation. Agricultural equipment made with metal rested on the availability of iron ore in which this history of geo-extraction is embedded in the fabrics of these technologies and infrastructures.

The collections that museums acquire are oftentimes complicit in creating new ecological circumstances. It is now well known that the destruction of ecosystems and large-scale industrial farming brought humans and non-humans together and led to the exchange of microbes, an exposure to new ones, as well as the arrival of new forms through spontaneous mutations in humans. When ecosystems are destroyed, non-humans are exposed to humans at the edges of their habitat.

Viral contagion multiplies through the routes, flows, and sequences of neoliberal capitalism and mobility. The material culture of global circulation, travel, and global supply chains, such as aircraft, trains, and ships represent spreading mechanisms. The "Marine Navigation" exhibition at the Deutsches Museum, according to the exhibition text, "allowed mankind to open up the world, encouraging communications and trade." Many of the objects on display showcased advancements in steel hulls and mechanical propulsion, all of which are the trademark of the industrial age representing the most important technological periods: sail, steam, and the diesel engine.[44]

The Smithsonian Air and Space Museum collection includes more than 30,000 aviation and 9,000 space objects and is best known for its rare and historically significant aircraft and spacecraft such as 1903 Wright Flyer, representing the beginning of flight. The Boeing Milestones of Flight Hall celebrates some of the most significant airplanes, rockets, and spacecraft in history and includes information on Charles Lindbergh's solo trip across the Atlantic in his plane the "Spirit of St. Louis," the first American jet aircraft, the Bell XP-59A Airacomet, and the Bell X-1 in which "Chuck" Yeager first broke the mythical "sound barrier." These machines, and the milestones they represent, according to the Museum, "have made our planet smaller and the universe larger."[45]

Museum displays of the Anthropocene, such as the landmark exhibition *Welcome to the Anthropocene: The Earth in Our Hands* (2014) at the Deutsches Museum in Munich, and the *Anthropocene* exhibition at the Museum of Tomorrow in Rio de Janerio (2016-), focus on human-centred narratives in regard to the human mastery of the planet and humanity's destructive forces. But what we witnessed in recent years is not human mastery but viral power. Covid-19 has now become a more influential agent than modern humans. The Viroscene and its potential destructive forces brought capitalism to a standstill. Human populations emerged as a new object in the Viroscene, one that serves as exhibition content for the future.

While viruses can be recognised by cell receptors deep inside bodies, Lowe explains, "they are not accessible to perception, proprioception, or interoception." Viruses can only be inferred through symptoms or recognised artificially, or in a transformed state through scientific investigation. They are only able to be seen under powerful microscopes.[46] This is what makes viruses different from other organisms that have a discernible presence able to be put on display. For these reasons, they are given very little attention in museum space.

The pandemic disrupts existing species relations between people, museums, buildings, staff, and audiences and remakes these ontologies in new ways. It therefore gestures towards new opportunities for re-making fundamental museum forms. Museums are complicit in the spread of capitalism on one hand and have themselves thus become subjects of its failure.

Viral inhabitations

The Covid-19 virus became a new organising force in the museum sector that has the potential to enable us to think and live differently but not in a normative sense of relating. The pandemic starkly illustrated the permeability of nature and culture and, indeed, the inadequacy of taxonomic categories in reflecting life itself. Perhaps we needed a viral contagion to become more ecologically minded and empathetic species-beings.

Historically, museums comprise separate departments and promote separate domains of life based on disciplinary categories. Our lifeworlds and bodies, however, have always been profoundly more than human – deep, complex, and dynamic enmeshments of socio-economic, political, cultural, technological, biological, and all manner of planetary domains and forces, all of which the museum sector have failed to adequately acknowledge and represent.

New humanisms emerge and forms of human agency come to the fore in becoming viral communities, all of which are inhabitations of a distinctive kind that museums must promote. The social and the cultural, as a distinct domain of life, and as a specific type of human-centred ecological thought, are therefore held open again and become a subject of critique in museums. The notion of the social as explanation, and of representation, becomes one of profound ecological enmeshment.

Here feminist theorist Donna Haraway's use of the biological concept of sympoiesis, as a way of explaining multispecies enmeshments, is useful for considering Covid-19 as a more-than-human embodied subject and curatorial agent in museum spaces. For Haraway, sympoiesis means "making with." That is, everything is co-constitutional and important for each other's survival. Nothing makes itself, nothing is really autopoietic or self-organising: "sympoiesis enfolds autopoiesis and generatively unfurls and extends it." Critters therefore, as Haraway explains, "become-with each other, compose and decompose each other, in every scale and register of time and stuff in sympoietic tangling, in ecological evolutionary developmental earthly worlding and unworlding."[47]

We are profoundly sympoietic. We become with viruses. We are viral bodies, viral carriers, and hosts creating new viruses within ourselves. As part of the human "metagenome," Lowe explains "viruses inhabit every corner of our bodies, vastly outnumbering human and bacterial cells alike, and are arguably responsible for life as we know it."[48] Multiple viral fragments come together as an entangled genealogy with other species spanning 300 million years.[49] Viruses exist invisibly within and around us.[50] We are part of the viral mix in which new viruses form within us and across us in recombination with other species.[51] It is likely that Covid-19 was not one single event but a multiple recombination of events in which viral strands come together and mix and then become part of other events.[52] With Covid-19, multispecies' shared vulnerability comes to the fore. For example, we share H2 receptors, on the outside of our cells, with cats in meshes of sympoietic becoming.

The virus is inside us and asserts itself; its intention is to survive and multiply. New humanisms arise as fundamental viral inhabitations and reposition the humanist figure in viral clouds. Our sympoietic natures rest on the similar cellular machinery we share with viruses. Covid-19 invades us as its host cell, binding with our molecular machinery, producing its progeny, and then exits the cell and invades others.[53] As literature scholar James Baumlin observes, the signs and alarms are inside us as Covid-19 takes hold and latches onto our cells, to infect, to disrupt, and to shed, coursing through our organs, blood stream, and respiratory system,[54] leading to cytokine storms and potential death. The idea of the sovereign individual and the human body as structurally bounded is no longer tenable. Human hubris is brought into question. The virus highlights our permeability and the porous processes in which we continue to be made and remade.

Haraway invites us to make kin with multispecies.[55] It is not a case of making kin with Covid-19, however, as the virus is already kin with a long genealogy of familial, sympoietic relations. There is no division between us and our viral kin. This also illustrates how the virus itself curates not in a humanist sense medically, socially, and politically but also simultaneously as multiplying rhizomic viral curating agencies. Viral museologies become an organising logic within museum spaces and out in the world

It is acknowledged that humans are embodied in elemental and chemical processes and through the exchange of DNA in evolution. As Nigel Clark explains, the historical embodiment of past experiences, which inscribes the efforts and trials of other bodies and aptitudes in dealing with past environmental limitations, is stored in complex chemistry memories as a deep material force. These chemical memories materialise the distribution of fat stores as a result of environmental stresses and food shortages and skin tone as a response to solar flux.[56] Similar sympoietic processes, in becoming with viral kin both inside and outside our bodies, help us ward off other more deadly viruses and are therapeutic agents in place of antibiotics. They destroy cancer cells and hold genetic innovation as a mode of evolution. Viral kin are also deep material forces in our bodies. Viral code is inserted into DNA, in which some strains are linked to long-term memory function. Others encode proteins that move information between cells. Viruses have infected eggs and sperm inserting their genes into human populations for many millennia.[57] Human's ability to have live births is attributed to a bit of genetic code that was co-opted from ancient retroviruses that infected our ancestors more than 130 million years ago.[58] We are radical viral kin interfaces within ourselves, binding and inhabiting as practices of co-making in tighter, deeper enmeshments fundamental to our prospects of living and dying. The perceived unity of the sovereign individual is disrupted.

Haraway's interpretation of the sympoietic is biological and focused on lifeforms. Sympoiesis, for Haraway, is synonymous with the compost, a mesh of decaying organic matter and aerobic processes in becoming. Charles Darwin's theory of evolution, through the exchange of DNA between species, is just one sympoietic thesis. Haraway's multispecies entanglements are another. Media theorist Jussi Parikka's geomaterial thesis for digital media[59] and my explication of digital cultural heritage as profoundly ecological and material are others. All these theses are examples of more-than-human co-making with, in, and through different genetic, geological materials, biological processes, and digital data and infrastructures. Haraway explains that, biologically, "all things beget other things."[60] The viral sympoietic thesis is, however, one of planetary worlding, a co-constitutive material enmeshment on a planetary scale, dramatically shaping an emerging milieu in which human populations come to exist and as one of complex uncertainty. This confluence of viral forces and their unintended consequences is what Lowe calls "viral clouds."[61]

The sympoietic evolutionary thesis extends to all domains across cultural, political, social, technological, and economic realms of life, including earthly processes. The thesis is one of a complex co-making and becoming with planetary histories through different and complex registers of interrelatedness and enactions. All things beget other things in dense more-than-viral eco-curating processes.[62] Covid-19 curating, for example, has usurped the geopolitical in the same way that climate change has done. The global geopolitical returns to the national, to the regional, to the local as situated practices of sympoietic making with others.

We are bound into sympoietic enmeshments of becoming, not just biologically with viruses, but also through complex associations with, between, and across global capitalism and its failures. These enmeshments and their consequences included the unravelling of economies and industries, the resurgence of nationalism and the composing of new forms of sovereignty, of border regulation and restrictions, shifts in atmospheric relations and the drop in CO_2 emissions, environmental destruction, toxic environments, deforestation, viral infection and its threats unfurling and reshaping a "new normal" of cultural and political life as folded and enfolding material processes. These enmeshments represent viral and eco-systemic processes of sympoietic tangling and becoming with others[63] as community, and also as a deeply material process from the profound, intimate sharing of cell machinery, molecular, DNA, chemical, elemental, metal, mineral, metal elements and animal and plant life to complex technological and cultural infrastructures, social, political institutions, community organising, digital transformations at multiple scales, as situated and dynamic histories of co-making and becoming of shaping and reshaping. The virus exists in co-making relations with human and animal evolution, but also as a social, political, and technological development in which earthly life is curated by all manner of entities through sympoiesis. This is what I call ecological compositional design, or eco-curating processes, in which one action begets another and another and so forth, as a complex mesh of begetting.[64]

These viral entanglements, enfoldings, interdependences, and inhabitations are much more deeply bonded and fundamental to life prospects than we could have ever imagined. More-than-viral curating, through sympoietic evolution in more-than-human settings, is co-making and materialising diverse economies, brought neoliberal capitalism to its knees, reduced greenhouse gas emissions, instituted regimes of control, co-made huge amounts of waste from PPE gear and masks to supermarket packaging, and led to human and more-than-human systems becoming fossil in landfills, to become radical viral inhabitations in the archaeological record. None of these events preexist our enmeshment with viral kin, although preconditions activate such enfoldings. These preconditions include, for example, prehistory and the emergence of coronaviruses, the rise of neoliberal capitalism and its hubristic, extractivist, colonialist, and destructive logics. The invention of aircraft and shipping routes, the development of the biological sciences, taxonomy, even the orientation of the museum form towards resource extraction and laboratory investigation, and racial ideologies all lay the foundations for viral emergence. The emergence of the modern state and the notion of borders, the advent of the steam machine and mechanical, industrial-scale farming, environmental collapse, biodiversity loss, geological extraction, and the rise of alchemy and so forth are all sympoietic processes that connect with vital contagion in direct and indirect ways.

The war against the virus is erroneous. In debunking the myth of human hubris, feminist posthuman theorist Rosi Braidotti explains that the human subject never masters, nor possesses, but rather inhabits, crosses, and is always community with others.[65] Indeed, how will this viral kin co-make our bodies in the long term and what residue markers will remain? Viral kin in and with us has the potential to express what Braidotti calls an affirmative, ethical dimension as a form of new world building, and, for Haraway, as a new ethics of care and response-ability. What appeared to emerge was a sympoietic thesis of care and empathy, new social institutions, political structures, and community organising. As the pandemic normalised, hope for change was never realised.

We are sympoietic, we have shared material histories and have knitted together, at a molecular level, with the elements of water, fire, and air, with wood, metals, minerals, chemical and viral histories and, therefore, with museum buildings, digital programs and all manner of collections in museum spaces. Viral lifeforms have always inhabited museum spaces with us and in us. Viral curating is implicated in museum histories from the materials building were made from to the design and flow of spaces, to the cultural, technical, and biological collections through sympoietic making with different registers of inanimate, animate things and processes. Museum surfaces, stainless steel door handles, railings, and toilets are inhabited by viral lifeforms. Doors and handles made from extracted materials, iron, chromium, and carbon contribute to environmental destruction and biodiversity loss, and their human culturing (design, engineering, and production) into these forms using carbon-intensive production methods are places for viral droplets to inhabit for up to 72 hours. Their contagion is co-made with these metals and surfaces, in which people pass and disperse droplets alongside temperate and humid air-conditioning systems.

Museum staff, visitors, and collections themselves live with and comprise coronaviruses. All collections comprise non-humans and their entanglements with human designs. Biological collections as possible hosts of coronaviruses become entangled with museum practices of collection and documentation deployed for sympoietic, rather than for taxonomic, research. The Natural History Museum in London, for example, is using its collection of frozen bats, rodents, and many other host animals that embody preserved pathogens, to link them to new Covid research that seeks to connect disease outbreaks to pathogen hosts. Using genetic analysis, museum curators are using DNA sequences of pathogens in animals and humans to identify and track pathways of transmission.[66] Similarly, curators in the natural sciences and mammology, researching bat biology, ecology, evolution, and conservation at the American Natural History Museum have established a Covid-19 research initiative identifying unrecognised host species for SARS-CoV-2 using the institution's natural history collections and related data. This will contribute to future sampling for emerging viruses.[67]

The power of viral eco-curating, in its ability to rearrange the social, ideological, and material, refigured the very DNA of museum institutions and is akin to what I call rhizomic liquid museum forms.[68] These movements represented efforts to curate beyond the digital solution, sparked by instituted viral contagion in respect to museum closures, the inability to admit visitors to enclosed spaces, and the way in which museums adapted to these circumstances in recognition of the mechanisms of viral spread and its social constraints. The routes, sequences, and material properties and behaviour of the virus became scripted into museum physical forms. At the same time, the virus unfurled or decomposed the material and spatial inhabitations of physical museum buildings and lead to the development of outdoor and immersive exhibits. "Free the Museum" seeks to challenge the museum world to rethink definitions of the museum as large physical edifices, to reconsider their location, format, and the type of objects on display with the intention of activating new museum models. The aim is to transform everyday places into the museum experience "out in the world" reflection, healing, activism, and informal learning.[69] In this respect, any space out in the world can be considered a museum space. By thinking outside building edifices, "Free Museum" seeks to embrace the notion of ubiquitous museum practice and museology as a series of locational activations embedded in community. Retail experiences in a thrift shop became museum-ified. QR codes are, for example, were linked to stories of specific businesses and their original owners. Sound galleries became integrated into construction canopies around Chicago. A sound scape as a pop-up installation in a London Street interpreted the influence of Bauhaus furniture in the design world.[70]

From dis-embeddedness to sympoietic design in museums

In the collection titled *Pandemic Solidarity* Rebecca Solnit predicted that

> the pandemic is the end of something, a version of postwar prosperity for the global north predicated on exploitation of other regions, of other human beings and of nature itself, of a set of assumptions about our capacity to control that nature, and of many orders that are about to become history.[71]

The many lockdowns Solnit suggested made a lot of people question what they considered to be "normal," "valuable," or "productive."[72] We witnessed empathy for doctors, nurses, care workers, and waste disposal workers. For a moment, people's wellbeing usurped economic growth in many places. Homelessness for many essentially ended overnight due to lockdowns. The post-Covid rebuild has the potential to lead to the emergence of a different kind of economy and politics. For those on the left, the crisis therefore opens up opportunities to build a more caring community, to usher in post-carbon and post-capitalist economies, and to inspire an ideological shift away from the economy to a new community form based on mutuality, generosity, empathy, inclusion, and hope.[73] The Contagion Design: Labour, Economy, Habits symposium, convened by the Institute for Culture and Society, Western Sydney University (22 October to 12 November 2020) explored the opportunity for political regeneration born out of contagion that has the potential to build a different kind of economy and politics.[74]

Museums in all their figurations have important roles to play in the post-Covid rebuild, not just in narrating or representing new relations of viral inhabitations or changing physical structures but also in shifting structures, thoughts, ideologies, and assisting in the incorporation of new forms of design for ethical re-worlding and responsibility.

The ethical dimension of a post-Covid rebuild must signal the end of museum practices that promulgate the thoughts and actions that have got modern societies into this mess. Museums have the capacity to shift these forms of thinking, hubris, exploitative mindsets, capital as aspiration, racial and gender inequalities, and practices and thoughts that promote dis-embeddedness. Indeed many museums are already doing this work, such as the Museum of Capitalism, an institution based in Oakland California, New York City and Boston, through its mission to educate people about the ideology, history, and legacy of capitalism through artefacts of capitalism, exhibitions, and programs.[75] Pedagogical work is directed towards promoting justice for victims of capitalism, promoting those who have resisted this mode of economy and those who have helped to develop alternatives in an effort to inspire future generations. We have also seen an ethical commitment to highlight and address structural inequalities and racial prejudice through connections between museum collecting and documenting the Black Lives Matter movement.

Since the mid-2000s I have been pioneering a new field of research and practice centred around the emergence of posthuman museum practices and museologies by conducting ecologising experimentations across domains of museum practice from climate change narratives to institutional forms, to collections and documentation, and to digital cultural heritage and the new museology.[76] These ecological experiments and reconfigurations have been strategic, and directed towards promoting a new ethics of care that might encourage respect for various forms of animate and inanimate things.

They have sought to promote multispecies flourishing and connections, intercultural relations, inclusiveness, and interaction and seek to return modern societies to their rightful place as part of a larger dynamic that ultimately has the potential to support the continuation of life itself. I've sought to lever museums' expertise, knowledge, authority, trust, and power to achieve this.[77] Ecological thinking is no longer the sole domain of the biological sciences. Modern populations need to embed ecological thinking across museum practice by instituting a sympoietic thought and design agenda across all aspects of museum work.[78]

In returning to the pandemic and its representation, museums have been preoccupied with viral design in control mode, a design directed to preventing viral dispersal. But as institutions museums have important roles to play, beyond human-centred forms of biosecurity and social control, in generating different models of life.

In a post-pandemic world, museums, for example, can showcase and mobilise different ways of living with contagion as a constitutive form, not in quarantine mode but rather as contagion mutuality as a form of life. Exhibitions and programming can promote ideas and practices of multispecies flourishing, for example, or foster collaborations between humans and their viral kin, or showcase human destructive practices that encourage and prevent pathogenic viral emergence and humans as viral beings. They can also exhibit histories of colonialisation and exploitation towards the non-human world, commonly figured as exploitable resources to serve human needs. Institutions, in a critical mode, can challenge capitalism as a blueprint of global and technocratic control.

Furthermore, institutions can challenge the spatial and temporal assumptions that sit behind contagion. They can make explicit that human populations are not taxonomic, we are not held apart; rather everything begets everything else in sympoietic relations and as profoundly deep histories of humans and non-humans in relational sympoietic terms. Human agency must be recast as sympoietic in which *making with* is linked to ecological communitarian design. In doing so, museums can refigure taxonomies from forms of relational patterning alongside hubris, the notion that the viral as dangerous and so forth to that of sympoietic patterning of enfolding, begetting, composing, and decomposing. Institutions can promote "friendly" forms of agriculture and habitat and can illuminate re-wilding to support multispecies flourishing. New genetic and information technologies could be used to make viruses visible in museum spaces so they become a lever to help rethink new polities of confederation in a more-than-human sympoietic world.

Sympoietic thinking and ecological communitarian, or compositional, design could be instituted across domains of museum curatorial practice. New narratives have the potential to emerge based on the relational refiguration of human and viral agency and by ecologising human hubris as something that has always been ecological. Thinking with language, text, and speaking in a sympoietic way performs a distinctive sympoietic emergence through the way others respond. This can result in tangles of speaking, tangles of narrating, and as ways to illustrate and represent tangles.

Institutions could showcase the idea of contagion seepage as one in which the past, present, and future are co-implicated as multiple rhizomic histories. Narrating and documenting contagion, or sympoietic histories, for example in an expanded thesis, involves the folding and enfolding, begetting with, unfurling, decomposing, and composing processes through and across social, cultural, biological, terrestrial, planetary, economic, and technological domains. Their enfoldings with viral power do not precede their relatings but rather emerge out of previous complex sympoietic processes as mutating and new mutant forms through a long and deep history of enmeshment and emergence.

Such narratives and forms of documentation are productively focused on relationality, interconnectedness, and becoming, where everything is connected to many things which are then connected to other things. In this process, transforming, becoming, and affecting come to the fore as complex webs of interconnecting and collaborating sympoietic entanglements. Most importantly these forms of ecological experimentation, communitarian design, and narration allow reflection and foreground methods of composition. A focus on specific types of connections in terms of who, how, and in what ways we are bound up with the proximity and specific nature of such sympoietic enfoldings can provide some direction on how humans bind themselves up with others and how our worlding practices might unfold.

Virospheric collections and their interpretation could productively focus on what Haraway calls processes of begetting. Museum collections are viral. That is their embedding of the Virocene in the earth's crust in the form of waste such as plastic shields, billions of potentially contaminated facemasks, and vaccine vials may be recovered later from the geological record as radical sympoietic interfaces.

Vaccines, for example, are conventionally viruses appropriated for human uses. They are fundamentally more-than and other-than-human and at the same time elemental, chemical, mineral, and viral. Vaccines have long histories beyond the sympoietic biological and human culturing and crafting occurring in the present in their co-making. Narratives that highlight begetting as a method for interpreting vaccine samples, for example, foreground who, how, and what are we bound up with alongside the notion of proximities. The Astro Zeneca vaccine is a radical human-viral sympoietic interface made from an altered virus, it is a modified version of a common cold virus, with its origins in chimpanzees, designed to carry a gene from the novel coronavirus' spike protein. The insertion of a new virus into human bodies variously begets and triggers an immune response in humans to manufacture antibodies. This trains human bodies how to respond should you become infected, a sympoietic process linked through millions of years of viral sharing and development of common cellular machinery. Through Covid-19 as vaccination, we become bonded at a cellular and biochemical level with the virus.

Sympoietic living together operates through intra-actions across and through all manner of registers and entities and not just biological systems. The close interaction between humans and viruses is made evident through tracing and mapping the sequences, interrelatedness, temporalities, and affects of viral curating across domains of life. This includes the networks of sympoietic begetting and their long non-linear histories and extended dynamic enmeshments of becoming. The genomic sequencing of the virus, used in the identification of viral unfurling as it absconded hotel quarantine like for example the Holiday Inn hotel case in Melbourne in 2021 where the virus escaped a nebuliser in a corridor, are examples of sympoietic enaction and the tracking of viral contagion and its affects. The use of imaging technologies makes visible the radical interfacing of the virus as it latches on. It also makes clear the unfolding of its capacities and affordances within us.

Covid collections are connected to a broader viral cloud and its extended, dynamic, and ecological composition. Viral contagion in Australia, for example, begets economic collapse, and contagion control. Lockdowns, border control, and hotel quarantine sought to contain the virus's unfurling as its mutations seeped into the community. Becoming one with the virus through vaccines offered hope, of openings, of economic growth, of survival, but also renewal. Museum work has the potential to make the sympoietic visible. Medical research processes offer up narrative forms in which to research, document, and interpret collections.

Co-opting embodied and embedded sympoietic exhibition, documentation, and design principles enables the enaction of Haraway's "thick presence" and offers up different spatial, and also ethical, relations of thinking, acting, movement and exchange in museum spaces, programs, and exhibitions in a post-Covid world. Thinking with other epistemologies, such as Indigenous knowledge about biodiversity and through narrative directed to representing the "in – worlding" formation of identity and concepts of agency, provide new methods for dwelling in the world directed to articulating and co-making inhabitations of a different kind.

Conclusion: Contagion mutuality – The post-Covid rebuild and museum agency

As activist academic Neil Howard observed in the UK, there was a nationwide eruption of localised mutual aid groups connected via digital platforms, such as Facebook, to ensure the provision of basic needs for those unable to secure them.[79] This form of mutuality often as casual networks and organised groups did not exist before the crisis and was amplified because of the failure of governments to limit the impact of the virus, to support people in crisis, and to provide social services.[80] Agency, communion, and connection became amplified in a Covid world. Empathetic projection towards others strengthens not just the human but also the non-human world.

In the modern world, human populations are disciplined to be a sovereign individual. The body of the strong sovereign individual is impermeable and intact, free of diseases of hosts. But through viral curation sovereign individuals, states and cultural institutions are foregone. Political organisation becomes viral compositions and forms of social organisation. We witnessed a fundamental shift from a focus on me the sovereign individual to us in the broadest sense. Linked to the design of sympoietic communities, institutions must reconsider how they might work outside neoliberal capitalism in which museums have co-opted themselves. Colleagues Declan Kuch and Stephen Healy coined the term "contagious mutuality" to express what these new forms of community signify in the context of food security where non-profits and social enterprises focus on getting food to people by building on existing practices of mutuality.[81] Contagion mutuality is not just of the human social but has a resonance of more-than-humankind as new forms of sympoietic relations.

Museums in all their forms, from lofty edifices to embedded and ubiquitous liquid institutions are places for developing models of future societies, and which we glimpsed emerging in the midst of the crisis. That is, places promoting practices of mutuality and solidarity, empathetic attunement, ecological communitarian forms of thinking, and acting with others in close sympoietic relations. Interventions could include assisting in the development of mutual economies and cooperatives that focus on nature as commons rather than in the promotion of neoliberal forms of exploitation and appropriation. This can be achieved through concepts such as degrowth, the formulation of carbon-neutral economies, and the promotion of health and wellbeing as forms of sympioetic community building and design. We have seen the way new museum forms activated the outdoors and community at a distance as forms of viral curation. The sympoietic relations of these rhizomic institutional forms have the potential to generate new ideas, structures, community, and economy to better cope with future crises, as well as the potential to create new social, political economic wealth and resources that might be distributed in different ways. These concepts of mutuality can travel conceptually and practically across the museum sector, forming their own distinctive constituencies and constituent forms as they become embedded in communities as centres for both speculation and renewal.

Co-opting embodied and embedded sympoietic exhibition, documentation, and design principles enables the enaction of Haraway's "thick presence" and offers up different spatial and also ethical relations of thinking, acting, movement and exchange in museum spaces, programs, and exhibitions for a post-Covid world.

Notes

1 Fiona R. Cameron, "Viral Agencies and Curating Worldly Life Differently in Museum Spaces," in *Postmodern Pandemics*, ed. Saul Newman and Tihomir Topuzovski (London: Bloomsbury Academic, 2021), 219–46.
2 Celia Lowe, "Viral Clouds: Becoming H5N1 in Indonesia," *Cultural Anthropology*, 25, no. 4 (2017): 625–49, 626; Celia Lowe, "Viral Ethnography: Metaphors for Writing Life," in *Troubling Species: Care and Belonging in a Relational World*, ed. Multispecies Editing Collective (RCC Perspectives: Transformations in Environment and Society, 2017), 91–96.
3 Cameron, "Viral Agencies," 219; Fiona R. Cameron, *The Future of Digital Data, Heritage and Curation in a More-than-Human World* (Abingdon: Routledge, 2021), 9.; Fiona R. Cameron, "Theorizing Digitizations in Global Computational Infrastructures," in *The International Handbook of New Digital Practices in Galleries, Libraries, Archives, Museums and Heritage Sites*, eds. Hannah Lewi, Wally Smith, Steve Cooke and Dirk vom Lehn (Abington: Routledge, 2019), 55–67, 66; Fiona Cameron, "Posthuman Museum Practices," in *Posthuman Glossary*, eds. Rosi Braidotti and Maria Hlavajova (London: Bloomsbury Academic, 2018), 349–51, 350; Fiona R. Cameron, "Ecologizing Experimentations: A Method and Manifesto for Composing a Post-humanist Museum," in *Climate Change and Museum Futures*, eds. Fiona R. Cameron and Brett Neilson (Abington: Routledge Museum Research Series, 2014), 16–33; 3.
4 Cameron, "Viral Agencies," 220.
5 Cameron, "Viral Agencies," 220.
6 Cameron, "Viral Agencies."; Cameron, "Ecologizing Experimentations: A Manifesto."
7 Cameron, "Viral Agencies."
8 Cameron, "Ecologizing Experimentations: A Manifesto," 22; Cameron, "Theorizing Digitizations," 60.
9 Carl Zimmer, "Welcome to the Virosphere," *New York Times*, March 24, 2020. accessed February 25, 2023, https://www.nytimes.com/2020/03/24/science/viruses-coranavirus-biology.html.
10 Fiona R. Cameron, "Liquid Governmentalities, Liquid Museums and the Climate Crisis," in *Hot Topics, Public Culture, Museums*, eds. Fiona R. Cameron and Lynda Kelly (Newcastle: Cambridge Scholars, 2010), 112–28; Fiona Cameron, "The Liquid Museum: New Ontologies for a Climate Changed World," in *Museum Theory: An Expanded Field*, eds. Andrea Witcomb and Kylie Message (Oxford: Blackwell Publishing, 2015), 345–62.
11 NEMO (Network of European Museum Organisations), "Corona Survey Results," (2020), accessed February 25, 2023, https://www.nemo.org/fileadmin/Dateien/public/NEMO_documents/NEMO_Corona_Survey_Results_6_4_20.pdf, 1, 2, 5.
12 NEMO, "Corona Survey Results," 13.
13 UNESCO, "Museums around the World in the Face of COVID-19," (2020), accessed February 25, 2023, https://unesdoc.unesco.org/ark:/48223/pf0000373530, 8, 14.
14 Hastings Contemporary "Robot Tours," 2020, accessed February 25, 2023, https://www.hastingscontemporary.org/exhibition/robot-tours/; Amy Levin, "Isolation as a Collective Experience: Museums' First Responses to COVID-19," accessed February 25, 2023, *Museum and Society* 18, no. 3 (2020): 295–97. https://journals.le.ac.uk/ojs1/index.php/mas/issue/view/202.
Hastings Contemporary "Robot Tours." Levin, "Isolation as a Collective Experience," 2.
15 Getty Animal Crossing Art Generator. 2020, accessed February 25, 2023, https://experiments.getty.edu/ac-art-generator.

16 Anna Guboglo, "Digital Museology Under Test: Challenges and Opportunities for Russian Museums," *Museum and Society* 18, no. 3 (2020): 311–13. 311. accessed February 25, 2023. https://journals.le.ac.uk/ojs1/index.php/mas/issue/view/202.
17 Jade French, Nic Lunt and Martin Pearson, "The MindLab Project. Local Museums Supporting Community Wellbeing Before and After UK Lockdown," *Museum and Society* 18, no. 3 (2020), accessed February 25, 2023, https://journals.le.ac.uk/ojs1/index.php/mas/issue/view/20.
18 Levin, "Isolation as a Collective Experience."
19 Areti Galani and Jenny Kidd, "Hybrid Material Encounters – Expanding the Continuum of Museum Materialities in the Wake of a Pandemic," *Museum and Society* 18, no. 3 (2020): 298–301, accessed February 25, 2023, https://journals.le.ac.uk/ojs1/index.php/mas/issue/view/202.
20 Balfour, L. A., "Ground Zero Revisited – Museums and Materiality in an Age of Global Pandemic," *Museum and Society*, 18, no. 3 (2020): 295–97, accessed February 25, 2023, https://journals.le.ac.uk/ojs1/index.php/mas/issue/view/202.
21 Guboglo, "Digital Museology Under Test," 311.
22 Marina Sitrin, ed. *Pandemic Solidarity: Mutual Aid during the Covid-19 Crisis* (London: Pluto Press), xvii.
23 Museum of Chinese in America, "One World Covid 19 Collection." (2020), accessed February 25, 2023, https://moca40.mocanyc.org/oneworld/.
24 Nadja Sayej, "Their Stories Should be Told Right: How Museums are Documenting the Protests," *The Guardian*, July 7, 2020, accessed February 25, 2023, https://www.theguardian.com/artanddesign/2020/jul/07/museums-document-black-lives-matter-protests.
25 Brett Neilson, "Migration and Labour," *Contagion Design: Labour, Economy, Habits, Data, Symposium*. 2020, accessed February 25, 2023, https://www.westernsydney.edu.au/ics/events/contagion_design (audio recording).
26 Emma Thorne-Christy, "Rethinking Where We Exhibit in Light of COVID-19," *MuseumNext*, November 15, 2020, accessed February 25, 2023, https://www.museumnext.com/article/rethinking-where-we-exhibit-in-light-of-covid-19/.
27 Sarah Laurenson, Calum Robertson and Sophie Goggins, "Collecting COVID-19 at National Museums Scotland," *Museum and Society* 18, no. 3 (2020): 334–36, accessed February 25, 2023, https://journals.le.ac.uk/ojs1/index.php/mas/issue/view/202.
28 Museum of Home, "Stay Home Collecting Project." 2020, accessed February 25, 2023, https://www.museumofthehome.org.uk/explore/stay-home-collecting-project/.; Tim Deakin, "A New Kind of Challenge: Curating in the COVID-19Crisis," *MuseumNext*,https://www.museumnext.com/article/a-new-kind-of-challenge-curating-in-the-covid-19-crisis/.
Deakin, "A New Kind of Challenge: Curating in the COVID-19 Crisis."
29 Museum of Home, "Stay Home Collecting Project"; Deakin, "A New Kind of Challenge."
30 Deakin, "A New Kind of Challenge."
31 Powerhouse, "Coronovirus Collection," accessed February 25, 2023, https://collection.maas.museum/set/7574.
32 Cameron, *The Future of Digital Data,* 255.
33 Cameron, "Posthuman Museum Practices," 350–52.; Cameron, *The Future of Digital Data,* 139–40. Cameron, "Theorizing Digitizations," 60.
34 Donna Haraway, *Staying with the Trouble: Making Kin in the Chthulucene* (Durham, NC: Duke University Press, 2016), 2.
35 John Dupré, and Stephan Guttinger, "Viruses as Living Processes," *Studies in History and Philosophy of Biological and Biomedical Sciences* 59 (2016): 109–16, 109.

36 Natural History Museum, London, "Museum Specimens Could Help Fight the Next-Pandemic: Why Preserving Collections is Crucial to Future Scientific Discoveries," December 16, 2020, https://theconversation.com/museum-specimens-could-help-fight-the-next-pandemic-why-preserving-collections-is-crucial-to-future-scientific-discoveries-148293.
37 Eben Kirksey and Stephan Helmreich, "The Emergence of Multispecies Ethnography," *Cultural Anthropology* 25, no. 4 (2010): 545–76, 545.
38 Smithsonian National Museum of Natural History, "The Art and Science of Bats," 2020, accessed February 25, 2023, https://www.si.edu/spotlight/bats.
39 Intergovernmental Science-Policy Platform on Biodiversity and Ecosystem Services (IPBES), accessed February 25, 2023, https://www.ipbes.net/pandemics-media-release.
40 Fiona R. Cameron, Ben Dibley and David Ellsworth, "Climate Collections, and Photosynthetic Fossil-Fuelled Atmospheres," *Environmental Humanities*, in press (2023): 62–84.
41 Paul Crutzen and Eugenie Stoermer, "The Anthropocene," *IGBP Global Change Newsletter* 41 (2000): 17–18.
42 Powerhouse. Wool specimen 2021 accessed 10th June 2023 http://ma.as/370865
43 National Wool Museum collection, accessed June 23rd 2023 https://www.geelongaustralia.com.au/nwm/collection/article/item/8d1c1779f0f1a0d.aspx
44 Deutsches Museum, "Marine Navigation Exhibition," accessed February 25, 2023, https://www.deutsches-museum.de/en/exhibitions/transport/marine-navigation/.
45 Smithsonian National Air and Space Museum, "Boeing Milestones Flight Hall," accessed February 25, 2023, https://airandspace.si.edu/exhibitions/boeing-milestones-flight-hall.
46 Lowe, "Viral Ethnography," 93.
47 Haraway, *Staying with the Trouble*, 58, 58, 97.
48 Lowe, "Viral Ethnography," 93.
49 David Cyranoski, "Profile of a Killer: The Complex Biology Powering the Coronavirus Pandemic," *Nature* 581 (May 2020), 22–26, 24, accessed February 25, 2023, https://www.nature.com/articles/d41586-020-01315-7.
50 Lowe, "Viral Ethnography," 92.
51 Lowe, "Viral Clouds: Becoming H5N1 in Indonesia," 625–49, 625.
52 Eben Kirksey, "The Emergence of COVID-19: A Multispecies Story," *Anthropology Now* 12, no. 1 (2020): 11–16.
53 Cyranoski, "Profile of a Killer," 26.
54 James Baumlin, "From Postmodern to Posthumanism: Theorizing Ethos in an Age of Pandemic," *Humanities* (2020): 9–46.
55 Haraway, *Staying with the Trouble*, 58.
56 Nigel Clark, *Inhuman Nature: Sociable Life on a Dynamic Planet* (California, Sage, 2011), 209.
57 Lowe, "Viral Ethnography," 93.
58 Rachel Nuwer, "Why the World needs Viruses to Function," *BBC Future*, June 18, 2020, accessed February 25, 2023, https://www.bbc.com/future/article/20200617-what-if-all-viruses-disappeared.
59 Jussi Parikka, *A Geology of Media* (Minneapolis: University of Minnesota Press, 2015).
60 Haraway, *Staying with the Trouble*, 58.
61 Lowe, "Viral Ethnography," 92.
62 Cameron, *The Future of Digital Data*, 139–40.
63 Fiona R. Cameron, "Theorizing More-Than-Human Collectives for Climate Change Action in Museums." *L'Internationale Online Climate Change Special Issue* to coincide with the UNFCCC meeting in Paris, November 30–December 11, 2015, accessed February 25, 2023, http://www.internationaleonline.org/research/

politics_of_life_and_death/48_theorising_more_than_human_collectives_for_climate_change_action_in_museums.

64 Cameron, *The Future of Digital Data*, 9; Cameron, "Theorizing Digitizations," 60–61.; Cameron, "Posthuman Museum Practices."; Cameron, "Ecologizing Experimentations: A Manifesto."

65 Rosi Braidotti, *The Posthuman* (Cambridge: Polity, 2013), 193.

66 Pamela Soltis, Joseph Cook and Richard Yanagihara, "Museums Preserve Clues that can Help Scientists Predict and Analyze Future Pandemics," *The Conversation*, June 24, 2020, accessed February 25, 2023, https://theconversation.com/museums-preserve-clues-that-can-help-scientists-predict-and-analyze-future-pandemics-141175.

67 American Museum of Natural History, "The Science of Covid 19," accessed February 25, 2023, https://www.amnh.org/explore/covid-19-science.

68 Cameron, "Liquid Governmentalities."; Cameron, "The Liquid Museum."

69 Emma Thorne-Christy, "Rethinking Where We Exhibit."

70 Free the Museum, accessed February 25, 2023, https://www.freethemuseum.org/.

71 Rebecca Solnit, "Pandemic Solidarity," in *Pandemic Solidarity: Mutual Aid during the Covid-19 Crisis*, ed. Marina Sitrin (London: Pluto Press, 2020), 178.

72 Solnit, "Pandemic Solidarity," 178–79.

73 Neil Howard, "A World of Care," in *Life after Covid 19: The Other Side of the Crisis*, ed. Martin Parker (Bristol: Bristol University Press, 2020), 21–30.

74 *Contagion Design: Labour, Economy, Habits, Data, Symposium*, accessed February 25, 2023, https://www.westernsydney.edu.au/ics/events/contagion_design.

75 Museum of Capitalism, accessed February 25, 2023, https://www.museumofcapitalism.org/events.

76 Cameron, "Theorizing Digitizations."; Fiona R. Cameron, "Stirring Up Trouble: Museums as Provocateurs and Change Agents in Polycentric Alliances for Climate Change Action," in *Addressing the Challenges in Communicating Climate Change Across Various Audiences*, eds. Walter Leal Filho, Bettina Lackner and Henry McGhie (New York: Springer, 2019), 647–73; Cameron, "Posthuman Museum Practices."; Cameron, "The Liquid Museum."; Fiona R. Cameron, "Theorising More-Than-Human Collectives for Climate Change Action in Museums," in *L'Internationale Online Climate Change Special Issue* 2015, accessed February 25, 2023, http://www.internationaleonline.org/research/ politics_of_life_and_death/48_theorising_more_than_human_collectives_for_climate_change_action_in_museums.; Fiona R. Cameron, "From 'Dead Things' to Immutable, Combinable Mobiles: H.D. Skinner, the Otago Museum and University and the Governance of Māori Populations," *History and Anthropology* Special Issue 25, no. 2 (March 2014): 208–26; Fiona R. Cameron, "We Are On Nature's Side? Experimental Work in Rewriting Narratives of Climate Change for Museum Exhibitions," in *Climate Change and Museum Futures*, eds. Fiona R. Cameron and Brett Neilson (Abingdon: Routledge Museum Research Series, 2014), 51–77; Cameron, "Ecologizing Experimentations: A Manifesto."; :Fiona R. Cameron, "Climate Change, Agencies and the Museum and Science Centre Sector," *Museum Management and Curatorship* 27, no. 4 (October 2012): 317–39; Cameron, "Liquid Governmentalities," 127; Fiona R. Cameron, "Object-orientated Democracies: Conceptualising Museum Collections in Networks," *Museum Management and Curatorship* 23, no. 3 (2008): 229–43.

77 Cameron, "Ecologizing Experimentations: A Manifesto."

78 Cameron, "Ecologizing Experimentations: A Manifesto."; Cameron, "Posthuman Museum Practices."

79 Howard, "A World of Care," 21–30.

80 Sitrin, *Pandemic Solidarity*, xi.

81 Stephen Healy and Declan Kuch, "Mutuality by Design: Designing Food and Mobility Futures," in *Contagion Design: Labour, Economy, Habits, Data*, eds. Gay Hawkins and Ned Rossiter (London: Open Humanities Press, 2021), 68–78, 77.

References

American Museum of Natural History. *The Science of Covid 19*, 2020, accessed February 25, 2023, https://www.amnh.org/explore/covid-19-science.

Balbour, Lindsay. "Ground Zero Revisited – Museums and Materiality in an Age of Global Pandemic." *Museum and Society* 18, no. 3 (2020): 295–97, accessed February 25, 2023, https://journals.le.ac.uk/ojs1/index.php/mas/issue/view/202.

Baumlin, James, S., "From Postmodern to Posthumanism: Theorizing Ethos in an Age of Pandemic." *Humanities* 9 (2020): 46. https://doi.org/10.3390/h9020046

Braidotti, Rosi. *The Posthuman*. Cambridge: Polity, 2013.

Cameron, Fiona R. "Object-orientated Democracies: Conceptualising Museum Collections in Networks." *Museum Management and Curatorship* 23, no. 3 (2008): 229–43.

Cameron, Fiona R. "Liquid Governmentalities, Liquid Museums and the Climate Crisis." In *Hot Topics, Public Culture, Museums*, edited by Fiona Cameron and Lynda Kelly, 112–28. Newcastle: Cambridge Scholars, 2010.

Cameron, Fiona R. "Climate Change, Agencies and the Museum and Science Centre Sector." *Museum Management and Curatorship* 27, no. 4 (October 2012): 317–39.

Cameron, Fiona R. "Ecologizing Experimentations: A Method and Manifesto for Composing a Post-humanist Museum." In *Climate Change and Museum Futures*, edited by Fiona R. Cameron and Brett Neilson, 16–33. Abingdon: Routledge Museum Research Series, 2014.

Cameron, Fiona R. "We Are On Nature's Side? Experimental Work in Rewriting Narratives of Climate Change for Museum Exhibitions." In *Climate Change and Museum Futures*, edited by Fiona R. Cameron and Brett Neilson, 51–77. Abingdon: Routledge Museum Research Series, 2014.

Cameron, Fiona R. "The Liquid Museum: New Ontologies for a Climate Changed World." In *Museum Theory: An Expanded Field*, edited by Andrea Witcomb and Kylie Message, 345–62. Oxford: Blackwell Publishing, 2015.

Cameron, Fiona R. "Theorizing More-Than-Human Collectives for Climate Change Action in Museums." *L'Internationale Online Climate Change Special Issue* to coincide with the UNFCCC meeting in Paris, November 30–December 11, 2015, accessed February 25, 2023, http://www.internationaleonline.org/research/politics_of_life_and_death/48_theorising_more_than_human_collectives_for_climate_change_action_in_museums.

Cameron, Fiona R. "Posthuman Museum Practices." In *Posthuman Glossary*, edited by Rosi Braidotti and Maria Hlavajova, 349–51. London: Bloomsbury Academic, 2018.

Cameron, Fiona R. "Stirring Up Trouble: Museums as Provocateurs and Change Agents in Polycentric Alliances for Climate Change Action." In *Addressing the Challenges in Communicating Climate Change Across Various Audiences*, edited by Walter Leal Filho, Bettina Lackner and Henry McGhie, 647–73. New York: Springer, 2019.

Cameron, Fiona R. "Theorizing Digitizations in Global Computational Infrastructures." In *The International Handbook of New Digital Practices in Galleries, Libraries, Archives, Museums and Heritage Sites*, edited by Hannah Lewi, Wally Smith, Steve Cooke and Dirk vom Lehn, 55–67. Abingdon: Routledge, 2019.

Cameron, Fiona R. "Viral Agencies and Curating Worldly Life Differently in Museum Spaces." In *Posthuman Pandemics*, edited by Saul Newman and Tihomir Topuzovski, 219–46. London: Bloomsbury Academic, 2021.

Cameron, Fiona R. *The Future of Digital Data, Heritage and Curation in a More-than-Human World*. Abingdon: Routledge, 2021.

Cameron, Fiona R., Ben Dibley and David Ellsworth. "Climate Collections, and Photosynthetic Fossil-Fuelled Atmospheres." *Environmental Humanities* 15, no. 2 (July 2023): 62–84.

Clark, Nigel, *Inhuman Nature: Sociable Life on a Dynamic Planet.* California: Sage, 2011.

Contagion Design: Labour, Economy, Habits, Data, Symposium, accessed February 25, 2023, https://www.westernsydney.edu.au/ics/events/contagion_design.

Crutzen, Paul and Eugenie Stoermer, "The Anthropocene." *IGBP Global Change Newsletter* 41 (2000): 17–18.

Cyranoski, David, "Profile of a Killer: The Complex Biology Powering the Coronavirus Pandemic," accessed February 25, 2023, *Nature* 581, no. 7806 (May 2020): 22–26. https://doi.org/10.1038/d41586-020-01315-7. https://www.nature.com/articles/d41586-020-01315-7.

Deakin, Tim, "A New Kind of Challenge: Curating in the COVID-19 crisis," (2021) accessed February 25, 2023, https://www.museumnext.com/article/a-new-kind-of-challenge-curating-in-the-covid-19-crisis/.

Deutsches Museum, "Marine Navigation Exhibition," accessed February 25, 2023, https://www.deutsches-museum.de/en/exhibitions/transport/marine-navigation/.

Dupré, John and Stephan Guttinger. "Viruses as Living Processes," accessed February 25, 2023, *Studies in History and Philosophy of Biological and Biomedical Sciences* 59 (2016): 109–16.

Free the Museum. 2020. https://www.freethemuseum.org/.

French, Jade, Nic Lunt, Martin Pearson. "The MindLab Project. Local Museums Supporting Community Wellbeing Before and After UK Lockdown," accessed February 25, 2023, *Museum and Society* 18, no. 3 (2020) https://journals.le.ac.uk/ojs1/index.php/mas/issue/view/202.

Galani, Areti and Jenny Kidd. "Hybrid Material Encounters – Expanding the Continuum of Museum Materialities in the Wake of a Pandemic." *Museum and Society* 18, no. 3 (2020): 298–301, accessed February 25, 2023, https://journals.le.ac.uk/ojs1/index.php/mas/issue/view/202.

Getty Animal Crossing Art Generator. 2020. https://experiments.getty.edu/ac-art-generator.

Guboglo, Anna, "Digital Museology Under Test: Challenges and Opportunities for Russian Museums." *Museum and Society* 18, no. 3 (2020): 311–13, accessed February 25, 2023, https://journals.le.ac.uk/ojs1/index.php/mas/issue/view/202.

Haraway, Donna, *Staying with the Trouble: Making Kin in the Chthulucene.* Durham, NC: Duke University Press.

Hastings Contemporary. 2020, accessed February 25, 2023, https://www.hastingscontemporary.org/exhibition/robot-tours/.

Healy, Stephen and Declan Kuch, "Mutuality by Design: Designing Food and Mobility Futures." In *Contagion Design: Labour, Economy, Habits, Data,* edited by Gay Hawkins and Ned Rossiter, 68–79. London: Open Humanities Press, 2021.

Howard, Neil. "A World of Care." In *Life after Covid 19: The Other Side of the Crisis,* edited by Martin Parker, 21–30. Bristol: Bristol University Press, 2020.

Intergovernmental Science-Policy Platform on Biodiversity and Ecosystem Services (IPBES). "IPBES Report Executive Summary." 2020, accessed February 25, 2023, https://www.ipbes.net/pandemics-media-release.

Kirksey, Eben and Stefan Helmreich. "The Emergence of Multispecies Ethnography." *Cultural Anthropology* 25, no. 4 (2010): 545–76.

Kirksey, Eben "The Emergence of COVID-19: A Multispecies Story." *Anthropology Now* 12, no. 1 (2020): 11–16.

Laurenson, Sarah, Calum Robertson and Sophie Goggins. "Collecting COVID-19 at National Museums Scotland," *Museum and Society* 18, no. 3 (2020): 334–36, accessed February 25, 2023. https://journals.le.ac.uk/ojs1/index.php/mas/issue/view/202.

Levin, Amy, "'Isolation as a Collective Experience': Museums' First Responses to COVID-19." *Museum and Society* 18, no. 3 (2020): 295–97, accessed February 25, 2023, https://journals.le.ac.uk/ojs1/index.php/mas/issue/view/202.

Lowe, Celia. "Viral Clouds: Becoming H5N1 in Indonesia." *Cultural Anthropology* 25, no. 4 (2010): 625–49.

Lowe, Celia. "Viral Ethnography: Metaphors for Writing Life." In *Troubling Species: Care and Belonging in a Relational World*, edited by The Multispecies Editing Collective, RCC Perspectives: Transformations in Environment and Society 2017, no. 1, 92: 91–96.

Merrit, Elizabeth, "No Walls, New Ways: Giving up the Building to Connect and Create," accessed February 25, 2023, https://www.aam-us.org/2017/08/31/no-walls-new-ways-giving-up-the-building-to-connect-and-create/.

Powerhouse. "Coronavirus Collection." 2021, accessed February 25, 2023, https://collection.maas.museum/set/7574.

Powerhouse. "Coronavirus Collection, Bush Billboards." 2021, accessed February 25, 2023, https://collection.maas.museum/object/609057.

Powerhouse. 2021. "Wool specimen", accessed June 10, 2023 http://ma.as/370865

Museum of Capitalism. 2021, accessed February 25, 2023, https://www.museumofcapitalism.org/events.

Museum of Chinese in America. "One World Covid 19 Collection." 2020, accessed February 25, 2023, https://moca40.mocanyc.org/oneworld/.

Museum of Home. "Stay home Collecting Project." 2021, accessed February 25, 2023, https://www.museumofthehome.org.uk/explore/stay-home-collecting-project/.

Museum of London. "Collecting Covid." 2021, accessed February 25, 2023, https://www.museumoflondon.org.uk/discover/museum-for-london-collecting-covid.

National Wool Museum collection, accessed June 23, 2023 https://www.geelongaustralia.com.au/nwm/collection/article/item/8d1c1779f0f1a0d.aspx

Natural History Museum, London, "Museum Specimens Could Help Fight the Next-Pandemic: Why Preserving Collections is Crucial to Future Scientific Discoveries," December 16, 2020, accessed February 25, 2023, https://theconversation.com/museum-specimens-could-help-fight-the-next-pandemic-why-preserving-collections-is-crucial-to-future-scientific-discoveries-148293.

Neilson, Brett. *Migration and Labour Contagion Design: Labour, Economy, Habits, Data, Symposium*. 2020, accessed February 25, 2023, https://www.westernsydney.edu.au/ics/events/contagion_design (audio recording).

NEMO (Network of European Museum Organisations). 2020, accessed February 25, 2023, https://www.ne-mo.org/fileadmin/Dateien/public/NEMO_documents/NEMO_Corona_Survey_Results_6_4_20.pdf.

Nuwer, Rachel. 2020. "Why the World needs Viruses to Function." 2020, accessed February 25, 2023, https://www.bbc.com/future/article/20200617-what-if-all-viruses-disappeared.

Parikka, Jussi. *A Geology of Media*. Minneapolis: University of Minnesota Press, 2015.

Rasha Aridi, "To Prevent Future Pandemics, Protect Nature." October 30 2020, accessed February 25, 2023, https://www.smithsonianmag.com/smart-news/protecting-nature-will-protect-us-how-prevent-next-pandemic-180976177/.

Sayej, Najda "'Their Stories Should Be Told Right': How Museums are Documenting the Protests." July 7, 2020, accessed February 25, 2023, https://www.theguardian.com/artanddesign/2020/jul/07/museums-document-black-lives-matter-protests.

Sitrin Marina, ed. *Pandemic Solidarity: Mutual Aid during the Covid-19 Crisis.* London: Pluto Press, 2020.

Smithsonian National Air and Space Museum. 2021, accessed February 25, 2023, https://airandspace.si.edu/exhibitions/boeing-milestones-flight-hall.

Smithsonian National Museum of Natural History. "The Art and Science of Bats." 2021, accessed February 25, 2023, https://www.si.edu/spotlight/bats

Soltis, Pamela, Joseph Cook and Richard Yanagihara. "Museums Preserve Clues that can help Scientists Predict and Analyze Future Pandemics." *The Conversation*, June 24, 2020, accessed February 25, 2023, https://theconversation.com/museums-preserve-clues-that-can-help-scientists-predict-and-analyze-future-pandemics-141175.

Thorne-Christy, Emma. "Rethinking Where We Exhibit in Light of COVID-19," accessed February 25, 2023, https://www.museumnext.com/article/rethinking-where-we-exhibit-in-light-of-covid-19/.

UNESCO. "Museums around the World in the Face of COVID-19." 2020, accessed February 25, 2023, https://unesdoc.unesco.org/ark:/48223/pf0000373530.

Zimmer, Carl, "Welcome to the Virosphere." *New York Times*, March 24, 2020, accessed February 25, 2023, https://www.nytimes.com/2020/03/24/science/viruses-coranavirus-biology.html.

7

CURATING SUSTAINING PRACTICES IN AND FOR MORE-THAN-HUMAN WORLDS

Introduction

The sun is setting and the men of a village on the island of Iripiv, off the coast of Malekula and in the Vanuatu archipelago in the Pacific Ocean, have gathered in the Nukamal (kava hut) to drink shells of kava, a nightly ritual. Women are excluded from this men's-only space, nor do they drink kava, but Chief Numa Fred, also curator of the Malekula Kaljoral Senta, has kindly invited me to talk about climate change. Earlier that day we had walked around the island and he had pointed out how sea level rise and king tides were eating away at their island. I drink a shell of kava as a mark of respect and converse about the immediate threat of climate change and what it means to the Iripiv community.

Back in 2008, Iripiv was without electricity, internet, cars, and roads. Salt water was rising through the water table and compromising the availability of fresh water and the villagers' ability to grow crops. Others in the group explained how the seasons were changing which was affecting their ability to predict, grow, and harvest fruit and how rising temperatures were leading to an increase in cyclones.

One man in his mid-40s who I call Jeremiah, who had previously been working in the capital, Port Vila, in the heritage and tourism industry, said to the group,

> I don't want to be pulled into a Western cash economy and be stuck working for a wage forever. What sort of life is that? We have everything here – we have an idyllic lifestyle, we have plentiful food, catch our own fish, grow our own crops and have time to spend with our families and the community.

DOI: 10.4324/9781315212067-7

Numa Fred points to an island just offshore sinking under the waves. He says, "we are losing our island – we have to build a sea wall or move." Another elder interjects and claims, "carbon from the sky is compromising our livelihoods and our home." This discussion took place in 2008. Five years later, I returned to work with a group of ni-Vanuatu people on a policy framework for the use of traditional knowledge for climate change adaptation, "Save Blong Climate Change Mo Disaster."

Fast forward to 2019. All manner of institutions, including the International Council of Museums (ICOM) and other professional bodies, are adopting Agenda 2030 and the Sustainable Development Goals (SDGs), a set of 17 goals developed to frame global initiatives for developing sustainable futures based on sustainable development. Through an expanding network of museum institutions and alliances, these policy frameworks are going viral across the sector. In this chapter, I am interested in unpacking the dominant thoughts, frames, and systems of value inherent in museums' engagement with the SDGs and exploring the types of futures these frameworks materially enact.

Agenda 2030 and its 17 SDGs is a relatively new global agenda and indicator framework linked to climate change action outlined in the Paris Agreement (PA). This policy framework seeks to foster principles and measures that promote a better, more equitable and just world for humanity and the planet, and to simultaneously foster prosperity as well as a protected environment.[1] Agenda 2030 and the SDGs have been set up by nation-states as a framework of global governance by government officials and policy makers. It is formulated as a series of universal standards on the basis of technical skills and expertise in the fields of each of the goals. The shift to "sustainable development," geographer Jessica Hope explains, constitutes mainstream development actors' response to global warming and intensifying environmental change. The Agenda seeks to strengthen links to development banks, multilateral and bilateral institutions, states, international non-government organisations (INGOs), civil society organisations, and, increasingly, private sector actors. The aim of these connections is to balance social, economic, and environmental arenas within a "plan of action for people, planet and prosperity."[2]

While aspirations enshrined in the goals – such as ending hunger, alleviating poverty and violence, attaining gender equality, peace, justice, good health and wellbeing, clean water, and sanitation – are admirable, they are also profound in their failure to invoke a relational interdependence between humans and the non-human world in respectful, recuperative ways. The invocation of the SDGs by museums in unreflexive ways, therefore, potentially threaten and undermine planetary habitability, as I will now explain.

The idea of sustainability, and therefore Agenda 2030 and its SDGs, emerged in the 1980s out of the Brundtland Report[3] and both were based on the assumption of preserving the planet at the same time as meeting human needs, both without compromising the ability of future generations to meet their needs. Sustainable development, within which the SDGs are framed, is presented as an aspiration that is multidimensional and universal and that incorporates societal requirements alongside environmental protection, economic growth, and prosperity, equity in living conditions between the rich and the poor, a capitalist framework for present and future generations, and the maintenance of natural capital so that future growth is assured. The SDGs frame a series of indicators from which progress on the implementation of the goals can be measured based on empirical data from the field through which the performance of each goal and its impacts might be calibrated locally and globally.

The SDGs promote a neoliberal ecological modernisation model for social transformation through the construction of a series of so-called socially and economically viable relationships between society and "nature." This policy framework articulates the ways humans should act towards the non-human world under conditions of growth so that resources continue to support human life. While the SDGs and their indicators are articulated as a horizon to aspire to, and the monitoring of progressive steps towards their achievement is cast as vitally important for the future of human societies, these universal indicators are human-centred and intended to support human, social, economic, and technological progress. This people-centred neo-capitalist agenda, therefore, describes the non-human world in human-centred terms and needs.

The way in which human–non-human relations are framed in respect to the economy is outlined by UN Secretary-General António Guterres in the publication, *Making Peace with Nature*. In his foreword, Guterres states the following:

> Humanity is waging war on nature. The consequences of our recklessness are already apparent in human suffering, towering economic losses and the accelerating erosion of life on Earth. Ending our war does not mean surrendering hard-won development gains. Human prosperity relies on the wise use of the planet's finite space and remaining resources, as well as on the protection and restoration of its life-supporting processes and capacity to absorb waste.[4]

Climate change and environmental degradation, therefore, are seen as events that undermine past economic successes and future prospects. The philosophy behind the SDGs, in the face of socio-ecological collapse, is to strengthen and prioritise the economy, avert threats to economic loss, and support development as the central pillars. These are at the expense of ecological integrity. In the future, human prosperity is set to be built on the service value and the utilisation of nature's capacities and agencies.[5] Human ingenuity is believed to be able to harness and craft these accordingly. In regards to development, valuing the role of the non-human world, or what feminist environmental activist Ariel Salleh calls the "raw materials warehouse,"[6] is a fundamental principle of the SDGs. But the belief that nature is a resource to support economic development is an erroneous one, particularly when it overrides other local and/or Indigenous knowledges. Furthermore, the superior articulation of living standards made possible through such modes of thinking and acting, and as aspirations for so-called developing nations under the logic of capitalism is both neo-colonial and patriarchal.

Each of the goals incorporates forms of social scientific ecological thinking which privilege human needs, economic growth, and wellbeing as being dependent on the environment, and through the regular incorporation of the word "nature" in specific SDGs shows the necessity of the non-human world for continuing development. The aim of the goals is to maintain essential ecological processes, preserve genetic diversity, and sustain the use of species and ecosystems.

Each of the SDGs is an ecosystem in itself and is described in terms of resources and the provision of service functions and to shift thinking from unminded resource extraction to the maximisation of nature's resources for the production of material outputs. Low carbon and climate resilient development, for example, requires protecting and restoring the health, productivity, and resilience of oceans, seas, and coastal areas.[7] Despite this, though, all life becomes financialised with all human–non-human transactions and activities becoming questions of monetary exchange. The term "harmony with nature" articulates actions that assume natural capital is preserved. Human agency in the SDGs remains linked to economic development, and an extractivist logic in which thoughts and acts of control and exploitation are articulated but within limits.

The links between the PA and the SDGs are through science and a rights-based agenda. Whereas chemistry and physics modelling is used to establish emissions and temperature scenarios in the PA, these same disciplines inform the SDGs. For example, a 1.5-degree scenario operates as a threshold for achieving the SDGs of maintaining human health, wellbeing, and ecosytem function.

SDG setting assumes a direct linear relationship between setting the goal to exact and mobilise tasks and its achievement through a series of empirical indicators. Goal setting is not an ecological framework that interrogates relations of interdependence in order to strive for a more holistic system. This is because the SDGs are built around technocratic aims of uninterrupted production, the maximisation of efficiency, and the profitability of resources, all while seeking to impose less harm on the environment. Significantly, decisions made are based on Western modern worldviews, with information derived from technical methods and data that often disregards, overlays, and relinquishes other worldviews and practices. The way the SDGs are framed currently, therefore, is a form of neo-colonial imposition because they appear undemocratic and are insufficiently nested in local worldviews and practices.

In many fields, research is being conducted to refine the goals, particularly in terms of the development of new measures and indicators tailored to particular contexts. Researchers at the Stockholm Environment Institute, for example, are integrating knowledge about the whole supply chain – from production to consumption – to develop new indicators of effectiveness for Goal 12, "Responsible Production and Consumption."[8]

Many museums have for some time been working with aspirations that are now enshrined in the goals, in particular questions of equity and inclusion. The British Museums Association, for example, states in its mission statement: "The best museums use their position of trust to encourage people to reflect on society's contemporary challenges. They promote social justice and human rights, challenge prejudice and champion fairness and equality." Professor Richard Sandell and colleagues at the University of Leicester Museum Studies program have for many years been conducting significant research on how museums might combat inequality in regard to sexuality and disability. Other academics, artists, and curators are doing remarkable work in regard to gender and racism. An example is the photo essay exhibition by photojournalist Bradley Secker. It was titled "*Kütmaan*" and was held at Leighton House Museum (2012), and it took a stance against inequality and injustice by exploring the plight of LGBTQIA+ people who have been forced into exile from Iraq, Iran, Syria, and Libya because of their sexuality or gender identity.[9]

Presently, the SDGs, and commitment to them, have become a global campaign across all sectors. In the museum and more recently in the heritage sector, this movement has been led by consultants Jasper Visser and Henry McGhie, the ICOM Working Group on Sustainability, the International Centre for the Study of the Preservation and Restoration of Cultural Property (ACCROM), Museu do Amanhã (Museum of Tomorrow), We are Museums among others. The Museum for the United Nations (UN Live) is an example of an independent institution that has been created to connect people everywhere to the work and values of the United Nations by dramatically increasing the number of people and museum institutions who work to achieve the goals of the UN. This institution has a specific agenda to implement the SDGs across all societies. The museum's cofounders – Olafur Eliasson, Jan Mattsson, and Henrik Skovby – observed that in order for these goals to be successful the world needs to mobilise billions of people to take action. This birthed the idea of a wholly new kind of organisation that strives to bring the work of the UN closer to its people and to link the power of the UN to a people-driven connected global movement. UN Live is an example of what I call museological globalism at its most ambitious. Furthermore, the Canadian Museums Association's Sustainable Development Guide for Canada's museums which was written to help Canadian museums develop and implement the SDG's policies is an example of museological nationalism. Through the SDGs, in this case, the Canadian Museum Association explicitly supports the enrolment of museums in development theory. Development theory and strategy are based on the protection of natural resources and the environment under the condition of stimulating economic development and improving the quality of human life.[10] By supporting the SDGs museums, through museological globalism, introduce values and behavioural norms that present an idealised view of the world and one based on economic prosperity as sustainable futures rather than desirable futures in situated contexts.

Back in 2018, consultant Jasper Visser advocated that museums should lead the way in the implementation of the SDGs. He enrolled museums in neoliberal economic strategies by making museum resources and influence available to entrepreneurs and local initiatives supporting the SDGs.[11]

As cultural theorist and colleague Brett Neilson suggests, development, even in a sustainable form, means economy and growth in a capitalist constellation.[12] The articulation of sustainable development, and its promotion through museums, enrols all communities and the non-human world in a trajectory of capitalist advancement using the latter as resources to fuel and achieve growth. Through these policy instruments museums, sustainability, and their communities are firmly located within a western economic neoliberal capitalist model.

For example, Goal 8, "Decent Work and Economic Growth," calls for a 7% increase in annual GDP in the least developed countries and higher levels of economic productivity.[13] The British Council, and its program "Cultural Heritage for Sustainable Growth," enrols heritage institutions in Goal 8 through the promotion of an increase in GDP through sustained, inclusive, and sustainable economic growth in the museum and heritage industry, and through the expansion of employment opportunities.[14]

Furthermore, Jasper Visser, The Canadian Museums Association, the "Cultural Heritage Counts for Europe" Consortium, and others see museums as having a critical role in implementing Goal and Target 8.9 by devising and implementing policies to promote sustainable tourism that generates economic growth, local culture, and creates jobs and products. In doing so cultural institutions, through the implementation of Goal and Target 8.9, support a narrow suite of economic measures based on paid labour and monetary exchange, rather than other forms of work outside capitalist economies.

Goal 12, "Responsible Production and Consumption," is premised on the idea of achieving economic growth and sustainable development without the overextraction and degradation of environmental resources. This is proposed to be done through policies that promote efficiency, reduce waste, and streamline sustainability practices. At the same time, this goal seeks to promote urgent action to decrease ecological footprints by changing the way we produce and consume goods and resources, and their accompanying waste, that conserves and manages natural resources of future generations. Similarly, both the British Council and the Canadian Museums Association support

Goal 12, "Responsible Production and Consumption." This organisation and museum peak body and the museums they represent are therefore obliged to support Goal and Target 12.2, "more sustainable patterns of consumption and production through the sustainable management and efficient use of natural resources." And in addition, these museums, through Goal and Target 12.8, are required to promote a universal understanding of sustainable lifestyles by ensuring that people everywhere have the relevant information and awareness for sustaining development and lifestyles in harmony with nature.[15] This goal and target supports a third industrial revolution and a methodological globalism in which museums promote international laws and the values of the SDGs in regards to sustainable production and consumption. The non-human world is therefore conceptualised as a "critical economic asset," legitimising commodification (green capitalism) and unbridled consumerism. Here museums, through Goal 12, come to support objectives around growth and consumption through pedagogical programs that instil specific values and lifestyles. In this way, the destructive mindset that led to the climate crisis in the first instance is made regulatory as a form of ecological thought in which nature provides vital life-supporting resources for humans.

Goal 11, "Sustainable Cities and Communities," is the only goal that mentions natural and cultural heritage in Target 11.4: "Strengthen Efforts to Protect and Safeguard the World's Cultural and Natural Heritage." More specifically Goal, Target, and Indicator 11.4.1, "Safeguarding Natural and Cultural Heritage," proposes the measuring of total per capita expenditure on the preservation, protection and conservation of all cultural and natural heritage, by source of funding (public, private), type of heritage (cultural, natural), and level of government (national, regional, and local/municipal). Therefore, natural and cultural heritage protection is gauged in terms of financing, that is then put to work to attract people (inhabitants, workers, tourists, etc.), create jobs, increase GDP and financial investments, and to ultimately enhance the total amount of expenditure.

References to culture in Agenda 2030 and the official SDGs documentation are virtually non-existent. As a consequence, the initiative, titled "Culture 2030 Campaign for the Goals," seeks to acknowledge the many ways in which cultural values influence and can contribute to sustainable development.[16] The omission of culture as a pillar in Agenda 2030 is due to the fact that the implementation and reporting process does not necessarily include the communities, leaders, and cultural networks who are most engaged in culture and who understand cultural impacts best.[17] Culture 2030 puts cultural practice – forms of expression, creativity, and identity building – to work in the SDGs as a means to strengthen and enhance people-centred, sustainable development. The addition of culture through Culture 2030 is mobilised under Goal 12 "to support new sustainable consumption and production patterns that contribute to the responsible use of resources and address the adverse impact of climate change."[18] Culture therefore acts as a frame and an endorsing rationale for bolstering consumption and production values in a neoliberal mode.

As a member of Culture 2030, UNESCO articulates Goal and Target 11.4 by highlighting how the cultural value of natural and cultural heritage can act as a lever for its protection rather than just being indicators based on expenditure. While sustainable development remains a central feature of goal implementation such aspirations are invoked through the tourism, cultural, and creative industry sectors by raising awareness of the value and educational purpose of heritage. According to UNESCO,

> Museums have great potential to raise public awareness of the value of cultural and natural heritage and of the responsibility of all citizens to contribute to their care and transmission. Museums also support economic development, notably through cultural and creative industries and tourism. This Recommendation draws the attention of Member States to the importance of the protection and promotion of museums and collections, so that they are partners in sustainable development through the preservation and protection of heritage, the protection and promotion of cultural diversity, the transmission of scientific knowledge, the development of educational policy, lifelong learning and social cohesion, and the development of the creative industries and the tourism economy.[19]

Furthermore, the implementation of Goal and Target 11.4 by Norway, in respect to the preservation of Sami indigenous culture, places cultural knowledge, as opposed to technical or financial concerns at the centre.[20]

Museums support goals around financial measurements, production, consumption, and economic growth, and thus endorse the notion that neoliberal capitalism is the driving force for a habitable planet through minor adjustments to the way we use planetary resources, and through the invocation of culture in developmental strategies. Cultural practices are therefore activated to drive economic development rather than values and practices central to diverse economies and cosmologies based on other forms of consumption, production, and exchange. The new mantra, "most good, less harm" underlays all this. The aim supported through cultural institutions and capital is to find a sustainable scale in which the drawdown of nature's resources, and the use of new technologies in industrialisation and its transfer, seeks to be less wasteful and therefore does not exceed nature's replenishment. While it is acknowledged that natural capital decreases as capital grows, nature, or the non-human world, is still regarded as a continuing resource for human purposes and as a supplier of ecosystem services in terms of the natural income it provides.

Across all the SDGs, however, there is a recognition that there is something profoundly wrong with endless industrial growth and the threat social-economic practices lead to. Museums are thus complicit in promoting a profound contradiction between the logic of capitalism, on which the SDGs are founded, and a desire to inspire sustainable economic growth and a consideration of the vitalities of metabolic systems to which viable futures depend.

Henry Mcghie, from Curating Tomorrow, states that museums are fundamental to the achievement of Agenda 2030, along with other collections-based institutions, cultural sites, monuments, and natural heritage. Mcghie also articulates the role of museums across a range of SDGs and has developed a comprehensive handbook on how the SDGs and their specific indicators might be implemented through and across the sector,[21] and at the same time expands the target and indicator parameters. For example, Mcghie reinterprets Goal and Target 11.4 as closely linked to the achievement of many other SDGs because collections, he argues, serve as the basis for a wide range of activities, notably education (Goal 4) and research and innovation (Goal 9). Mcghie also believes that natural history collections operate as heritage science research infrastructures that can contribute to Goal 2 ("Sustainable Agriculture"), Goal 13 ("Climate Action"), Goal 14 ("Life in Water"), and Goal 15 ("Life on Land"), by ensuring the protection and safeguarding of natural heritage in nature. As forms of heritage science, natural history collections are enrolled in a scientific research agenda. They enact a technocratic and industrial complex of development directed towards a range of initiatives including support of investments in scientific and agricultural research, the advancement of vaccines, the establishment of sustainable production and consumption patterns, and the backing of developing countries to enhance their research and technological strategies and technological transfer.

In terms of environmentally directed goals, museums have been engaged in Goal 15 ("Life on Land"), which is directed to "Protecting, restoring and promoting the sustainable use of terrestrial ecosystems, the sustainable management of forests, initiatives directed to combating desertification, land degradation and biodiversity loss."[22] Targets of specific interest include 15.1, "Conserve and Restore Terrestrial and Freshwater Ecosystems in Line with International Agreements," and 15.9, "The Integration of Ecosystem and Biodiversity into Govermental Planning." The Climate Tool Kit for museums, gardens, and zoos,[23] for example, developed specific indicators and forms of measurement for natural history collections. That is by integrating scientific ecosystem and biodiversity values into national and regional planning and by establishing processes, alongside purpose-built research and collections facilities, for the effective conservation and restoration of terrestrial ecosystems. While a positive development such conservation science initiatives form part of a methodological globalism that does not sufficiently acknowledge local practices and ecological knowledge including that of First Nations that is derived from understanding metabolic processes.

Goal, "Life on Land" and Target and Indicator 15b "Finance and Incentivise Sustainable Forest Management," seeks to provide incentives to developing countries to advance forest management based on western conservation and reforestation practices. As described in Chapter 4, such practices require the enclosure of lands and absorb self-reliant diverse economies and long-standing practices of conservation and stewardship into a western remit while privileging the practices of global capitalism, finance, and its extractive logic of resource use. Western scientific measurements and global environmental assessments, therefore, are central to measuring goal success.[24] In such a circumstance human prosperity depends on the life-supporting capacities of ecosystems and forests that need restoration to support such needs.[25] Ecosystems and forests become natural capital and therefore an economic measurement of sustainability.

When SDG implementation involves using economic terms, such as GDP, to think through the relations between human populations, forests, and terrestrial ecosystems, it is necessary to quantify and balance the sustenance of human communities with impacts such as erosion and microclimate damage, as well as a consider all forms of ecosystem services such as biodiversity and habitat retention in an effort to avoid ecosystem services collapse. Measuring this sustainable scale involves linking the rate of resource depletion with the reduction of natural capital's ability to provide the level of natural income it did in the past.[26]

Climate change engagement by museums is now central to many institutional activities and strategic plans. Goal 13, "Take Urgent Action to Combat Climate Change," is therefore one of the goals taken up by institutions and their networks. Goal 13's aim is to maintain global temperature rise to 2 degrees. This target, however, is founded on a business-as-usual logic of capitalism, and the continuation of industrial and economic growth as the main indicator of progress.[27]

To maintain this temperature target, this goal is directed to improving climate resilience and adaptation capacity through Goal and Target 13.1, "The Integration of Climate Change Measures into National Policies," and Goal and Target 13.2, "The Reduction of Emissions Through Nationally Targeted Mitigation and Adaptation Plans and Policy, Bolstering Responses to Climate Impacts and Capacity Building in Partnership with Civil Society and Marginalized Groups." Central to these goals and targets is the commitment of developed country parties to the United Nations Framework Convention on Climate Change (UNFCCC) to annually commit $100 billion to address the needs of developing countries to mitigate climate change through the capitalisation of the Green Climate Fund outlined in Goal 13A. These funds are directed to building capacity for effective climate change-related planning and management in least developed countries and small island developing States. A focus is also placed on supporting women, youth, and local and marginalised communities, as described in Goal 13B. This goal endorses the UNFCCC as the primary international, intergovernmental forum for negotiating the global response to climate change. Typically museum commitments are directed to Goal and Target 13.3, "Improve Education, Awareness-Raising and Human and Institutional Capacity on Climate Change Mitigation, Adaptation, Impact Reduction and Early Warning."[28] The Goal and Target is aligned with the PA and the educational implementation frameowrk, the Action for Climate Empowerment, as discussed in Chapter 2.

The thoughts and frames driving Goal 13 are modern, capitalist, technological, and technocratic. The development of technologies and their resourcing by nature capital privileges modern mitigation strategies over the nurturing of living and vital ecologies in which communities are embedded. Climate financing, for example, ensnares people into a dependency on neoliberal capitalism and modern visions for mitigation and, in doing so, disenfranchises them from their own economies of value and sustenance located within living ecologies and ecosystems built on practices of attunement and care.

Climate change mitigation, as outlined in Goal 13, supports the continuation of industrialisation through new technologies and processes, and through their economic provisioning. In doing so, technological solutions, such as renewable technologies that support the shift in energy sources and energy efficent transport, are presented as clean and green and, in doing so, the belief that the biosphere won't be destroyed by the mining or manufacturing processes by which they are made. Furthermore, while zero carbon emissions might be achieved in sectors such as electricity generation, transport, and agriculture, extraction continues along with a failure to account for the multifarious metabolic processes in which western societies and their emissions reduction schemes are enmeshed. Renewable technologies also have their own material consequences implicated in toxicity and environmental degradation. Chile, for example, has one of the largest reserves of lithium in the world holding a quarter of global markets.[29] The mining of lithium in the Atacama desert salt flat in northern Chile to contribute to efforts to reach zero emissions globally through a range of interventions, including the use of this mineral in batteries, for electric cars, and to store energy is exhausting and contaminating water resources, exacerbating climate-induced drought, and restricting the availability of fresh water to the Indigenous Atacameño communities.[30] In all processes, and the control of them, extraction is legitimated and continues to lead to intensifying climate consequences. Decarbonisation through renewable technologies should not become renewed opportunities for further growth.

Nature-based solutions (mitigation based on the agency of plants to store carbon) are directed to reducing emissions, increasing the sink capacity of soils and plants, enhancing resilience within and across forestry, agriculture, oceans and food systems, and to soil and biodiversity conservation. In all these efforts, harnessing nature's agency is deemed to preserve the atmosphere. Nature-based mitigation is integral to the implementation of Goal 13 in which nature acts as a regulatory service through the capacity of soils and plants to gather up and absorb carbon. Critically, reforestation directed to promoting soil and carbon sequestration transforms forests into carbon sinks that not only enable the continuation of carbon burning but also sideline the livelihoods of forest people to create carbon offsets. The non-human world thus becomes political and economic nature as the preservation of carbon sinks allows for the sustenance of human life now and into the future. Human–non-human relations therefore are defined by the former exploiting, controlling, and crafting the latter.

Through Goal 13 museums articulate change as being an improvement in environmental performance that leads to changes in economic performance and a reversal of the causal relationship between radiative forcing and heating. Museums support investments in nature performance, and a reduction in heating in the atmosphere, by substituting energy sources and thus avoiding loss of economic prosperity and in actual fact enhancing growth. In such a scenario, the atmosphere becomes commodified as nature capital and is directed to sustaining future life when fossil fuel resources are substituted by solar and wind energy. This sparks a green technological revolution that is supported through the management of nature-based assets. All these narratives are central to museum engagement and articulated through exhibitions such as "Atmosphere." Nature is capital preserved in service to human needs. While lucrative in generating dividends, emissions trading programs act against the maintenance of ecological metabolic pathways.

Goal 7, "Ensure Access to Affordable, Reliable, Sustainable and Modern Energy for All," seeks to provide access to electricity, promote energy efficiency and renewable energy for cooking, technology, transportation, and heating and encourages higher levels of ambition in renewable energy and its transfer. The Canadian Museums Association, for example, seeks to encourage action through Goal and Target 7.2, "Green Energy Use," and Target 7.3, "The Efficent Use of Energy for All and Specifically for the World's Poor." Similarly, the Climate Toolkit for Museums, Gardens, and Zoos embraces 24 goals and targets for addressing climate change, notably Goal 7, "Energy, and Investment Assisting Institutions to Reduce Emissions Within Their Own Operations and Generate Renewable Energy Onsite."[31] Engagement with this goal supports the transformation of energy sources from carbon-intensive to renewables and is a strategy inextricably linked to economic growth and social progress for museums, zoos, and gardens.

Museums' engagement with the goals supports the alignment of all communities into global economic activity, transforming economies through international trade, technology transfer, sustainable industrialisation, and energy efficiency, global supply chains and the financing of economic, resource and energy flows through technology, science, and innovation. Such an engagement signifies a shallow rather than a deep radical transformation. Furthermore, as explained in Chapter 2, the linking of consumption and production patterns to clean technologies as a strategy to avoid harmful activities is erroneous.[32]

When SDGs are implemented through museums, new concepts, international norms of human rights and environmental standards are given over to audiences. The indistinctness of the boundaries between local, national, and global frameworks leads to the emergence of supranational norms. This includes sustainable development and sustainable futures, and each of the 17 goals. The aspirations of people, the planet, prosperity, and peace, construct a new knowledge system of observation, a new way of interpreting the world, and an illustration of what livelihood and success look like and how it is to be measured. Through Agenda 2030, local communities are subjected to interventions based on international scales and norms whereby promoting the idea that "global problems require global solutions." This in turn undermines specific responses based on local knowledge. Museums play into this by enacting museological globalism in seeking to build a global citizenry through SDG commitment and by strengthening cooperation between museum institutions and coordinating policy implementation across networks such as We are Museums and FORMS. In doing so, museums and networks embed international norms and policy frameworks, promote international co-governance actions, and upgrade governance capacity through the values of Agenda 2030.

The so-called given right to development for developing countries, which museum engagement supports, is governed by the assumption, enshrined in Agenda 2030, that other knowledge and alternative economies are deficient, backward and vulnerable rather than as attuned stewardship practices of planetary life. These claims to development, and to green energy and energy efficiency, are made according to a liberal ideology of rights in which vulnerability is the basis of right attribution. Overall the SDGs invoke an environmental justice approach based on human beings and their rights relating to quality of life and health.[33] Equally, ecological justice is a rights-based approach attributed by human populations to the non-human world to enable the former's survival.[34] These rights-based approaches are erroneous because they are humanist and anthropocentric and are framed according to developed world population values; they position the non-human world as vulnerable rather than acknowledging their agenital capacities.

Another goal that maps onto museum pedagogy is Goal 4, "Inclusive and Quality Education for all in Both Formal and Informal Settings." This goal is directed to ensuring that all learners acquire the knowledge and skills to promote sustainable development, sustainable lifestyles, human rights, gender equality, a culture of peace and non-violence, global citizenship, and appreciation of cultural diversity and its capacity to contribute to sustainable development. The two key targets identified by the Canadian Museums Association guidelines, and cited by consultant Henry Mcghie in his handbook, are Goal and Target 4.4 "Increase the Number of People with Relevant Skills for Financial Success," and Goal and Target 4.7, "Education in the Service of Sustainable Development and Global Citizenship Through an Appreciation of Cultural

Diversity and of Culture's Contribution to Sustainable Development." In respect to both these targets, education is cast as a skill set that enables upward socioeconomic mobility as a key strategy for escaping poverty,[35] and in promoting the principles of sustainable development and economy. In comparison to the Canadian Museums Association, the Monash Sustainable Development Institute's interpretation of this goal is more expansive in considering how education improves life prospects more generally, thereby "provides people with the knowledge and skills to increase employment prospects, reduce poverty and improve health and wellbeing."[36]

Goal 4 is also closely tied to the promotion of gender equality. Millions of women across the world are either denied, or only given a limited, education and therefore this aspiration is commendable. A lack of education is also deemed to result in reduced health, wellbeing, ability to take part in decision-making processes and increased social isolation. Museum commitment through this target is laudable in upholding and encouraging gender equity, but unfortunately makes museums complicit in placing education, culture, and cultural diversity within a neoliberal capitalist and developmental mindset. That is whereby paid labour and socioeconomic advancement are considered the key to emancipation and alleviating poverty. Thus, museums again support the development of a top-down, one-world economy, universal educational values, and pedagogical imperatives without sufficient regard to local context.

An example of this approach is demonstrated by the museum, UN Live which seeks to roll out the SDGS across the world and foster a belief that this model is universally applicable to all cultures and contexts. In doing so, the Museum fails to acknowledge the diversity of economies, worldviews, and practices in which communities are embedded. Instead, culture and cultural diversity are harnessed to enhance and expediate sustainable development. In this way, museums such as this one are not only enrolled in the governance of natural and cultural heritage through the notion of sustainable development but also through education. Concerningly, museums thus use the goals to eradicate poverty by educating populations to support the expansion of economic growth and consumption, and governing climate change according to technocratic, technological methods, and market mechanisms. In terms of the environment, ecological problems are defined as a risk as opposed to an opportunity, as events or crises to manage and measure rather than map, attune, and regenerate. In doing so, the sector promotes problematic values and concepts such as the carrying capacity of the environment in respect to extraction, the substitutability of produced and natural capital, the transition of energy sources to reduce emissions and sustain natural capital, and the resilience of ecological systems, measured and calibrated, by sustainability science. Museums thus come to articulate habitability scenarios through market mechanisms, technology, and the better management of natural capital and at the same time promote further destructive practices by negating or rendering invisible the fundamental political, economic, and social changes the world requires to create a habitable planet. Through their enrolment in Agenda 2030, museum institutions mimic classic development models, ecological politics, and imaginings based on financing, technocratic measures, scientific and technical knowledge, moral rights in a modern form, the idea of nature as resource, the promotion of universal values at the expense of locally situated practices, and neoliberal capitalism. Change towards accomplishing goals, therefore, is achieved through a series of checks, balances, and updates on a global scale.

The concept of countries having different development levels is a foundational principle underlaying each goal. The PA and SDGs, and the classification of developing and developed countries that inspires them, is a neo-colonial form of ecological modernisation in which developing countries are, according to feminist academic and activist Ariel Salleh, viewed as "not yet fully capitalist," are "cast as victims and unsophisticated" and in need of being "inducted into neoliberal capitalism to progress."[37] To achieve this, the SDGs use euro-centric ideas of poverty, development, and subsistence economies to make these claims. Underlaying this is the belief that communities are unevenly linked to global capitalism based on their levels of development. Neoliberal capitalism also operates as a lever to pull developing countries out of their predicaments. These thoughts and frames of reference override and disregard developing countries' own knowledge and material practices, including different communities' relational, spiritual, and metabolic understandings of the non-human world and the efficacy of their often low-carbon livelihoods developed over thousands of years. Museums, through their commitment to and implementation of SDGs, perpetuate this modern form of progressive evolutionary thinking in respect to societal development.

So I now return to Jeremiah, mentioned on the first page of this chapter. Would he see the goals as being relevant to him and the Iripiv community? Is this the sustainable future they might imagine? A future that foregrounds the importance of economic and GDP growth as a sustainable paradigm, the bolstering of waged labour to promote growth, their enrolment in the production and consumption of goods as a life-sustaining strategy and their heritage as assets for generating income and jobs? Is this the good life? A world governed by neoliberal capitalism? Would Jeremiah and his community view themselves as poor and under-developed? Is neoliberal capitalism the answer to pull them out of their assumed predicament?

UNESCO acknowledges that the knowledge systems and environmental management practices of Indigenous, First Nations, and local people provide insights that enable better management of ecological challenges – preventing biodiversity loss, reducing land degradation, and mitigating the effects of climate change.[38] To a degree, this acknowledgement breaks the tight hold of economy and technical knowledge embedded in the SDGs, but even still

Indigenous knowledge systems are seen as an add on to the proposed aims and not integral to them.

Goal commitment, in regard to Indigenous communities, is often enacted by the adoption of principles enshrined in the goals such as equity, inclusion, cultural values, and participation in order to meet the requirement of cross-cultural engagement and research design, rather than to legitimately engage with First Nations knowledge systems. Such applications are often separate to the goals. The Australian Institute of Aboriginal and Torres Strait Islander Studies (Australian Institute of Aboriginal and Torres Strait Islander Studies) Code of Ethics for Aboriginal and Torres Strait Islander Research, for example, deploys the goals to ensure research is undertaken in appropriate, ethical, and culturally appropriate ways in collaboration with Aboriginal and Torres Strait Islander communities.[39] This supports a number of goals and targets relating to the fair sharing of benefits arising for example Goal and Target 1.4 on access to property and inheritance, Goal and Targets 2.5 and 15.6 on fair use of genetic resources and traditional knowledge, and Goal and Targets 4.4 and 4.7 promoting participation of Aboriginal and Torres Islander communities in research. Other Goals and Targets implemented in such an approach are Goal and Target 10.2 which seeks to empower and promote the full social, economic, and political inclusion of all, Goal and Target 10.3 focusing on reducing inequalities of outcomes, and Goal and Target 10.4, which aims to adopt policies for inclusion in order to promote inclusive participation. Goal and Target 11.4 enshrines principles of cultural rights in regard to the creation of cultural heritage during the research process. It endeavours to ensure it is properly cared for, and that respect is given to Australian Aboriginal and Torres Strait Islander people and their perspectives and rights.

On a global scale museums' engagement with the SDGs, through ICOM, for example, inducts millions of people into a capitalist economy of exchange, forcing them to abandon their autonomy. Furthermore, the mobilisation of the SDGs across museums globally is problematic because it supports the enlargement of economic development as an agenda, across the planet, penetrating deep into the more-than- and other-than-human, non-human worlds, and re-defining them according to human needs and a desire for growth. This process, in which museums are complicit, represents the potential production of new ecological commodities, markets, and their financialisation.

The goals are intended to fit all and to represent common interests. Each of the goals seeks to capture absolutes and a singular aspiration. At the same time, the development plan of each goal dissociates ecosystems from each other by delineating them into separate goals. For example, clean water is located within a goal separate from other systems, such as industrial agriculture, deforestation, manufacturing, urbanisation, and climate change, despite the obvious interconnectedness between them all. For example, museums advocating for Goal 6, "Clean Water and Sanitation," enrols communities and their water resources into a capitalist value of exchange. Salleh, for example, explains how under the United Nations Environment Program, communities in Mexico and Brazil's water sources fall under the eurocentric green model payment system for managing watersheds and how concerningly such an arrangement makes these water resources amenable for future commodification.[40] Such economic reasoning leads to the enclosure of these water resources for bottled water articulated as a human right to clean water. Really each of the goals is deeply contradictory because on the one hand they promote economic growth within a capitalist regime and on the other seek to encourage environmental sustainability. The goals incorporate the legacy of colonisation and its driver capitalism and the consequences that arise from these two agendas, political economist John Foster explains, is the creation of a metabolic rift damaging ecosystems and appropriating people's livelihoods for the manufacture of commodities, service industries, and so forth.[41] In driving goal commitment, museums become complicit in environmental destruction and in the deeply contradictory coupling of economic growth and environmental sustainability. While not connected to the goals specifically, the Australian Museum in Sydney has revised its mission statement to include a stronger emphasis on climate action through the promotion of First Nations knowledge. This is commendable for articulating a new vision of being "a leading voice for the richness of life, the earth, and culture in Australia and the Pacific."[42]

The problem with the SDGs and their implementation in context is well documented. For anthropologists, such as Eric Hirsch, sustainable development extends neoliberal capitalism into new contexts while proclaiming to remain conscious of economic and environmental constraints. This agenda is pitched in a way that suggests careful capitalist extension is potentially limitless and at the same time harmless.[43] The rationale behind this is that sustainably maintaining natural capital stock is the key to not letting resources further decline.[44]

The notion of a finite planet, central to Agenda 2030, emphassises that natural capital is a limited resource and thus development must be sustainable.[45] This logic supports the idea that maintaining stock is a prudent minimum condition for assuring sustainability. This paradigm, sociologist Paul James explains, remains the basis of human flourishing and is marginally offset by ameliorative efforts that take the environment into account.[46] In engaging Agenda 2030, museums promulgate the notion of the planet as finite and conflate the idea of capital extraction into a discourse for maintaining planetary stock together as a noble aspiration. In doing so museums, and professional bodies such as ICOM, insinuate that the non-human world is an object that needs to be socialised and financialised into new markets globally, and into new venues and sites for capital accumulation. It remains an extractive ecological thought.

Through museums, the SDGs and their implementation standardise growth models and ways of integrating the economy, society, and the environment, and thus the future, according to an ecological modernisation agenda. The position of the human also becomes agential in sustainability through the coupling of wage labour and GDP growth.

Eco-curating habitability practices

Ecological concepts are distinctive relations of coexistence that situate human populations in different ways in diverse material contexts. Even Cartesian philosophical binaries of nature/culture or subject/object are not binary but relations of entanglement but not in a conventional sense. One element, say, nature, must be compared to another such as culture in order to have efficacy as a system of relations and practices.

In museum practice, the predominant thought is that objects or things are placed in some type of relationship with human subjectivity. However, modern systems of relatedness in Agenda 2030 lack a fundamental attunement across each of the goals in respect to other-than-human, or more-than-human, relations. Rather such binaries re-count attunement in terms of human needs and the economy. Within this policy framework, museums make it their public mission to focus on human responsibility and strong human agency as a means of creating a positive future, of taking action, either overtly or implicitly, and in advancing the UN Global Goals for 2030.

Museological globalism, and its eurocentric ecological modernisation agenda, crosses all the SDGs and becomes an aspiration and method for framing a better future, or series of better futures. In an attempt to roll out the Agenda 2030 vision of an equitable global economic system while preserving the planet for our children and future generations, museums impose technocratic methods on others through targets, indicators, and monitoring. This is in stark contrast to envisioning a series of proactive, speculative, innovative habitability practices that engender the spirit of the goals but don't necessarily adhere to their valuations of economy and environment. Most concerningly museological globalism enshrines the idea of backwardness for those who do not live according to a capitalist agenda or refuse such an imposition.

While it is acknowledged that society and nature are entangled, in order to sustain needs the non-human world is deemed to lack its own agencies except in the case of nature-based solutions, for example, the capacity of plants to photosynthesise carbon as a climate mitigation strategy. People become human capital and nature becomes natural capital. In essence, Agenda 2030 breaks down local ecosystems and sustainable economies, locks communities into being controlled by overseas business interests and multilateral bureaucracies, and undermines cultural diversity and autonomy.[47] Salleh suggests that metabolic value, as opposed to commodity value, must be attributed to the non-human world in order to enable organic flourishing. Coupled with this, sustaining aspirations must promote the commoning of energy, land, water, and air and at the same time recognise the wisdom of low-carbon societies.

Anthropologist Marisol de la Cadena works in the Andes with the Quechua people and conveys problematic stories regarding people's efforts to resist development and the conversion of their relations with more-than-human entities into "resources."[48] For such communities, it is evident that engaging the spirit or some of the principles of the goals would be better implemented through local practices. For example, Goal 2, "Zero Hunger," when implemented according to local sustainable food production, as opposed to the SDG framework, uses less energy, eliminates dependence on imported goods, retains carbon in the soil, and supports biodiversity.[49] With such knowledge, the question for museums may be what other planetary-human configurations might be proposed and enacted for partial recuperation and regeneration so that museum projects are not just ones stabilising or sustaining the planet's resources for future generations in an anthropocentric mode but rather reconstituting, re-imagining and rethinking human agency and the planetary as praxis beyond the modern constitution in situated contexts. That is, through alternative material figurations of relationality, attunement, and economic practices to those presented by capitalism. An example of this in action is taking Goal 15, "Life on Land" and re-working it in contexts that ni-Vanuatu anthropologist Ralph Regenvanu explains as nature metabolism,[50] which rests on the expertise of farmers who live and are attuned with non-human rhythms.

Museums are venues pursuing a global agenda promoting a suite of new common interests, values, and characteristics for all human beings as sustaining and as neo-capitalist economic entities. In doing so, they enact a collective agenda that seeks to build a global community of people, nation-states, and territories through forms of museological globalism. But museums could pursue a different agenda whereby the SDGs become communitarian design projects combining context-specific ecological thoughts alongside a consideration of material metabolic rhythms and systems.

This can be achieved by learning to live differently by being inspired by other knowledge practices, including indigenous ones. Through such knowledge, different relations of, and to, water, land, forests, and the non-human can exist and be enacted that don't aim to turn "nature" into a "resource" and humans into capital accumulators.

Having said this different groups interpret the SDGs in their own ways and direct actions to strengthening their own cultural indicators of relevance to specific contexts in an attempt to rectify the imbalance between economic and environmental priorities or pillars. Some groups are actually developing living standards frameworks to address the deficiencies in the SDGs that privilege economic growth over community wellbeing. For example, the *Living Standards Framework* in Aotearoa/New Zealand engages a broader set of measures rather than appraising progress solely in economic terms. The mantra of leaving no one behind, evident in the SDGs, is replaced by the Māori proverb, "He waka eke noa – 'we are all in this together'." The special status of Māori as the *tangata whenua*, Indigenous people of New Zealand, is central and the concept of *kaitiakitanga*, or guardianship, of the natural environment puts sustainable intergenerational wellbeing at the centre of policy-making and the management of resources.[51] The *People's Plan* connects with the government's Living Standards Framework but articulates a clearer alignment with the SDGs' universal vision of the 2030 Agenda and the global framework of the SDGs, with shared targets and measurable accountability. Involving consultations with civil society, including tangata whenua (the Indigenous Māori people of Aotearoa New Zealand), community groups, non-government organisations, educational institutes, unions, and the private sector acknowledges the place of indigenous knowledge, culture, and contributions to solutions, targets, and indicators written in response to specific Aotearoa/New Zealand concerns.[52] The targets and indicators, however, adhere to an economic pillar and neoliberal capitalism as the founding principle. Using the term "capital" is a way of articulating human and nature "assets."

Another example of a group not simply adopting the SDGs is the ni-Vanuatu which also developed their own version, *Vanuatu 2030 The Peoples Plan (The Plan)*, based on their own cultural values, traditional knowledge, and Christian beliefs. With a balance between the social, environmental, and economic pillars of sustainable development. ni-Vanuatu cultural heritage is placed as the foundation.[53] In both the Aotearoa/New Zealand and Ni-Vanuatu examples while both groups interpret the SDGs in their own ways each is driven by universal values of economy, sustainability, and development. Culture nonetheless remains in the service of sustainable development and involves the translation of the universal language of the SDGs into the language of specific communities, cities, and regions.[54]

In respect to Agenda 2030, museological globalism is increasingly directed to using the SDGs to strengthen cooperation across territories through a series of common interests determined by universal values. Even hybrid arrangements, such as those described above, are, at their foundation, neoliberal in both conception and implementation. Furthermore, museological globalism homogenises diverse cultural values, economic practices, and futurisms into a series of universal goals, indicators, and measures around neo-capitalist economy and sustainable development.

This is incorrect, however, as the global order is characterised by the coexistence of multiple knowledge practices rather than a singular authority. Directed towards the assessment of human concerns and needs first and foremost, the systems, notions of economy, and technology on which the SDGs are framed, as a form of methodological globalism, are designed to enhance human abilities, capacities, wellbeing, and enconomic growth to survive and thrive rather than address an emergency borne out of a failure to acknowledge and work with entanglements. Human-centred thinking in the west has blindly constituted the world through the use of neoliberal capitalism as a strategy to harness, and now protect, the non-human world. In such thinking, nature is an entity that is under human control to support growth.

The SDGs, therefore, posit relationships and differences of the way of valuing the non-human world between and across communities, as a set of values and practices that must be converted for the purposes of development rather than as many different practices of world making. No singular economic, knowledge, value system, or idea of human agency, in relation to the more-than or other-than-human world or universe, exists in which all life and diverse cultural perspectives can be accommodated.

The convergence of these different living frameworks, including the perspectives of the ni-Vanuatu on Iripiv, and Agenda 2030 on the other hand, demonstrates that pluralism as the only viable proposition. However, how these pluralities are imagined different implications result. In order to explore this notion further, and in the context of sustainability and Agenda 2030, I return to cultural theorist John Law's[55] concept of "fractiverse," formulated through his analysis of the Australian post-colonial context and the stark differences between colonial and Aboriginal Australian relationships to land. Law argues that cultures express and enact different reals or universes or possible worlds as a series of fractiverses[56] rather than as one world, a universal framework, or even as a pluriverse (a plural one-world framework). In such a framing, cultural knowledge, practices and their respective value systems offer a different interpretation, role, significance, relevance, and priority to living circumstances and their future prospects; they formulate different habitability practices. Operating in this way a fractiversal optic opens up a space to deeply interrogate different habitability realities as the effects of contingent and multifarious culturally inflected norms, values, politics, histories, enactments, and sets of relations. The term "fractiverse" also better explains why and how different concepts of human agency, human–non-human, ecological relations, and material attunement, as well as different economic practices, and why groups respond to the SDGs, their narratives, indicators, and measures, or reject them and enact futures differently.

In respect to curating for planetary habitability, fractiverse is a concept that can be put to work to explain how diverse concepts and practices operate in disciplinary contexts and how different knowledge practices, that lie behind the notion of habitability, require distinctive responses and seemingly attract competing interests. Overall, it explains why global management schemes, based on a universal technocratic frame, are also difficult to devise. Habitability thoughts and acts are a fractiversal composition and herein lies the problem with the SDGs. Indigenous relations and cultural practices, in respect to climate mitigation, for example, are acts that have their own histories, politics and cultures, and ways of working with earth others.

The "Environmental Injustice – Indigenous Peoples' Alternatives" exhibition, at the Musèe d'ethnographie, Genève (24 September 2021–21

August 2022) was framed around the climate emergency and presented the viewpoints, perspectives, and knowhow used by Indigenous people to deal with the environmental damage caused by climate change, in particular the protection of biodiversity, soil, water, and ecosystems.[57] It operates as a space through which First Nations people talk about how they mobilise their ancestral knowledge to deal with climate change impacts, live sustainably, and change modern relationships to ecosystems in order to weave a common future based on the values of care, protection, repair, respect, and responsibility for the environment.[58]

Applying the fractiverse concept, as a way of framing global diversity, comes with its difficulties. Rejecting the idea of one universe, inhabited by thoughts and practices, also means refusing the possibility of any overarching standardising logic such as the SDGs in order to adjudicate between different sustaining practices. These different realities, as per the fractiverse concept, might be thought of as different ways of being, doing, acting, conceptualising, or objectifying phenomena, as domains of knowledge such as law, science politics, or indeed different heritage concepts or museologies. When thinking fractiversally sustaining, the world is enriched and enlarged beyond a one-world framework. Each world has its own unique system of economic, cultural, and living practices, and ways of working relationally within different ecological circumstances that must be taken into account when implementing sustaining policies and projects.

While the SDG framework acknowledges that each goal is a system of ecological thought and a series of actions in their own right, it enacts a limited range of coordinates. In considering this, the sector must come to terms with the implications of these goals in respect to what they mean and, most importantly, what they materially enact. They need to be conceptualised not only in terms of their vital cultural engagements but also in regard to the underlaying values, assumptions of agency, sense of community, economic development and finance, and the types and forms of human ingenuity that mobilise tasks, measure and monitor performance, and affirm achievement. The values embodied in the SDGs include culture as economic assets, intergenerational equity management, and the passing on of resources that have the potential to provide for the needs of the Earth's future human populations.

Development and a neo-capitalist economy remain at the centre of the SDGs obviating a deeper engagement and critique of the relations of culture and economy that is required, and indeed exists. In doing so, the SDGs undermine culturally diverse, situated, and sustaining knowledge and practice, as well as material practices of economy and ecology that might arise from these engagements. In Agenda 2030, and museums' embeddedness in policy, culture enters, if at all, in instrumental terms. For example, museums are recognised in both developed and developing countries as cultural institutions and important sources of income, employment, and economic growth, and their work is directed to the sustainability of local communities through such avenues as cultural tourism.[59]

In a fractiversal constellation, however, Agenda 2030 is just one mode of existence that can be put to work, not as a universal policy framework but directed to achieve the spirit of the goals such that they are interpreted and implemented in many ways.

The Buen Vivir movement, for example, imagines and practices an alternative, community-oriented future, unshackled from the demands of capitalism that are driving the planet to disaster. Buen Vivir literally means "well-being" and the movement emphasises the intrinsic connection between humans and nature in an extended community that is envisioned as the goddess Pachamama. Such frameworks, and other alternatives, are not economy-centred in a capitalist sense but can still be economically directed within local frameworks by seeking to find a balance between environmental and economic concerns while rejecting capitalism and economy-centred development.

For First Nations Indigenous communities in Australia, sustainability has an entirely different meaning, although there are some intersecting lines with eurocentric frameworks. Aunty Fran Bodkin of the D'harawal people and Bidiagal clan from southwestern Sydney views sustainability as a concept in which non-human resources can be harnessed for human flourishing but in a way that is based on respect and care through sophisticated understandings of the vitality of Country, animals, and plants as a holistic concept.

> In the dreaming it was realised that in order to live, we needed to ensure the availability of those resources we needed to sustain ourselves. Over generations we observed and experienced those conditions on This Land, recording in story and song what we had learned, how the times of day were important for certain duties, how the changes in the weather were rhythmic, recurring year after year, and how other, longer cycles either lengthened or shortened the pulse of the rhythms. We learned that the availability and sustainability of those resources upon which our life depended could be extended if we respected the Land rather than used it.[60]

In contrast, sustainability concepts in Agenda 2030 have neoliberal capitalism at their core: a waged economy, resource extraction, growth, education for waged labour, and so forth. Furthermore, Agenda 2030 – and its sustainable development agenda and measurements of efficiency, quantification, data analysis, extraction, and GDP growth –is incompatible with understandings of vitalism, attunement, and respect.

Sustainable development is presented as a universally applicable strategy that transcends cultural, economic, and political lines. Environmental scholars David Thorsby and Ekaterina Petetskaya argue that a fully realised model of sustainability, applied in non-Indigenous societies, will only be possible if political institutions acknowledge the importance of the insights that have accumulated over generations in Indigenous knowledge systems.[61] This is beginning to happen in Australia in the aftermath of the devastating 2019–2020 bushfires on the east coast of Australia people when the land management and cultural burning practices of First Nations were increasingly acknowledged as critical adaptive strategies attuned to Country that needed to be considered more seriously.

The Brundtland Commission's definition of sustainable development published in 1987 is presented as "development that meets the needs of the present without compromising the ability of future generations to meet their own needs."[62] Agenda 2030 and museums embrace the core idea of sustainability, and particularly the implication of long-term viability and a capacity of systems to reinforce and maintain themselves over time. A similar concern with the long-term underpins Indigenous cultures whose very existence is grounded on the inheritance of traditional knowledge and the transmission of cultural values to subsequent generations. In such cultures, the nature of sustainability is understood and experienced in terms of relationships to kinship, land, language, and spiritual knowledge as a holistic system for interpreting the world.

While Indigenous and modern sustainability concepts share a concern for the efficient and equitable management of resources so that present and future human needs are met, how these are managed, and to what ends, and the values underlying them, vary greatly. The Agenda 2030 sustainability concept abandons holistic ideals in favour of economic or sectoral gain at the expense of environmental or social principles. In contrast, Indigenous societies place a stronger focus on the absolute necessity of a whole systems approach and a more balanced interpretation of priorities.[63]

Agenda 2030 and Indigenous holistic frameworks are motivated by long-term concerns for intergenerational transmission of knowledge, whether that is natural or cultural capital or more complex phenomena such as human capacities for satisfying economic, social, or cultural needs. The origins of such motivations are, however, different. For Indigenous First nations, these concerns are founded in the recognition of the continuity of human life, and interpreted in both material and spiritual terms.[64] With Agenda 2030, on the other hand, the idea of maintaining harmony with nature means to work within a capitalist mindset rather than to attune and be responsive to the vitality of the non-human world. As Thorsby and Petetskaya state Indigenous First Nations' holistic approaches have been very successful and existed long before the concept of sustainability and Agenda 2030. Furthermore, these knowledge practices have proven to be successful in achieving, for their societies, the sorts of sustainability goals enshrined in the Agenda 2030. The insights of Indigenous communities cannot be ignored or overlaid by Agenda 2030 or by museum practices, or indeed any development agendas.

Furthermore, Agenda 2030 is a direct outcome of environmental destruction wrought by the capitalist machine. Interestingly, through the goals the non-human world becomes natural capital amendable to extraction comprising renewable and non-renewable resources and on the other comprising ecosystems and species that support and maintain life in the biosphere.[65]

In contrast, Indigenous holistic frameworks are more likely to focus on a steady state with an emphasis on maintenance rather than development.[66] For example, the Yolngu people of Eastern Arnhem Land in Australia see their role as maintaining capital stocks as they are. That is, their country, their culture, their traditional knowledge, and their creative practices. However, this does not imply cultural or economic stagnation. Indeed, remote communities in Eastern Arnhem Land are pursuing new methods for creative expression and are looking for new economic opportunities that can be interpreted within an overall framework of cultural maintenance.[67] This includes new avenues by which the Yolngu can earn revenue – including new media and the utilisation of copyright law – in order to ensure the continuation of Yolngu culture.[68]

Moreover, Indigenous holistic frameworks are predominantly location and society-specific, in contrast to Agenda 2030 sustainability concepts that were developed to be applied worldwide. This difference reflects the knowledge systems on which these frameworks are based. Indigenous knowledge originates from particular locations and regions, while Western scientific knowledge is thought to be universal and applicable irrespective of location or scale.[69]

Indigenous cultures are built on a system of shared responsibility, whereas the predominant value system underlying economic and social policy in the West, and that promoted through museums, is one of individualism that supports private ownership of resources and ideals of unrestricted economic growth. In these instances, local perspectives can be seen to be important in defining the political, social, ecological, and cultural contexts within which sustainability is sought. Indigenous holistic frameworks can not only complement the Agenda 2030 model but serve as alternatives to neoliberal growth.

In regard to museums, they can engage cultural values not only in terms of showing how they can be coupled with industrialisation, economies, technologies, and the production of goods and services but also in showing how they can embody more-than-human ways of knowing, being, and acting that is more successful for sustaining life forms and ecological health through low emissions practices. Although not articulated in conventional mitigation terms, or through Goal 13, museums can make significant contributions towards more effective and equitable policies in relation to terrestrial carbon stores, low-energy agricultural and agroforestry systems, and the regeneration of Country and marine and terrestrial landscapes. Furthermore, museums can engage with Agenda 2030 and its hybrid re-envisioning in multifarious ways.

The "fractiverse" concept opens up possibilities for discussing sustainability as multiple reals, as well as considering fractiversal living, or habitability practices, as governing projects. The connections between these different reals can be thought of as modes of existence within a "fractiverse," each with their own system coordinates with intersecting lines and relations that can be composed in different ways to form sustaining projects and interconnected ecological formations. Fractiversal positions become, and are, aligned to situated contexts rather than mobilised in support of a return to the modern values and concerns of Agenda 2030 and museological globalism.

Re-working the SDGs

In refiguring the SDGs, and incorporating practices that exist outside these frameworks as a fractiversal constellation, many different sustaining habitability practices come into view that are situated and place-based. When investigating the SDG framework and the way the goals might be reworked in modern neoliberal contexts, three problems become evident. The first is the centredness of the figure of the modern human in the form of the male, the entrepreneur, the technocrat, and the policy maker. The second is the system of accumulation, which provides the impetus for neoliberal capitalism and its development agendas, enshrined in each of the goals and the reinstitution of non-human agencies beyond their human-centred utilisation. The anthropocentric illusions of the SDGs must be debunked and at the same time development agendas must be removed as the drivers for change. Generally, the SDGs must become post-anthropocentric concepts and policies in which humans are put back into their ecological place.

In this section, I draw inspiration from my previous work and scholarship in the critical posthumanities and other ontological frameworks in order to rework the SDG framework. I work with these optics to formulate more-than-human habitability practices as an alternative to the existing human-development-centric approach, and then I consider how this novel framework might be enrolled in museums.

What I call sustaining, living, or habitability practices not only refer to humans and future generations but take account of all inhabitations. Posthuman scholars, Olga Cielemecka and Christine Daigle, claim that sustainability and, as I will argue here, the SDGs must be reconceived to envision ways of common co-surviving rather than foregrounding the survival of human populations.[70]

The conceptual reinstatement of the relationality of vital material systems and knowledge practices across cultural, social, economic, political, and biological spheres is central to achieving this aim. Such an approach follows the insights provided by Indigenous ontologies but does not appropriate them. Sustaining practices, as described by Aunty Bodkin of the D'harawal people of the Bidiagal clan in Sydney, for example, are situated and based on relational thinking in which all spheres of life are considered interconnected and vital. Such thinking results in actions successful in caring for Country, and maintaining community, both economically and culturally.

The first step in overcoming the privileging of the human over the non-human in the SDGs, and the divisions that keep relational connections as separate goals, entities, and indicators, is to reinstate non-human agency as vital actors in implementation projects. To do this, and frame concepts and practices of co-survival, the first step is to draw on the "ecologizing method" I developed in the pioneering 2014 work, in my co-edited collection, *Climate Change, Museum Futures*, and in particular the chapter "Ecologizing Experimenttions: A Manifesto for a Posthumanist Museum." This practice was further developed in Rosi Braidotti and Maria Hlavajova's, 2018, *Posthuman glossary*, my entry, "Posthuman museum practices" and more recent monograph, *The Future of Digital Data, Heritage and Curation in More-than-Human World.*

All these texts make visible and fold out the many coordinates and entities, as well as the multifarious agencies, involved or as many that can be identified in order to work with them in interdependent co-surviving arrangements.

In such a configuration the vitalism of non-human agencies are, and can be, refigured to work together with human agencies in respectful, rather than exploitative, ways, as well as link to other agencies and entities in eco-curatorial processes of becoming. Vitalism and stewardship as a process ontology can then become core ideas and practices that are mobilised conceptually, materially, and metabolically to more closely align with the realities of life itself.

Fundamental to this, as Cielemecka and Daigle argue, is that sustainability is conceived according to conventions of human environmental management and seeks to secure access to natural resources for future generations of humans. Sustainable human activity, therefore, is one that ensures that "our children" will be able to inhabit the world as we do.[71] Modelled on the logic of intergenerational inheriting, however, Cielemecka and Daigle rightly contend that "sustainability is limited in its ability to help think through the entangled, multi species temporalities in which responsibility for the past, present, and future converge."[72]

Agenda 2030, as a sustainability agenda, follows the same principles of intergenerational inheritance as a human-centred form. Museums, through their complicity with Agenda 2030, promote an anthropocentric focus on human responsibility for future generations rather than on an expansive scope of community and future generations, including all inhabitations – that being all instances of life and earthly processes. Museum institutions, therefore, promote a narcissistic view of the future in which generations to come are seen as extensions of both current and future humans. When re-working human bodies and agency as forms of humanness radically embedded in compositional processes, with non-human others and earthly processes, the "I" becomes "us" and human agency, and indeed community, is always more-than-human.[73]

The agential configuration of the SDGs must be pluralised in their composition and distribution and in respect to who is involved, both human and more-than-human. Central to this is a consideration of how each of the goals – their coordinates and the non-humans they stand for and their metabolic processes – become entangled with others and how their temporal durations might be reworked beyond human time.

Indeed, empirically in the field an understanding of the relational interconnections between the coordinates involved – including their figurations, durations, and metabolisms – is foundational to any implementation strategy to counter strong human agency. Human time, like intergenerational inheritance, is foundational to Agenda 2030. For example, the year 2030 is the agenda's implementation date. This timeframe to achieve the aspirations of the goals reflects the human time of the United Nations and also includes museum time in terms of policy implementation timelines. The Working Group for Sustainability and Climate Change Action plan proposes the modern human time of 2024 as a goal for implementing the plan for ICOM.

Rather than a strict adherence to human time the SDGs, in all their complexity, need to comprise multiple temporalities and durations of human, non-human, and more-than-human coordinates, and implementation strategies need to take account of these multiplicities. Goal 15, "Life on Land," the sustainable use of terrestrial ecosystems, the sustainable management of forests, initiatives directed to combating desertification, land degradation, and biodiversity[74] not only comprises strong human agency in respect to sustainable management, but also unique, multiple entanglements of rhythmic and metabolic temporalities of plant times with soil times, entangled with multiple human times of planting and harvesting, and the deep geological temporalities of minerals, chemicals, rocks, atmospheric durations of climate change and multispecies. Sustaining practices are therefore cuts into the webs and knots of these unfolding processes and durations. Rather implementation strategies need to be crafting and composing procedures that work with ecological complexity, become acts of radical attunement, and recognise the entanglement of many different agencies, and the plurality of their temporal dimensions and different scales of emergence. In the museum laboratory, and through pedagogies, for example, such implementation strategies would become multilayered and rhythmic material practices of composing that are distinct to past configurations that foreground strong, and human agency, scientific research, and capital accumulation. Through reframing the SDGs as material, rhythmic, and involving many non-human agencies a less anthropocentric figuration of the goals might be instituted in context.

I will now return to the Boulton and Watt "Lap" Engine discussed in Chapter 3 and explore how it might be reinterpreted within the SDG framework, and specifically a refiguration of Goal and Target 11.4, "Strengthen Efforts to Protect and Safeguard the World's Cultural and Natural Heritage" that takes into account a more-than-human framework.

When considering the Boulton and Watt "Lap" Engine conventionally in regard to this goal, the objective becomes safeguarding the Engine and its material fabrics so it does not deteriorate. This would be measured according to how much money needs to be spent on its restoration compared to the income it generates and the jobs it creates in education, curation, and engineering sectors to achieve these objectives. Thus, the Engine would be interpreted as a cultural asset directed to sustaining community, economy, waged labour, and GDP growth. In framing the Boulton and Watt "Lap" Engine in terms of Goal and Target 11.4 alongside UNESCO's interpretation of this same goal, it is evident that a limited view of sustainable futures as heritage protection is created rather than a consideration of the role the Engine might play in sustaining earthly communities. My question, therefore, is how can the Engine be put to work to enact sustaining thinking and act beyond the concerns of GDP and development?

One way the Boulton and Watt "Lap" Engine can be deployed for enacting sustaining thinking is through its figuration as an ecological composition, that is through the folding out of all the human and other-than-human coordinates, and their histories, involved in its emergence. As an ecological composition, a fuller picture emerges of its expanded histories of entanglement, co-survival, and, most importantly, the specific ways modern populations might reimagine themselves as more-than-human. While the Boulton and Watt "Lap" Engine gestures towards the troubling futures in which it is complicit beyond heritage protection, it offers new lessons in sustaining practices through an inversion of hubris, and the potential for thinking about populations through the new humanisms it embodies as dynamic biochemical lifeforms bonded into its wooden fabrics.

On the one hand, the embodiment of human-material climates, illustrated through the Engine's histories, mechanical function, and material fabrics, gesture towards its complicity in capitalist accumulation, and carbon-intensive industries such as mining, industrial and production processes. In doing so, it also signals what Kathryn Yusoff calls, Black anthropocenes,[75] because the Engine embodies the histories of sugar plantations and the slave trade, the factory system and labour exploitation, and its dispersal through settler colonialism, Indigenous colonial legacies, the dispossession of land and climate injustice.[76] The Boulton and Watt "Lap" Engine, its materials and engineered crafting and design, bear witness to its role as a carbon emitter. While the folding out of its histories, materials, design, and embodied fabrics makes visible the interconnected processes of its emergence as troubling consequences, in doing so, it can also incite contemporary practices of decolonisation, decarbonisation, recuperation, and alternative economic practices.

The folding out of the deep geological times of coal, chemicals, soils, and photosynthesis, to name a few, enmeshed within the Engine's fabrics and as new humanisms link back to earlier discussions on the SDGs and the invocation of attunement as a provocation in respect to the emergence of a new suite of disciplines and research agendas to understand agency and ontological processes. These disciplines include more-than-human humanities, biological, earth sciences, physics, chemistry, and medical sciences. Thus, the Boulton and Watt "Lap" Engine provides an opportunity to explore the entanglement of non-human and human agencies for use in future speculative work on ecological recuperation.

Through an analysis of the Boulton and Watt "Lap" Engine's history and composition, it becomes clear many non-humans and processes become entangled with human designs. Community enmeshed in the Engine and its contexts, therefore, extends to include earthly processes, multispecies, and the Earth's strata made up of minerals and chemicals; these are all critical figurations in respect to what is required for planetary co-survival. By illustrating how non-humans are entangled with human designs, the Boulton and Watt "Lap" Engine illustrates how this knowledge acts as a starting point for considering sustaining futures taking account of the vitality of human–non-human enmeshments in which it embodies in respect to formulating post-carbon speculations, and thoughtful and respectful crafting in attunement.

The Boulton and Watt "Lap" Engine also illustrates the potential for activating more-than-human habitability practices in which nature and the non-human world are no longer a resource but as being vital having its own agencies entangled with modern populations in emerging eco-curating processes. Therefore, the Boulton and Watt "Lap" Engine is an object to think with, act with, and formulate sustaining habitability practices through the use of the layered materials from which it was made and operationalised to the multifarious human and non-human relations it embodies, all of which can be used to question extractivism, hubris, capitalist economy, human-centred engineered design, and so forth.

This consideration of the Boulton and Watt "Lap" Engine shows that museums can operate as experimental sustaining laboratories by exploring their collections in novel ways to consider new types of pedagogies around vitalism, non-human agency, embodied human agencies, and more-than and non-human durations on which sustaining living practices are contingent. More-than-human living habitability practices therefore can emerge for planetary praxes.

In supporting Goal 13, "Take Urgent Action to Combat Climate Change," museums are also involved in endorsing the reduction of atmospheric heating through the substitution of energy sources in order to avoid loss of economic prosperity and to ensure continued economic growth.

Technically, Goal 13 supports the continuation of industrialisation through new technologies and processes and through the economic provisioning of climate change mitigation and adaptation. Museums support investments in nature performance most directly through the promotion of nature mitigation, as discussed in Chapter 3, in which non-human agencies of the photosynthetic are exploited to support the continuation of fossil fuel combustion. When the Boulton and Watt "Lap" Engine is explored using a ^{13}C analysis, discussed in Chapter 3, alongside the folding up of its human–non-human relations, as done in this chapter, it becomes a lesson in carbon combustion and its consequences.

Under these lenses, the Boulton and Watt "Lap" Engine can be seen to be linked to Goal 13 in new ways and be understood not only as being a component of material climates but as having the potential to act as a cultural carbon mitigation strategy that can demonstrate the bonded human–non-human relations as enmeshed biochemical life forms. This biochemical convergence is also made visible through a range of bonded humanisms with coal, metals, chemicals, minerals, and human–non-human engineered and crafting processes. The Boulton and Watt "Lap" Engine, as embodied agencies of climate change, can be put to work to rethink human situatedness in climate change and be directed to expanding worldviews by making visible the distinctive and destructive humanisms bonded with carbon economies as material climates. Therefore, composing ethical habitable worlds cannot be achieved through changing energy sources and the measurement of ecosystem functioning with extractivism at its core, but through understandings of the distinctive humanisms that emerge and are made tangible and implicated in carbon economies as bonded material climates. Viewing the Boulton and Watt "Lap" Engine composition as materially interconnected and made manifest by its distributed composing agential forces and domains of influencing, not only embodies lessons of the past on destructive practice and what may result from these but also on the emergence of human–non-human climates and the potential for acts of recuperation.

Decarbonisation is no longer merely achieved through the substitution of energy sources and behavioural change. Nature mitigation and the exploitation of the agencies of the photosynthetic can no longer be put to work for human purposes. Rather it is time to frame new forms of humanism based on the notion of humans as biochemical lifeforms, and as being the product of plant, human histories and durations in which the temporalities of plant times and human times converge. Humanism, thus, becomes a question of wood bonding and intimacies in human, coal, and plant times in the atmosphere's composition. An understanding of modern humans as biochemical lifeforms make visible the entangled and distributed histories in which fossil-fuelled populations are bonded through the photosynthetic in dynamic ways, an understanding on which humanity's co-survival within an expanded sense of community depends. Such a figuration promulgates a deeper sense of co-survival and intimate bonding than greenhouse gas thinking. The deep material embeddedness of human and non-human climates activates Goal 13 in new ways as both a conceptual and material activation for the establishment of post-carbon futures.

Heritage science collections as research infrastructures can contribute to Goal 13, "Climate Action," as discussed in this chapter, as well as Goal 2, "Sustainable Agriculture," Goal 14 "Life in Water," and Goal 15, "Life on Land." Such a process would not seek to support development or commercialisation, but to compose worlds respectfully in regard to human–non-human relations through an account of their vitalism, enmeshments, interrelatedness, and crafting and at the same time took into account the different temporalities required for co-survival. All the coordinates, including earthly processes embedded in these respective Goals, would thus become political agents in habitability practices as more-than- and other-than-human domains of influencing.

Sustaining practices activated through the Boulton and Watt "Lap" Engine and its heritage interpretation as an ecological composition has the potential to actualise an inclusive future where all non-human systems and earthly processes thrive. Through material and materialising lessons about the Engine's fabrics and deep histories of interconnectedness and begetting sympioetic relations can be understood through eco-curatorial processes by highlighting the Engine's componentry and alliances, and their entanglements in processes. The Boulton and Watt "Lap" Engine would thus become an exemplar of the eco-curatorial processes of life itself, a lesson on how humans harness the non-human world for their own ends, how modernity might redress the injustices evident in Boulton and Watt's distribution methods, and the un-minded damage and inhuman futures the Engine is complicit in that require recuperation.

In its current manifestation, standing in a museum, the Boulton and Watt "Lap" Engine represents a past aspiration of human futures fuelled by neoliberal capitalism and one inseparable from capitalist extractivism. At this level, the Engine seems to challenge the SDGs until one recognises that they too are based on development, capital, and natural resources.

As cultural theorist Claire Colebrook suggests,

> rather than see the humanity that faces an unjust and catastrophically tragic end as the agent of history whose survival is required in order to achieve a more just future, one should see the end of the world as the beginning of other worlds.[77]

The Boulton and Watt "Lap" Engine frames the beginning of other worlds, that of human–non-human futures of co-survival. Modern societies cannot see these relational contingencies because populations are addicted to fossil fuel and capital accumulation and are caught up with themselves and their needs. Museums thus have an important role in envisioning sustaining practices in a broader framework to protect planetary communities. Their task is one of rethinking the concept of sustainability and conceiving it in non-anthropocentric terms.

So, what might the SDGs look like beyond capitalist economies and the invocation of strong human agency? Thinking and acting must be activated as relational figurations across political, economic, biological, and geological durations, and in dynamic ways.

Mapping and imagining habitability, or living practices, through a critical posthumanism framework draws inspiration from diverse economies frameworks. Diverse economies research challenges the eurocentric nature of economic theorising. The motivation for diverse economies scholarship is to critique the status quo enshrined in the goals of business as usual, of capitalist practices that are exploitative and extractive and to explore how multiple forms of concentrated power exert undue influence on trajectories of change. Diverse economies scholarship offers new theorisations of livelihood and economic interconnection, it illuminates multiple trajectories of economic change that are not captured by unidimensional and unidirectional studies of economic practice and change, as I have described previously in respect to Agenda 2030. The starting point of diverse economies is imagining and enacting radically different, sustainable, non-anthropocentric, postcapitalist futures, to work with *what we have here at hand.*

Central to such a focus is the inventorying of economic diversity, of what exists and how it can become a strategy for opening up the economy, and indeed the SDGs, to new kinds of examination and to different kinds of economic subjectivity. In terms of the SDGs, inventorying economic diversity involves building ethical community beyond development and extractive agendas, and detailing low-carbon practices of livelihood. At the centre of my analysis is a reframing of the more-than-capitalist economy, towards ethical interdependence, or what might be called "*becoming* community economies."

When investigating implementation strategies for specific goals, in this case Goal 13, the first task is to de-prioritise human needs[78] and shift the focus away from capital, extractivism, and economic growth. Museums can play a role working with communities to map and establish different habitability practices based on low carbon practices for example, and explore economy in its multifarious forms beyond capitalist agendas through dialogic processes.

In respect to each goal and target in question, such a mapping exercise can also be put to work to identify each of the coordinates involved, both human and non-human, to achieve low-carbon economic practices. Such a mapping exercise can consider a number of factors, including the relational connections across the SDGs as interconnected processes, the nature of human–non-human relations, how human agency might be re-framed, how non-human agencies are considered and valued, the nature of different living principles and practices and the specific types of ecologies, practices of care, respect, recuperation, and regeneration that each goal enact or might enact. Central to this is an understanding of the different principles that exist in respect to the relational connections between human and non-human agencies, and the vitalism evident in their interconnectedness, the rhythms and durations of human and non-human time, and how they work together or impact implementation strategies. Further to this, a consideration must be made of how attunement, respect, and care for both human and non-human coordinates might be enacted in goal implementation and what forms of economy can support and promote these values. This process therefore comprises a thick description of and mapping of the eco-curatorial arrangements within and across each Goal. Refiguring the SDGs is a case of changing fundamental social relations and reframing what elements mean, such as land, human agency, and trees, for example. Such a reconceptualisation also involves considering what economy means, what technology means, how agencies are enacted, and so forth. In relation to Goal 11 and Target 4 heritage is not just things of value but involves questions of how Maori taonga, for example, become treasured possessions and what it means to have an understanding of the vitalism of heritage in specific contexts, such as being the embodiment of ancestors. For example, with Goal 13, "Climate Action," questions about another goal, "Life on Land," are no longer solely about nature mitigation as service providers but an attuned consideration of forest, plant histories and durations, and how these are folded into human times in respectful ways. In doing so museums, through their community connections, extend community, and intergenerational inheritance beyond human populations to establish place-based habitability or living practices, or regimes and cultures of care, recuperation, and regeneration. This is also a case of changing social relations between the elements. Central to such a scenario is embracing new thinking in diverse economies to comprehend the more-than-human nature of livelihood and interdependence, the work of non-humans (Earth Others) as well as humans, and the dynamics of resilience rather than growth.[79]

Conclusion: from sustainability and sustainable development to living habitability practices

Through all the approaches described in this chapter, sustainability becomes living habitability practices of care towards, and respect for the rhythms and vitalism of the non-human world and other entities entangled with human designs and the sustaining of all resources in respectful ways. Considering the non-human world as flourishing not just for human ends involves taking only what humans need and can use rather than adhering to an exploitive logic that historically has been fundamental to many cultures.

Sustainability and sustainable development must be re-framed as sustaining frameworks always in transformation, as fractiversal, as comprising a broad range of habitability living practices rather than being the imposition of one agenda. Revisioning sustaining living practices can be enacted by mapping and conducting dialogic processes relevant to different communities, suitable for building low carbon practices of livelihood, and directed to building ethical community beyond development and extractive agendas. Museums could champion what I call *becoming* habitability practices and living frameworks for developing sustaining futures.

Museums as stakeholders on the PA and within the SDGs are actually passive disseminators of these agendas in a top down, one-way deficit model of communication and implementation. Both represent modern and technocratic forms of governance. The alternative is for museums to work with communities and grassroots organisations to develop new strategies based on diverse knowledge systems and practices. Through museological habitability methods, museums can be put to work to rethink livelihoods as place-based and situated. Furthermore, through a fractiversal collective of sustaining practices, the idea of sustainability extends beyond the linear-human-centric concerns of Agenda 2030. An engagement with Indigenous struggles First Nations scholar Kyle Whyte encourages has the potential to support and legitimise life-affirming economic and ecological practices from which the world has much to learn.[80]

As diverse economies pioneers Katherine Gibson and Julie Graham once put it, changing our understanding of the world *is* to change the world, if only partially and locally.[81] As the discursive and material reconfiguration of the Boulton and Watt "Lap" Engine illustrates, by participating in the act of discourse making, coupled with material analyses, museums can actively invite shifts in bodily being, habitability, and subjectivity.

Notes

1 United Nations, Department of Economic and Social Affairs, "Transforming Our World: The 2030 Agenda for Sustainable Development," A/Res/70/1, accessed August 3, 2022, https://sustainabledevelopment.un.org/content/documents/21252030%20Agenda%20for%20Sustainable%20Development%20web.pdf.

2 Jessica Hope, "The Anti-Politics of Sustainable Development: Environmental Critique from Assemblage Thinking in Bolivia," accessed August 3, 2022, *Transactions of the Institute of British Geographers* 46, no. 1 (2021): 208, https://rgs-ibg.onlinelibrary.wiley.com/doi/10.1111/tran.12409.

3 United Nations, Department of Economic and Social Affairs, *Report of the World Commission on Environment and Development: Our Common Future* (*Brundtland Report*) 1987, accessed August 3, 2022, https://sustainabledevelopment.un.org/content/documents/5987our-common-future.pdf.

4 United Nations Environment Programme, *Making Peace with Nature: A Scientific Blueprint to Tackle the Climate, Biodiversity and Pollution Emergencies*, UN Secretary General's Foreword, António Guterres, Secretary-General of the UN,

February 18, 2021, accessed August 3, 2022, https://www.unep.org/resources/making-peace-nature, 4.

5 Ariel Salleh, "Climate Strategy: Making the Choice between Ecological Modernisation or Living Well," *Journal of Australian Political Economy* 66 (January 2010): 121.

6 Salleh, "Climate Strategy," 118–143.

7 United Nations, Department of Economic and Social Affairs, "Addis Abada Action Agenda of the Third International Conference on Financing for Development (Addis Abada Agenda)," July 13–16, 2015, accessed August 3, 2022, https://sustainabledevelopment.un.org/content/documents/2051AAAA_Outcome.pdf, 29–31.

8 Stefanie Chan, Nina Weitz, Åsa Persson and Caspar Trimmer, *SDG 12: Responsible Consumption and Production – A Review of Research Needs*, Technical annex to the Formas report Forskning för Agenda 2030: Översikt av forskningsbehov och vägar framåt. Stockholm Environment Insitute, Stockholm, 2018, accessed August 3, 2022, https://www.sei.org/wp-content/uploads/2018/11/sdg-12-responsible-consumption-and-production-review-of-research-needs.pdf.

9 Manuel Charr, "Museums and Sustainable Development Goals," *Museum Next*, March 27, 2022, accessed August 3, 2022, https://www.museumnext.com/article/museums-and-sustainable-development-goals/.

10 Xinyuan Wang, Hongge Ren, Pu Wang, Ruixia Yang, Lei Luo and Fulong Cheng, "A Preliminary Study on Target 11.4 for UN Sustainable Development Goals," *International Journal of Geoheritage and Parks* 6, no. 2 (December 2018): 18–24.

11 M. J. Villanueva, "Sustainable Development Goals for Museums, Selection by Jasper Visser," Abstract, *EXARC.net*, 2018, accessed August 3, 2022, https://exarc.net/sustainable-development-goals/museums-selection-jasper-visser.

12 Institute for Culture and Society Seminar Series – The sustainable development goals, accessed August 3, 2022, https://www.westernsydney.edu.au/ics/events/ics_seminar_series/seminars_in_2020/21_may_seminar_-_the_sustainable_development_goals.

13 United Nations, Department of Economic and Social Affairs, "Goal 8: Promote Sustained and Inclusive and Sustainable Economic Growth, Full and Productive Employment and Decent Work for All," accessed August 5, 2022, https://sdgs.un.org/goals/goal8.

14 British Council, *Cultural Heritage for Inclusive Growth*, 2018, accessed August 5, 2022, https://ocm.iccrom.org/documents/cultural-heritage-inclusive-growth.

15 Henry A. Mcghie, *Museums and the Sustainable Development Goals*, Curating Tomorrow, August 2019, accessed August 3, 2022, https://curatingtomorrow236646048.files.wordpress.com/2019/12/museums-and-the-sustainable-development-goals-2019.pdf 2019, 49.

16 Culture2030Goal campaign, *Culture in the Implementation of the 2030 Agenda. A Report by the Culture 2030 Goal Campaign, ICOMOS*, September 24–25, 2019, accessed August 3, 2022, https://www.icomos.org/images/DOCUMENTS/UN_SDG/culture2030goal_low.pdf, 5.

17 Culture2030Goal campaign, *Culture in the Implementation of the 2030 Agenda*, 6.

18 Culture2030Goal campaign, *Culture in the Implementation of the 2030 Agenda*, 34.

19 UNESCO, Culture for the 2030 Agenda UNESCO, accessed August 3, 2022, https://unesdoc.unesco.org/ark:/48223/pf0000264687.

20 Culture2030Goal campaign, 28–31.

21 McGhie, *Museums and the Sustainable Development Goals*, 49.

22 The Global Goals for Sustainable Development, "Goal 15, *Life on Land*, Protect, Restore and Promote Sustainable Use of Terrestrial Ecosystems, Sustainably Manage Forests, Combat Desertification, and Halt and Reverse Land Degradation and Halt Biodiversity Loss," accessed August 5, 2022, https://www.globalgoals.org/15-life-on-land.

23 International Centre for the Study of the Preservation and Restoration of Cultural Property, "The Climate Tool Kit," 2021, accessed August 3, 2022, https://ocm.iccrom.org/documents/climate-toolkit.
24 United Nations, "Transforming Our World."
25 United Nations Environment Programme, *Making Peace with Nature*, 4.
26 The Sustainable Scale Project, accessed August 5, 2022, http://www.sustainablescale.org/.
27 Hamid Jamali, "Critical Perspectives on the Sustainable Development Goals: Are Universal Indicators Meaningful?" *FLOWs: The Water Governance Blog at THE Delft Institute for Water Education*, 25 June 2018, accessed August 3, 2022, https://flows.hypotheses.org/1460.
28 McGhie, *Museums and the Sustainable Development Goals*, 49.
29 Thea Riofrancos, "The Rush to 'Go Electric' Comes with a Hidden Cost: Destructive Lithium Mining," *The Guardian*, Monday, June, 14, 2021, accessed August 3, 2022, http://www.theguardian.com/commentisfree/2021/jun/14/electric-cost-lithium-mining-decarbonasation-salt-flats-chile.
30 Riofrancos, "The Rush to 'Go Electric' Comes with a Hidden Cost."
31 International Centre for the Study of the Preservation and Restoration of Cultural Property, "The Climate Tool Kit."
32 United Nations, "Addis Abada Action Agenda of the Third International Conference on Financing for Development (Addis Abada Agenda)."
33 Eduardo Gudynas, "Value, Growth, Development: South American Lessons for a New Ecopolitics," *Capitalism, Nature, Socialism* 30, no. 2 (2019): 234–43, 237.
34 Gudynas, "Value, Growth, Development," 237.
35 United Nations, Department of Economic and Social Affairs, "Goal 4: Quality Education," accessed August 5, 2022, https://www.un.org/sustainabledevelopment/education/.
36 Monash Sustainable Development Institute, "Goal 4: Quality Education," accessed August 5, 2022, https://www.monash.edu/msdi/sustainable-development/sustainable-development-goals/quality-education.
37 Salleh, "Climate Strategy," 124.
38 UNESCO, "Culture: At the Heart of SDGs," April–June 2017, accessed August 5, 2022, https://en.unesco.org/courier/april-june-2017/culture-heart-sdgs.
39 Australian Institute of Aboriginal and Torres Strait Islander Studies, Canberra, Australia, *A Guide to Applying the AIATSIS Code of Ethics for Aboriginal and Torres Strait Islander Research*, 2020, accessed August 5, 2022, https://ocm.iccrom.org/documents/guide-applying-aiatsis-code-ethics-aboriginal-and-torres-strait-islander-research.
40 Salleh, "Climate Strategy," 121–22.
41 John Bellamy Foster, *Marx's Ecology: Materialism and Nature*. New York: Monthly Review Press, 2000, ix.
42 Laura McBride, "*The 2020 Project* First Nations Community Consultation Report," *Australian Museum*, 2021, accessed August 3, 2022, https://australian.museum/learn/cultures/the-2020-project/.
43 Eric Hirsch, "Sustainable Development," *Anthropology*, Oxford Research Encyclopedias, July 30, 2020, accessed August 3, 2022, https://oxfordre.com/anthropology/view/10.1093/acrefore/9780190854584.001.0001/acrefore-9780190854584-e-155.
44 Robert Costanza and E. Daly Herman, "Natural Capital and Sustainable Development," *Conservation Biology* 6, no. 1 (1992): 37–46, 37.
45 Costanza and Daly, "Natural Capital and Sustainable Development," 37.
46 Salleh, *Climate Strategy*, 124.
47 Salleh, *Climate Strategy*, 124.
48 Marisol de la Cadena, *Earth Beings: Ecologies of Practice across Andean Worlds* (Durham, NC: Duke University Press, 2015).

49 Salleh, *Climate Strategy*.
50 Ralf Regenvanu, "The Traditional Economy as Source of Resilience in Vanuatu," *Aid/Watch*, 2010, accessed August 3, 2022, http://milda.aidwatch.org.au/sites/default/files/Ralph%20Regenvanu.%20Traditional%20economy%20as%20a%20source%20of%20resistance%20in%20Vanuatu.pdf.
51 Deloitte NZ, "The Four Capitals," *Deloitte NZ*, accessed August 5, 2022, https://www2.deloitte.com/content/dam/Deloitte/nz/Documents/public-sector/sots5-fourcapisinfog.pdf.
52 SDG.org.nz, *The People's Report on the 2030 Agenda and Sustainable Development Goals: An Alternate Report for Aotearoa New Zealand (2019), SDG.org.nz*, 2019, accessed August 3, 2022, https://www.sdg.org.nz/peoples-report/.
53 Department of Strategic Policy, Planning and Aid Coordination, *Vanuatu 2030. The People's Plan: National Sustainable Development Plan 2016 to 2030*, Republic of Vanuatu, 2016, accessed August 3, 2022, https://www.gov.vu/images/publications/Vanuatu2030-EN-FINAL-sf.pdf.
54 Culture2030Goal campaign, *Culture in the Implementation of the 2030 Agenda*, 6.
55 John Law, "What's Wrong with a One-World World," *Heterogeneities.net*, September 25, 2011, accessed August 3, 2022, http://www.heterogeneities.net/publications/Law2011WhatsWrongWithAOneWorldWorld.pdf.
56 For a discussion of cultural perspectives as different reals and the universe comprising multiple fractiverses see Law, "What's Wrong with a One-World World."
57 Musée d'ethnographie de Genève, "Environmental Injustice Indigenous Peoples' Alternatives," *Exhibition*, accessed August 5, 2022, https://www.ville-ge.ch/meg/en/expo31.php.
58 Musée d'ethnographie de Genève, "Environmental Injustice Indigenous Peoples' Alternatives."
59 David Throsby and Ekaterina Petetskaya, "Sustainability Concepts in Indigenous and Non-Indigenous Cultures," *International Journal of Cultural Property* 23, no. 2 (2016): 119–40, 129.
60 Fran Bodkin, "D'harawal Stories of Cycles and Seasons: Land, Water and Fire," in *Reimagining Sustainability in Precarious Times*, eds. Karen Malone, Son Truong and Tonia Gray, (Singapore: Springer, 2017), ix–xiv, x.
61 Throsby and Petetskaya, "Sustainability Concepts in Indigenous and Non-Indigenous Cultures."
62 *World Commission on Environment and Development, Our Common Future* (Oxford: Oxford University Press, 1987).
63 Throsby and Petetskaya, "Sustainability Concepts," 122.
64 Throsby and Petetskaya, "Sustainability Concepts," 122.
65 Throsby and Petetskaya, "Sustainability Concepts," 122.
66 Throsby and Petetskaya, "Sustainability Concepts," 122.
67 Throsby and Petetskaya, "Sustainability Concepts," 122.
68 Throsby and Petetskaya, "Sustainability Concepts," 122.
69 Throsby and Petetskaya, "Sustainability Concepts," 122.
70 Olga Cielemęcka and Christine Daigle, "Posthuman Sustainability: An Ethos for Our Anthropocenic Future," *Theory, Culture & Society* 36, no. 7–8 (2019): 67–87.
71 Cielemęcka and Daigle, "Posthuman Sustainability," 71.
72 Cielemęcka and Daigle, "Posthuman Sustainability," 74.
73 Fiona R. Cameron, *The Future of Digital Data, Heritage and Curation in a More-Than-Human World* (London: Routledge, 2021).
74 The Global Goals for Sustainable Development, "Goal 15, *Life on Land*."
75 Kathryn Yusoff, *A Billion Black Anthropocenes or None*. (Minneapolis: University of Minnesota Press, 2018).

76 Fiona R. Cameron, Ben Dibley and David Ellsworth, "Climate Collections, and Photosynthetic Fossil-Fueled Atmospheres," *Environmental Humanities* 15, no. 2 (July 2023): 62–84.
77 Cielemęcka and Daigle, "Posthuman Sustainability."
78 James, "Alternative Paradigms for Sustainability," 31.
79 Gibson-Graham, Julie Katherine, "Diverse Economies: Performative Practices for 'Other Worlds'," *Progress in Human Geography* 32, no. 5 (2008): 613–32.
80 Kyle Whyte, "Indigenous Climate Change Studies: Indigenizing Futures, Decolonizing the Anthropocene," *English Language Notes* 55, no. 1–2 (Fall 2017): 153–62.
81 Gibson-Graham, Julie Katherine, "Diverse Economies," 615.

References

Australian Institute of Aboriginal and Torres Strait Islander Studies, Canberra, Australia. *A Guide to Applying the AIATSIS Code of Ethics for Aboriginal and Torres Strait Islander Research*. 2020, accessed August 5, 2022, https://ocm.iccrom.org/documents/guide-applying-aiatsis-code-ethics-aboriginal-and-torres-strait-islander-research.

Bodkin, Frances. "D'harawal Stories of Cycles and Seasons: Land, Water and Fire." In *Reimagining Sustainability in Precarious Times*, edited by Karen Malone, Son Truong and Tonia Gray, ix–xiv. Singapore: Springer, 2017.

British Council. *Cultural Heritage for Inclusive Growth*. 2018, accessed August 5, 2022, https://ocm.iccrom.org/documents/cultural-heritage-inclusive-growth.

Cameron, Fiona R. *The Future of Digital Data, Heritage and Curation in a More-Than-Human World*. London: Routledge, 2021.

Cameron, Fiona R., Ben Dibley and David Ellsworth. "Climate Collections, and Photosynthetic Fossil-Fueled Atmospheres," *Environmental Humanities* 15, no. 2 (July 2023).

Chan, Stefanie, Nina Weitz, Åsa Persson and Caspar Trimmer. *SDG 12: Responsible Consumption and Production – A Review of Research Needs*. Technical annex to the Formas report Forskning för Agenda 2030: Översikt av forskningsbehov och vägar framåt. Stockholm Environment Insitute, Stockholm, 2018, https://www.sei.org/wp-content/uploads/2018/11/sdg-12-responsible-consumption-and-production-review-of-research-needs.pdf.

Charr, Manuel. "Museums and Sustainable Development Goals." *Museum Next*, March 27, 2022, https://www.museumnext.com/article/museums-and-sustainable-development-goals/.

Cielemęcka, Olga and Christine Daigle. "Posthuman Sustainability: An Ethos for our Anthropocenic Future." *Theory, Culture & Society* 36, no. 7–8 (2019): 67–87.

Costanza, Robert and Herman E. Daly. "Natural Capital and Sustainable Development." *Conservation Biology* 6, no. 1 (1992): 37–46.

Culture2030Goal campaign. *Culture in the Implementation of the 2030 Agenda. A Report by the Culture 2030 Goal Campaign. ICOMOS*, September 24–25, 2019, https://www.icomos.org/images/DOCUMENTS/UN_SDG/culture2030goal_low.pdf.

de la Cadena, Marisol. *Earth Beings: Ecologies of Practice across Andean Worlds*. Durham, NC: Duke University Press, 2015.

Deloitte NZ. "The Four Capitals." *Deloitte NZ*, accessed August 5, 2022, https://www2.deloitte.com/content/dam/Deloitte/nz/Documents/public-sector/sots5-fourcapisinfog.pdf.

Department of Strategic Policy, Planning and Aid Coordination. *Vanuatu 2030. The People's Plan: National Sustainable Development Plan 2016 to 2030.* Republic of Vanuatu, 2016, https://www.gov.vu/images/publications/Vanuatu2030-EN-FINAL-sf.pdf.

Foster, John Bellamy. *Marx's Ecology: Materialism and Nature.* New York: Monthly Review Press, 2000.

Gibson-Graham, Julie Katherine. "Diverse Economies: Performative Practices for 'Other Worlds'." *Progress in Human Geography* 32, no. 5 (2008): 613–32.

The Global Goals for Sustainable Development. "Goal 15, *Life on Land*, Protect, Restore and Promote Sustainable Use of Terrestrial Ecosystems, Sustainably Manage Forests, Combat Desertification, and Halt and Reverse Land Degradation and Halt Biodiversity Loss," accessed August 5, 2022, https://www.globalgoals.org/15-life-on-land.

Gudynas, Eduardo. "Value, Growth, Development: South American Lessons for a New Ecopolitics." *Capitalism, Nature, Socialism* 30, no. 2 (2019): 234–43.

Hirsch, Eric. "Sustainable Development." *Anthropology*, Oxford Research Encyclopedias, July 30, 2020, https://oxfordre.com/anthropology/view/10.1093/acrefore/9780190854584.001.0001/acrefore-9780190854584-e-155.

Hope, Jessica. "The Anti-Politics of Sustainable Development: Environmental Critique from Assemblage Thinking in Bolivia." *Transactions of the Institute of British Geographers* 46, no. 1 (2021): 208–22. https://rgs-ibg.onlinelibrary.wiley.com/doi/10.1111/tran.12409.

International Centre for the Study of the Preservation and Restoration of Cultural Property. "The Climate Tool Kit." 2021, https://ocm.iccrom.org/documents/climate-toolkit.

Jamali, Hamid. "Critical Perspectives on the Sustainable Development Goals: Are Universal Indicators Meaningful?" *FLOWs: The Water Governance Blog at THE Delft Institute for Water Education*, June 25, 2018, https://flows.hypotheses.org/1460.

James, Paul. "Alternative Paradigms for Sustainability: Decentring the Human Without Becoming Posthuman." In *Reimagining Sustainability in Precarious Times*, edited by Karen Malone, Son Truong and Tonia Gray, 29–44. Singapore: Springer Singapore, 2017.

Law, John. "What's Wrong with a One-World World." *Heterogeneities.net*, September 25, 2011, http://www.heterogeneities.net/publications/Law2011WhatsWrongWithAOneWorldWorld.pdf.

McBride, Laura. "*The 2020 Project* First Nations Community Consultation Report." *Australian Museum*, 2021, https://australian.museum/learn/cultures/the-2020-project/.

Mcghie, Henry A. *Museums and the Sustainable Development Goals.* Curating Tomorrow, August 2019, https://curatingtomorrow236646048.files.wordpress.com/2019/12/museums-and-the-sustainable-development-goals-2019.pdf 2019.

Monash Sustainable Development Institute. "Goal 4: Quality Education," accessed August 5, 2022, https://www.monash.edu/msdi/sustainable-development/sustainable-development-goals/quality-education.

Musée d'ethnographie de Genève. "Environmental Injustice Indigenous Peoples' Alternatives." *Exhibition*, accessed August 5, 2022, https://www.ville-ge.ch/meg/en/expo31.php.

Parrotta, John A. and Mauro Agnoletti. "Traditional Forest-Related Knowledge and Climate Change." In *Traditional Forest-Related Knowledge: Sustaining Communities, Ecosystems and Biocultural Diversity*, edited by John A. Parrotta and Ronald L. Trosper, 491–533. World Forest Series, 12. Springer, Dordrecht, 2012.

Regenvanu, Ralf. "The Traditional Economy as Source of Resilience in Vanuatu." *Aid/Watch*, 2010, http://milda.aidwatch.org.au/sites/default/files/Ralph%20Regenvanu.%20Traditional%20economy%20as%20a%20source%20of%20resistance%20in%20Vanuatu.pdf.

Riofrancos, Thea. "The Rush to 'Go Electric' Comes with a Hidden Cost: Destructive Lithium Mining." *The Guardian*, Monday, June, 14, 2021, http://www.theguardian.com/commentisfree/2021/jun/14/electric-cost-lithium-mining-decarbonasation-salt-flats-chile.

Salleh, Ariel. "Climate Strategy: Making the Choice between Ecological Modernisation or Living Well." *Journal of Australian Political Economy* 66 (January 2010): 118–43.

SDG.org.nz. *The People's Report on the 2030 Agenda and Sustainable Development Goals: An Alternate Report for Aotearoa New Zealand (2019). SDG.org.nz*, 2019, https://www.sdg.org.nz/peoples-report/.

The Sustainable Scale Project, accessed August 5, 2022, http://www.sustainablescale.org/.

Throsby, David and Ekaterina Petetskaya. "Sustainability Concepts in Indigenous and Non-Indigenous Cultures." *International Journal of Cultural Property* 23, no. 2 (2016): 119–40.

UNESCO. "Culture: At the Heart of SDGs." April–June 2017, accessed August 5, 2022, https://en.unesco.org/courier/april-june-2017/culture-heart-sdgs.

UNESCO, "Culture for the 2030 Agenda." UNESCO, 2017, accessed August 5, 2022, https://unesdoc.unesco.org/ark:/48223/pf0000264687.

United Nations, Department of Economic and Social Affairs. "Addis Abada Action Agenda of the Third International Conference on Financing for Development (Addis Abada Agenda)." July 13–16, 2015, https://sustainabledevelopment.un.org/content/documents/2051AAAA_Outcome.pdf.

United Nations, Department of Economic and Social Affairs. "Goal 4: Quality Education," accessed August 5, 2022, https://www.un.org/sustainabledevelopment/education/.

United Nations, Department of Economic and Social Affairs. "Goal 8: Promote Sustained and Inclusive and Sustainable Economic Growth, Full and Productive Employment and Decent Work for All," accessed August 5, 2022, https://sdgs.un.org/goals/goal8.

United Nations, Department of Economic and Social Affairs. *Report of the World Commission on Environment and Development: Our Common Future* (*Brundtland Report*) 1987. https://sustainabledevelopment.un.org/content/documents/5987our-common-future.pdf.

United Nations, Department of Economic and Social Affairs. "Transforming Our World: The 2030 Agenda for Sustainable Development." A/Res/70/1, accessed August 3, 2022, https://sustainabledevelopment.un.org/content/documents/21252030%20Agenda%20for%20Sustainable%20Development%20web.pdf.

United Nations Environment Programme. *Making Peace with Nature: A Scientific Blueprint to Tackle the Climate, Biodiversity and Pollution Emergencies.* UN Secretary General's Foreword, António Guterres, Secretary-General of the UN, February 18, 2021, https://www.unep.org/resources/making-peace-nature.

Villanueva, M. J. "Sustainable Development Goals for Museums, Selection by Jasper Visser." Abstract. *EXARC.net*, 2018, https://exarc.net/sustainable-development-goals/museums-selection-jasper-visser.

Wang, Xinyuan, Hongge Ren, Pu Wang, Ruixia Yang, Lei Luo and Fulong Cheng. "A Preliminary Study on Target 11.4 for UN Sustainable Development Goals." *International Journal of Geoheritage and Parks* 6, no. 2 (December 2018): 18–24.

World Commission on Environment and Development, Our Common Future. Oxford: Oxford University Press, 1987.

Yusoff, Kathryn. *A Billion Black Anthropocenes or None*. Minneapolis: University of Minnesota Press, 2018.

8

CONCLUSION

More-than-human museologies

Introduction

While emerging as a concept from the second half of the nineteenth century, no one definition defines museology. Indeed, many museological and theoretical variants exist.

For some, it is the theory of museum work, a scientific discipline in itself, but one that is interdisciplinary. For others, museology has more practical dimensions as a set of practices underpinned by the subject perspectives and methods of many disciplines.[1] And for others museology is an aggregate of different theories and methods. The term museology in most figurations acts as a connecting structure across museum work and the multi-disciplines that currently drive curatorial work. From the Marxist–Leninist museology of Eastern Europe used as an ideological instrument of Communism prevalent into the early 1990s, to the postmodern museology in Germany and the United Kingdom, museum practices and museologies have always been multiplicitious.[2]

The aim of this chapter and indeed the book is not to define museologies as solely theoretical or practical. Rather my argument moves back and forth between museology and curation as a subject matter and the practical concerns of collecting, interpreting, exhibiting, and documenting collections. I consider how curatorial practices might be conceptually and theoretically reframed into a more-than-human framework and the new terms, methods, and theoretical research directions that necessarily arise. This approach follows and builds on ideas presented in my monograph, *The Future of Digital Data, Heritage and Curation in a More-Than-Human World* in reframing digital museology as more-than-human digital museologies.

DOI: 10.4324/9781315212067-8

The new museology – the dominant museological formation currently driving museum research in the Anglophone world – interrogates the history and social function of museums, their roles in society and the activities they engage in, including curating, conservation, public programming, and education. Evolved from the perceived failings of the so-called old museology, and its focus on the practical study of curatorial methods and procedures, the new museology was, and continues to be, a response to the perception, according to museologist Kenneth Hudson, that the role of museums in society needed to change. Museums, Hudson observed, had become disconnected from the concerns of the modern world, were elitist, and therefore obsolete, and accordingly a waste of public money.[3] Museum director Duncan Cameron's formative 1971 article, "The Temple or Forum"[4] sought to differentiate two distinct institutional forms: old museology, or the object-orientated reverential temple, and new museology, a community-centric postmodern forum with a new social commitment to society.

Since the 1970s, new museology, most widely publicised through Peter Vergo's 1989 collection, *The New Museology*,[5] has become a critical, reflexive, and theoretical study of museums, their activities and their roles in society. In doing so, this contemporary community-orientated museology has taken the socio-cultural foundations and philosophical basis of museums as an object of study and directed it to enhancing museums' relevance for the communities they are located within.[6] Such an approach has critiqued the traditionally human-centred institutions of museums using postcolonial, poststructuralist, and postmodernist standpoints. These theories have facilitated an analysis of the axis of power relations operating within and through institutions and the exclusions and inequalities these powers enacted. They have also considered the politics of representation and the potential for engaging plural cultural and disciplinary perspectives. Other optics of investigation include governmentality, feminism, gender studies, and ethics.[7] Through this new museological optic, institutions are cast as always being in a state of transition in response to changing political, societal, and economic circumstances. In museological debates, therefore, discussions are often directed to how institutions might reinvent themselves to remain relevant. During the 1990s, museums in the UK were tasked with orientating their work towards public service and implementing social policy directed to sustaining and developing economic, cultural, social, and educational wellbeing for their communities.[8] Indeed, museums and museology have always been in a state of transition but not only in the sense of a normative, human-centred societal optic used to investigate power and pluralism, or social policy. Such revisioning must go deeper to investigate and rework the powerful human-centred epistemologies that drive museological investigations.

As epistemological practices, the old and new museologies and the subsequent development of critical museology as a subset of the latter are underpinned by a human-centred disposition and focused on museum organisations as fixed structures made up of component parts. That is, whole museums and departments delineated by physically demarcated spaces and a division of labour, whole objects and disciplines and singular, rational, and self-directed audiences and users.[9] All these functional structures are directed to supporting the different methods, techniques, and procedures central to museum practice in its humanist form. Approaches to cultural diversity restrict social collectives to those of human communities. While critical museology reflects on practices of representation and seeks to empower previously marginalised people, cultural identity across all museological formations operates as a series of plural perspectives that make up a whole culture and then together a whole universe, all of which operate according to a one worlding principle. Even eco-museology with its strong place-based orientation is about people and their heritage practices located in their own environment[10] but is not about them as reciprocal enmeshed entities.

Curating is a suite of practices – comprising exhibition and program development, on and offline, collecting, documentation, archival, and interpretive practices – that operate according to the thoughts and frames that underpin their respective museological forms. In the case of the new museology, these concepts include positivism, modern humanism, material substances, structure, human agency, cultural pluralism, social constructivism, and dualisms such as subject/object, nature/culture, material/immaterial, and so forth. Within museological research, the human and the non-human worlds are figured, respectively, as separate cultural and biological entities that overlap but are not necessarily enmeshed.

Museology, in both forms, is governed by cognitive construals founded in teleological, essentialist, and anthropocentric thinking. Teleological thinking is the tendency to explain an event based on a goal, function, or outcome. This disposition is evident in institutions that are organised into functional divisions that are directed towards achieving certain goals. This, for example, includes curatorial specialisms based on discipline, and discrete functions such as exhibitions, programming, collection development, documentation, and education. Essentialist thinking, on the other hand, is the assumption that all things, including planetary processes, share underlying properties. For example, this conceptualisation may result in the belief that museological and taxonomic categories applied to collections are objective and, on some level, homogeneous or share common properties across a given data field of empirical reality. While worldviews and belief systems are culturally specific, on a fundamental level they are considered to fall into a universal schema often expressed through descriptive or classificatory terms. Anthropocentric thinking is the tendency to use explanations often used by modern societies to understand all other living things. It is mobilised from a superior, dominant, and comparative reference point.[11] Within this thinking process, modern knowledge practices and cultural means of reasoning are given authority as interpretive frameworks. So human bodies are considered in relation to other entities and phenomena, and cultural relations are positioned according to the concepts and biases of modern societies, and in the promotion of modern science. Human agency and accomplishments, often western and male, remain the overriding rationale and standpoint from which life and non-life are referenced.

Both the old and new museology are grounded in and mobilised by disciplinary knowledge such as anthropology, history, science, pedagogy, and art, and concepts such as culture and society. Each individually, and in collaboration, render practical influence on curatorial practice. Different knowledge practices promote other figurations of the human more connected to the non-human world. As vital relations of reciprocity, these perspectives are usually presented as anthropological or pre-modern.

Museums are facing a crisis of curation. This is because curatorial practices are routinely based on hubris, exceptionalism, and the cognitive interpretations that foreground teleological, anthropocentric, and essentialist thinking. This crisis of conceptualisation necessarily requires museums, practitioners, and scholars to ask new questions and frame novel methods in respect to how to curate with others, including the non-human world, in respectful ways and after modern museology in the context of impending socio-ecological collapse.

With reference to my previous studies, in this monograph, I rethink the concept of the human and human agency through a consideration of curatorial practice in historical and ecological transformations, anthropocentric climate change, socio-ecological planetary challenges, viral pandemics, and global policy frameworks.

In the last 200 years, the modern figuration of the human has, through curatorial practice, biological research, and concerns with technological advancement, been refigured and extended. For example, in science museums, Charles Darwin's theory of evolution rendered the human as entangled with other species through a common material signature, DNA. Similarly, in technology museums, the refiguration of the human is constantly changing in response to technological developments and the enhancement of human capabilities enabled through the former, such as the contingencies of artificial intelligence (AI) and biotechnology, which are rendered both as forms of the transhuman. These figurations, which still promulgate the idea of the western male as of superior intellect and agency, remain compelling.

In order to mitigate the effects of these troubling modern curatorial figurations, and to rework curatorial practices, in each of the chapters of this book, I took traditionally separate curatorial dimensions – technological, ecological, biological, and cultural – and enfolded them through a reconsideration of what it means to be human across and through disciplines, through different nature-cultures and in extended ecological circumstances. I promoted curatorial practice as novel enmeshed-relational approaches comprising new forms of humanisms, as involving objects as ecological compositions, and suggested that curating and museums themselves are enmeshed and processual. This includes, for example, the reworking of the figure of the human as situated humanisms bonded relationally and materially with other forces and entities, and not only through the sharing of DNA with other species. Viewed in this way these bonded humanisms can be observed as expanding into other realms, such as in the chemical composition of the atmosphere, in the processes of photosynthesis, within the forces of wildfire, in digital technological transformations and processes, and even within the composition of the coronavirus as a multispecies ethnography. The figure of the human is thus situated and at the same time becoming, for example, biochemical, fire, more-than-digital, and viral lifeforms.

In this respect, the arguments presented across the chapters represent the inversion of cognitive construals that have traditionally dominated curatorial practice, towards the emergence of "sym" populations and practices for more-than-human museological worlds. Museum and heritage time thus become multiple temporalities and durations enmeshed in complex ecological processes and circumstances thereby disrupting the notion of a modern linear and teleological past that museums traditionally promulgated.

Furthermore, the approaches foregrounded in this book both reference and look beyond the Anthropocene concept. The Anthropocene concept is mobilised by museums through exhibitions to represent environmental damage, inequality, and the treatment of the non-human world wrought by modern populations. More attention needs to be given to how museology, through the work of museums as human-centred curatorial practices, has wrought destructive environmental situations. To this end in this book, I proposed new methods and terms, challenged museums to rearrange their modes of expertise, and create new ones that are not disciplined or human-centric in a conventional sense. Curatorial practice and expertise thereby become creative practices of composition engaged with the empirical realities of embeddedness, and with enmeshment as emergent processes in and as multiple fractiversal constellations alongside, and with, different knowledge practices, figurations of the human and the agencies of the more-than and non-human world.

In respect to digital cultural heritage, like my previous monograph, *The Future of Digital Data, Heritage and Curation in a More-Than-Human World*, the arguments presented in this book reclaimed curatorial and museum practices as ecological, as of the earth, as more-than- and other-than-human, as a story of deep time and futures yet to come.[12] The notion of the past, present, and future is suspended. These distinctive temporal formations become acts of composing. And temporality becomes a series of creative acts between and across a range of time coordinates emerging as worlding formations in multiple and unexpected ways. Henceforth the agencies of a vast number of others, often more-than and other-than-human, become curating agencies in planetary processes.

I reframed modern discourse in respect to museums, curatorial practice, museology, and the fabrics of museums as ecological museologies building on my work on ecologising museologies in my chapter "Ecologizing Experimentations: A Method and Manifesto for Composing a Post-Humanist Museum," in my co-edited collection, *Climate Change, Museum Futures*.[13] That is as enmeshed and embodied processes, comprising different types and forms of humanisms. Thus, as this book demonstrated, curatorial practice is refigured as human, more-than-and non-human collaborations.

In reconsidering museologies as ecological, embedded, enmeshed, and embodied, I investigated new ways of organising knowledge and of representing it as material, embodied, as becoming with actions, thoughts, bodies, and all manner of other entities and eco-curatorial processes. The individual human becomes humanness in different forms and as agencies in the multiple that co-compose with others. Significantly, I don't use the term "subjectivity" as it is too humanist. Subjectivity instead becomes qualitatively different forms of cognition and intelligence, such as the sensing, calculative, discursive, and sympoietic interrelations in eco-curatorial composing processes.

Museums have a role directed to defining pedagogically what it means to be human, and what it means to be human and modern in relation to others. But now this role must entail what it means to be human entangled with others and earthly processes in co-making. I reworked the figure of the human, the concept of curatorial agency in order to promulgate creative capacities of entanglement in crises. Accordingly, in this book, I considered what human populations might become within diverse communities in planetary formations – as forming, transforming, and self-organising entities. I developed a suite of new forms of knowledge and curating practices that can be folded into existing ones in novel and interesting ways. In doing so, I inverted old interests in human development, progress, extraction, and speciesism to present new arrangements centred on attunement and new types and forms of more-than-human curatorial expertise.

All of this supports an understanding of situated human bodies, thoughts, and acts as enmeshed in the planetary, and as an empirical approximation of reality that museum practice must promulgate. Collections are foregrounded as embodied in deep histories beyond modern human time, as an empirical reality that has oftentimes been rendered invisible. Other materialisms and entities are no longer objectified according to humanist traditions of subject–object curatorial practice whereby they are positioned as passive. Rather they are no longer considered instrumental and categorical, but rather as compositions comprising unique coordinates and agencies with their distinctive affordances, temporalities, and durations.

I am not suggesting that museum practitioners completely reject humanist museum practice and curatorial work, and indeed the new and more recent critical museology, but rather think of this set of concepts and methods as just one ontological perspective, albeit a historical one that does a certain type of work.[14] At the same time, museum professionals and scholars must recognise the failings of modern museology as a contributor to humanist damage that unfortunately continues unabated.

Despite this, the contemporary museologies, I propose, can be put to work to address pressing planetary emergencies in the context of the crises of humanism that modern societies and their cultural institutions are now facing. In this book, I highlighted ways of thinking and acting through curatorial practice that engender recuperative measures are attentive and caring to all others, including earth others, and attune their audiences to the reality of material entanglement and accountability. Changing the types and terms of the relations and material enactments conceptually and empirically is the key to activating renewed forms of curating practice in a more-than-human world.

Also throughout this book, I've demonstrated how curatorial agency might be put to work to frame more attuned futures, and how planetary survival is performed with more-than and other-than-human others, conceptually and materially. To do this, I have focused on the intersections of the environmental as habitability praxis, as speculation, and as prospects in and for a more-than-human and other-than-human world. Human histories become the histories of many. Curatorial agency becomes ecological and deeply interconnected with everything. The notion of origin, and its authentic original, no longer exist but become dynamic, transforming, and connected to other things. Curatorial practices are therefore situated in the vast sprawling interconnectedness of life itself. But ecological curating practice, as praxis in recuperation, is more than consciousness. It must engender respect, care, and concern and, above all, mindfulness of our radical coexistence, but not as a paternal type of care, which is the case in nature conservation or the concept of caring for the environment as a holistic undifferentiated category in which non-human others become subject to human interventions based on command and control.

Following on from the arguments about digital heritage in previous publications, in this book I've examined the contradictions inherent in curatorial practices framed according to modern humanism. I've used this location as a starting point to turn my attention to explicating the newly visible enmeshments of the human and non-human in museums from the fabrics in which museums are made to the more-than and non-human entities embodied in collections. Museums themselves, in all their facets, are embodiments of human–non-human crafting. In reference to such an epistemology the actions of a singular human curatorial agent or agents at the centre of museum practice becomes another conceptual error. Rather with the conceptual, material extension of curatorial practice into the world curation becomes an embodiment of life itself. Curatorial practices therefore become those of human–non-human crafting and composing as new and conscious forms of attunement.

Museum curatorial agency, under the auspices of new museology, is founded on a misplaced idea of sovereignty and the notion of an autonomous, knowing self. Rather, enmeshed more-than-museologies and research become a seeking of encounters with vibrant matter, other entities, and things that agentially enable and exert influence as continual processes of transformation. Working within these empirical realities requires respectful and careful experimental forms of curation and research with an expanding understanding of "community," and this is explicated across the book's chapters. The human, or curatorial subject, and its object, or objects, are no longer separate or relationally connected in hubris ways but rather become part of embedded and embodied collectives and together activate new genres of research.

Like digital cultural heritage, all curatorial forms must be explicitly acknowledged as grounded material practices and technics of life itself, as out in the world and as contexts in which curation as a future-making capacity is borne out of other processes and made visible, at least in part.[15] Ethical and attuned curatorial practices also become dynamic expressions about what modern populations are becoming – more-than-human figurations and subjects – and express the futures curatorial practice might contribute to, through the convergence of conceptual-discursive, biological, technological, and material-geological crafting, and the transformative figurations such processes enact. Those futures, being an engagement with the planetary predicament on an intimate level, and with post-capitalist prospects, overall acknowledge the agencies of diverse communities in communion and their capacity to be agential in the creation of a common world.

As well as creating new ways of thinking, perceiving, and acting as novel curating formations, the new types of humanisms that emerge are more-than-humanist. These new humanisms, previously rendered as human subjects, are collective, material, cosmological, compositional, and porous subjects. They emerge as agential humanisms in situated ecological compositions. Here museums have a role in refiguring plural understandings of humanisms as diverse, multiple, and situated.

In curatorial practice, emerging more-than-human knowledges become specific types of situated ecological thoughts and practices that are distinct from eurocentric humanism. They exist alongside all other non-modern or non-human knowledges, all of which must be inducted into museum curatorial practice.

The new humanisms I proposed in this book are dynamic lifeforms composing with, and folding within, others and they signal a profound reorientation of the way modern populations can think about and live in the world. Therefore, significantly, museums and museology can be agential in breaking down the modernist humanist traditions of exceptionalism and hubris.

The Covid-19 pandemic made visible how deeply enmeshed modern populations are in a vibrantly creative material world comprising many entities.

The Covid case study, presented in Chapter 6, illustrated how enmeshed curation is with more-than-human agencies acting within, in, and through not only museum spaces but also staff and audiences to create profoundly influencing alliances. Viral curating, a bonded human–non-human eco-curating process, demonstrates how viruses can be embodied not only in our very DNA and organs but also how their influence is distributed beyond our fleshly bodies. That is in influencing the way we conduct our lives, curate social spaces (including museums and their closures), engage in economic and political decisions, and develop new technologies including social distancing and vaccines, the latter of which shape our immune systems. We think, act, and curate with and "through" viral worldings that both shapes us and "is" us at the same time.

Throughout the book, I not only refigured museology for a more-than-human world but also acknowledged museologies as fractiversal constellations. This necessarily involved the transformation of the new museology towards a focus on complexity and process.

Formulating more-than-human museologies

The new museology is based on a social constructivist approach to life and process and engenders a distinctive semantic of both terms, procedures, and actions. In contrast, more-than-human museologies are a distinctive museological disposition and set of curatorial methods that I have developed through a series of publications, beginning in 2008 with my article, "Object-orientated democracies: Conceptualising museum collections in networks." This article has been recognised as formative in establishing new currents of museological thinking.[16]

This work has been directed to rethinking and refiguring what a museum is, what comprises community, audiences, heritage, collections, curatorial practice, and modern museum time. Questions in relation to the roles and agencies of museums have routinely been framed within scholarly and professional circles according to neoliberal concerns, or the policies and keyworks of human-centred, modern disciplines, and political and cultural biases. What I sought to achieve in this book is a consideration of what the roles and agencies of museums within more-than-human figurations might become. Central to this was the reworking of past practices and, most centrally, the type and form of humanisms they embody.

In developing the idea of a distinctive museological form as more-than-human, I refigured the new museology according to new considerations that extend the notion of the curatorial to empirically embrace the agential and more-than and non-human worlds previously rendered invisible or lacking agency. This entailed illuminating new practices of acquisition, refiguring of museum spaces, engaging with the new humanisms embodied in collections, and considering planetary curating and curators as humanness, as domains of influencing in diverse communities through communions.

Accordingly, I used the term museologies in this book because I didn't wish to present yet another museological formation (singular) as a universal solution to the current worldly predicament. In an ecological and more-than-human framing, and by acknowledging the non-human in museological constellations, practice can shift from being focused on strong human agency to the recognition of our entanglements with others as socio-technical, cultural, and biological empirical realities. More-than-human museologies also encompass diverse communities, knowledges, and forms of cognition and sensing to form multifarious, agential, and eco-curatorial alliances.

Museological research and curatorial methods are read broadly, not as humanist or human-centred practices but as ontological processes. This represents a response to destructive planetary conditions where the western masculine figure viewed himself as the commander and controller of all others. But all manner of entities – animate and inanimate –have their own agencies and therefore, with others, become domains of influencing. For example, with the Boulton and Watt "Lap" Engine plants and circulating carbon in its various forms (as different expressions of humanness) are embedded in its very material composition. Together they explicate the domains of influencing wrought through the bonding of the photosynthetic and the actions of humans' as biochemical lifeforms to form a register of atmospheric change. Furthermore, multiple curatorial agencies are in action, including non-human elements such as the metals used to craft the Engine's boiler that enabled combustion.

As ontological processes, more-than-human museologies and curating comprise multiple temporalities and durations as multifarious time-related processes including their spatialities that cross over and converge – from the planetary and the earth's strata to the atmospheric and to deep time. New forms of curating integrate humans as humanness and the non-human world into curatorial constellations and at the same time rework curating as these multiple agencies. Curating becomes planetary as, really, it has always been.

In this context, more-than-human museologies become experimental and anticipatory and directed to sustaining futures. Conventionally, in a humanist framework, this is what might be termed as better futures. Here I advocate for futures, including museum futures, based on new principles. More-than-human museologies can do this work by realigning museological projects as conceptual, analytical, experimental, and deeply relational and as involving research directed to composing other worlds respectfully.

In this way, museological research becomes more-than transdisciplinary and involves new types of relations with diverse disciplines, agencies, and contexts. It becomes novel multiple histories in the making,[17] and more-than-human museologies a way of being in emergence in the world.

From substances to more-than-human heritages as ecological compositions

Collections are often classified by the dominant materials from which they are made. A collection item's material substance, its provenance or origin, or the human that made it if it is a cultural object, defines its humanist significance. Its significance is often defined as one point in historical, linear time when it was made or linked to an event or maker. Its non-human, or more-than-human crafting and making, is of little significance unless it is deemed to originate solely from the non-human world, such as geological specimens. It is only then that its non-human crafting is registered and made explicit. And even then, in the case of geological specimens, it is often their transactional value, its value to a neoliberal concept of economy and extraction, that is foregrounded.

But when considered as ecological compositions, these assumptions, that underlay collections and their substance definitions, are overturned. Furthermore, they are no longer obsolescent objects, nor in a normative sense rendered passive, and no longer relevant to the contemporary present only because they are representative of the past. They are not waiting to be curated by curators. They are already eco-curating themselves.

While this is comprehensible with geological specimens, it also applies to digital cultural heritage. It too is no longer immaterial. Donald Trump's tweets, for example, are a material record of life on Earth, and their histories are extended out to long time spans in earthly history through their reliance on the energy sources of the sun and fossilised carbon. Trump's tweets are actually exemplars of UNESCO's humanist material heritage and are enmeshed in the broader material and political history of human–non-human entanglements of planetary extent and impact, from minerals and rare earth elements to becoming techno-fossils rather than just digital memory traces of human expression. Furthermore, the recycling of component parts of his tweets as ecological compositions from e-waste such as cellphones, magnets, computer hard drives, alternative-fuel vehicles, and all manner of displays are returned to digital cultural heritage as ecological compositions becoming an ongoing material history of life through an endless cycle of obsolescence.

In a distributed refiguration, collections and museological research strategies become enmeshed curatorial methods. For example, digital cultural heritage has always included non-human coordinates; even forms of data storage on DVDs, for example, comprise minerals such as bauxite or aluminium, gold, and oil shale. In this regard, digital cultural heritage, as an ecological composition, becomes multiple material records of history, of duration, and of life itself. Furthermore, heritages become vital, enmeshed in, and subject to eco-curating processes in which no divisions exist that can be demarcated by a singular substance. Therefore, the main organising logic and its associated heritage definitions and in defining policy framed according to a substance or a lack of said substance such as – material, digital, or intangible – collapses. In addition, the acknowledgement of alternative heritage and data producers, such as machine intelligence, or indeed wood and the photosynthetic as a maker of the non-and more-than-human of cultural materials, overturns the idea of humanist heritage. Heritages are thus more-than-human and planetary.[18]

When conceptualised as ecological compositions, digital heritage, material heritage, and intangible heritage comprise different coordinates but they also have some commonalities. Accordingly, intangible heritage is not actually intangible but is profoundly material. In thinking beyond the modernist idea of substance, all collections are understood as material, discursive, more-than-human, social, and collective. Each has a different array of distributed coordinates. Distinctions based on substances can no longer be claimed. As an ecological composition, for example, digital heritage becomes entangled with natural heritage such as the geological deposits and the environment they exist within. Furthermore, digital heritage is enmeshed with built heritage such as the infrastructures of the digital economy, including disused electricity generation and nuclear power plants. In the case of intangible heritage while an act of preservation, the production and recording of, for example, traditions through digital technologies is positive but at the same time leads to environmental destruction, the expansion of capital, and the loss of land.

Through the emergence of more-than-human heritages, the bucket (discussed in Chapter 3), Trump's tweets, and the Boulton and Watt "Lap" Engine, as exemplars of humanist heritage and museological research, become a type of ontological perspective rather than a universal set of practices. They become one form of curatorial real, like humanist digital cultural heritage, operating among others in a fractiversial frame.

Comparison and difference to distributed social collectives

Museums and museological research are governed by principles of individual autonomy, sovereignty, rationality, and construals that promote mind and body separations, humanist concepts, and the division of the world according to distinctive substances and inert materialisms. Curatorial practice seeks to articulate the essence of the human – culturally and biologically – through forms of classification based on concepts framed around identification that are determined culturally and biologically in comparison with others. With a focus on species identification, including that of the human species, taxonomies act as the means through which differences are categorised and placed into relations with each other.

Differencing categorisations are activated from the position of hubris most distinctively rendered through the way museums present specimens as passive raw materials, compliant technologies, dead and domesticated organisms, or even "eco-system services" as seen through policy frameworks. But museums are also complicit in cultural differentiating. Museums and museological research create categories and define how communities belong and what they belong to, they outline the fundamental defining characteristics each embodies, and their relations with other communities and the non-human world. Modern populations, for example, belong to a culture, identify with a belonging category, ethically divided even further into cultures, genders, and species. Museums, through qualitative and quantitative methods, segment their audiences according to specific characteristics such as age, sex, income, and education in order to understand their essence.

As part of these differencing and categorising optics, museums always address the human as singular or the collective as groups of people that share common attributes. Representational and pedagogical strategies assume that individuals are discrete, singular entities. Such individualistic strategies evoke questions of how the world is represented, including the relations that people have with others and what individuals can do as part of broader human-centred communities. Even the projection of future imaginaries is routinely human-centred and concerned with how individuals become responsible for acting for themselves.

Agency, conventionally in curatorial practice becomes solely human, operationalised through fixed taxonomies, species categorisations, articulations of human identity, and divisions according to gender, culture, race, and substances such as the material and immaterial, the latter as productions of the mind. But human populations are not autonomous and sovereign. Agential capacity only happens with others. Substances do not have distinct demarcations that render or reveal objective categories that represent the empirical realities of life itself. Rather, substances are cultural constructs and are vital materialisms. Situated categories of race, gender, class, and species are signifiers and not the empirical realities of life itself. Climate action cannot be restricted to questions of climate precarity, strong human agency, and better science.

It is abundantly evident that museologies, museums, and curatorial practice exist in a world full of dynamic, transforming, creative, and even dangerous more-than and non-human entities and processes. Together they comprise and are in communion with, all manner of entangled agencies from the animate, the so-called inanimate, earthly processes, anthropocentric activities, and more-than and non-human modes of crafting. And while it may appear that many specimens and objects lay passive in display cases, these very displays are dynamic even if their durations and transforming processes are barely perceptible. Eco-curating processes illuminate this dynamism, and at the same time, it seems through this conceptual framework questions a belief in the existence of an authentic form and thus heritage value. This is clearly articulated in terms of conservation practice and the deployment of strong human agency to halt decay.

And yet while museums speak of individuals as anatomically, developmentally, culturally, and spatially as bounded bodies or non-human entities as distinct substances such as minerals, chemicals, cells and physical processes even museum buildings, collections, museum staff, and audiences are really all these things simultaneously. They are ecological, trans-corporeal, and compositional, and they share a deep intimacy with each other and the processes of life itself.

In my previous work, I refigured human populations as embodied and embedded in planetary processes within particular eco-compositional circumstances. Each population was shown to exist in radical interconnectivity on multiple levels through the sharing of DNA, and their mineral, chemical, and viral compositions, as agencies together influencing, culturing, and extracting. So rather than focusing on difference, museologies and curatorial practice must focus on interdependencies and modern populations and their relationalities as material belongings, and in becoming with and in communion. This new figuration of the situated human in museological research and curatorial practice is empirically operationalised through curatorial methods in more than situated knowledge constellations – materially and agentially. The curator, museum scholar, and indeed the visitor are situated and not autonomous. Therefore, the notion of strong human agency and the belief that we alone can achieve certain outcomes is erroneous. We must undertake research, curate, and design compositionally with others.

Museums and power

It is widely accepted that museological research, museums, and curatorial practice have historically been, and continue to be, complicit in narratives of progress. But what this means is an explication of the expansion of western masculine agency. Exhibitions, collections, and their narratives record a trajectory of increase in masculine power and agency in capitalist economies. This is clearly evident with the Boulton and Watt "Lap" Engine of the Industrial Revolution and its links to efficiency and capital accumulation, and to the profit-driven aspirations of Matthew Boulton and James Watt. Museums and curatorial practices are implicated in pedagogies of progress. The Boulton and Watt "Lap" Engine, for example, becomes something to marvel at in respect to man's genius and as illustrative of the intensification in power, agency, and capacity of fossil capitalism through its increased productivity in the expansion of its horsepower. The expansion of modern power is illustrated in a number of ways, for example through digital transformation, through Trump's attempts to influence using Twitter, or through efforts to overcome viral contagion through vaccines and other viral controls evident in emerging pandemic collections. The expansion of modern powers and agency is made manifest through Lappland's Kiruna Iron-ore Mine visitor centre and museum, which sits 540 metres underground, deep in the earth's crust. Developed to exhibit historical and contemporary mining activities to exploit the iron-ore substates of the planet with current mining activity at 1365 metres below ground, displays of mining equipment such as large-scale earth moving equipment and drilling machines, all illustrate hubris, the destructive and exploitative activities and the machines made to achieve this directed to enhancing capitalist accumulation.[19] Even exhibitions such as those in the Anne and Bernard Spitzer Hall of Human Origins at the American Museum of Natural History while demonstrating interconnections through molecular DNA, as a shared and embodied relation became a metaphor for modern power through a progressive logic in which hominds are placed in a chronological timeline of development.[20] Evolutionary thought conceptually articulated as Social Darwinism and represented in nineteenth-century museums was put to work to make visible racial hierarchies.[21] Dubai's Museum of the Future for example cast the future of the planet as technological imaginaries made by human, typically male prowess thereby enhancing human capacities over the non-human world.

Teleology or progress, as categorisation and a future-directed imperative, must be let go. In doing this, museums can, through an explication of the deep interdependencies of the agential interactions of all planetary things, shift from being complicit in an ideology of progress centred on modern masculine aspirations and achievements to becoming an ideology of life itself. In this way, evolutionary thoughts, as teleological thinking, can become a form of ecological thinking that can be put to work to consider how modern power can transform into a form of transversal humanness, and as domains of influencing.

Rethinking human agency in historical processes

Conventionally, human agency is figured as strong and all encompassing through museums and museological research and these forces are often presented, consciously and subconsciously, as appropriating, crafting, harnessing, and extracting things. Earth systems are typically viewed as separate from human activities and modern populations are typically shown to have an impact on earth processes, rather than the inverse. The exception is when the impacts that the non-human world has wrought on populations are shown in the case of natural disasters.

In reworking modern museum histories and humanisms, in this book I refigured the concept of the human from one of a static, bounded, whole fixed individual and biological entity, governed by and subject to forms of identification, categorisation, and representation on the basis of race, gender, age, cultural background, to the concept of dynamic humanisms of diverse and indeterminate forms.

In this way, human agency has been reworked to become an extended, transversal, and distributed form. The individual has become distributed agencies, as indeed it always has been. Strong human agency has become humanness and in the process rendered the individual human as categorically obsolete. This empirical figuration of the individual has become one of extended humanness within distributed ecological compositions.

Humanness is a figuration of situated and distributed human thoughts, intents, bodies, and activities in historical processes. It is a more dynamic and indeterminate figuration in which the humanist individual activities become agential capacities enrolled in extended historical durations. These multifarious humanisms emerge and are not predetermined entities. Rather they are dynamically embedded and enmeshed with earth others, chemicals, minerals, plants, animals, technology, digital infrastructures, biochemical processes, and viruses, for example. The use of the metaphor "life forms," in its many dimensions, becomes a figuration for thinking in terms of the distinctive humanisms and the specific bonds of different forms and qualities that emerge with others in historical, durational, and compositional processes as dynamic biochemical, viral, more-than digital, and as wildfire.

Humanness helps modern populations to think with others that are not necessarily human. Individuals and groups of people in climate action, for example, are framed as green citizens having attained all the information and practical knowledge to act for themselves and with others as a strong human agent of sustainable practice. The human, that being the educator, the curator, audiences, and, indeed, the designer are as central figures in museum programs directed towards greening themselves and others. Through this new figuration as diverse humanisms, modern populations are not only decentred in ecological and historical contexts but also become distributed in social collectives. This figuration opens up a space to reconsider the notion of green citizens as a form of engagement focused on individual action as strong human agency to instead be viewed as an individual enmeshed in a more-than-human world in which pedagogies are articulated around attunement and with a disposition for thinking and acting with all earth others.

The concept of humanness gestures towards different types and forms of relations at cellular, biochemical, biological, and cultural levels. And this figuration is expressed in a transversal sense as perception: as imagination, as creativity, as crafting, as hormonal, as viral, as biochemical, as habitual everyday practices, as extracting and as attuning, and so forth. That is, as multiple forms of agency.

Museum scholars and curators are actors of the enlightenment. History, including cultural history, is a humanist history according to modernist construals. Museum history is history written about humans and human agency. According to modern human-centred dispositions, natural history is the history of Nature. But histories and sciences are more-than-human histories.

More-than-human museological research and curatorial practice can explicate the many ways historical processes are more-than-human and other-than-human, and indeed always have been. Conventionally, other-than-human histories are humanist based and framed around the articulation of the agencies of the non-human world through modern science. Rather all histories become emergent as eco-curating processes, for example, with viruses, bacteria, chemicals, minerals, technologies, and many other entities. That is not a history involving the study of past events relating to the human or nature, but rather histories located within extended planetary becomings of intense complexity involving multiple coordinates and durations, including the socio-cultural, technological, biological, and geomaterial such as plants, soils, minerals and chemicals, and so forth.

Modern history in museums, and its demarcation into time periods of the past, present, and future, is reworked as diverse coordinates in composing processes, and as transforming in emergence. Accordingly, human populations are part of planetary ecologies and historical processes that are not just relational with multispecies but part of dynamic planetary ecologies. Strong human agencies become situated within historical processes.

The enaction of human agency in historical processes of planetary scale extends beyond that of geological agents of the Anthropocene, to include all agencies is best illustrated through the Boulton and Watt "Lap" Engine. The folding out of the Engine's elements, fabrics, and methods of production historically activates the Engine from being representative of male hubris and accomplishment and the advent of the Industrial Revolution to be part of embodied and multiplying histories involving human and non-human agencies. The Engine becomes forms of human agency such as carbon-burning practices, as well as more-than-human biochemical lifeforms entangled in its fabrics and contexts throughout its history This explicates the multiple ways the folding and enfolding of the Engine's coordinates have contributed to earthly historical processes of transformation.

Museum collection case examples – the green plastic bucket, the Boulton and Watt "Lap" Engine, viral contagion, Trump's tweets and policies, the Paris Agreement, and the Sustainable Development Goals (SDGs) – are not discrete objects but cuts into planetary ecologies and conventionally explicate strong human agency as command and control in multifarious ways. In this book, the bucket is used as an example of the exercise of strong human agency to control wildfire. The Boulton and Watt "Lap" Engine represents strong human agency as an engine integral to the advent of the Industrial Revolution and the desire on the part of Boulton and Watt to direct capital growth and accumulation through mechanic efficiency. Viral contagion and pandemic collections represent strong human agency as multifarious attempts to control the spread of the virus and its impacts on human populations. Trump's tweets, as a form of cultural heritage, are examples of the exercise of strong human agency mediated through and directed towards the control of North Korea's nuclear program, or to overturn the US presidential election result. And finally, both the Paris Agreement and the SDGs, through museum implementation, seek to command, control, and stabilise an out-of-control non-human world or render the latter's potential for continued capital growth.

At the same time reading these case examples as empirical and as interpretive cuts (discussed in Chapter 2) in respect to Trump's tweets and their interpretation and documentation in Chapter 3 in regard to the bucket and the Boulton and Watt "Lap" Engine demonstrate the newly visible, and broader and extended, social collectives in which each exists, is contingent on their materialising tendencies. Making as many coordinates as possible visible through my folding-out technique illuminates the distributed, eco-curating, and historical processes in which each is embedded and reliant.

The histories of all the collections' case studies are the outcome of the coordinates and their material agencies in which each is composed. Together they are in a constant state of sympoietic emergence. Each begets other things, becoming something else in enmeshment in multiple dimensions. For example, the Engine was a technology that engendered a sense of wonder when it was made. It became an Engine to be reviled as the concept and empirical reality of the Anthropocene emerged. Therefore, the historical processes by which the Boulton and Watt "Lap" Engine becomes a subject of emergence through the enmeshment of human bodies, thoughts, knowledge practices, and actions such as the fusions of carbon burning with photosynthetic processes, alternatively become ongoing sympoietic transcorporeal exchanges. Its histories emerge as multiple non-linear processes.

Thus, because histories are situated and become – with for example microbal and biochemical elemental exchanges – the discursive and human intent becomes part of distributed multiple enfoldings, practices of interpretation, and documentation as cuts into these multiple emergent processes. The human-centredness of conventional historical imaginaries also become cuts into ontological processes. Perhaps one of the most powerful innovcations of these new situated humanisms is illustrated by Bill Putt, and his green plastic bucket and house discussed in Chapter 3. The inversion of the human-centred stories around Bill, his house and bucket, towards an appreciation of the relations and interactions that explicate the distributed agencies and durations in which wildfire becomes agential, means Bill, his bucket and house, become a mechanism for considering the circumstances that led to the Black Saturday bush fires of 2009. Furthermore, the green plastic bucket can be considered an object that can be used to rethink these relational contingencies in order to understand human agencies in distributed collectives, who and what was involved, and alongside the agential capacities of the bucket itself and its fabrics, atmospheric change, the weather conditions, the combusted bush, human interventions and their enmeshments.

According to posthuman theorist James Williams, duration is one of continuous multiplicity in emergence.[22] Therefore, in the context of digital cultural heritage, and across all manner of collections, time and historicisation as a record for modern heritage become durational perspectives, or the variable time of ecological processes of the coordinates involved. In this way, collections and their temporal frames and fields of influence are greatly expanded, and collection time is understood to be the durations and composites of many different elements and processes.

Therefore, in museological research and curatorial practice, every collection item is an ecological composition because it is enabled and created by a wide range of coordinates no longer solely anchored in a singular event. Even though the green plastic bucket is framed around a human story and Bill Putt's attempts to save his house, the bucket comprises other-than and more-than-human coordinates, durations, and temporalities. The bucket is embodied, addressed to, and enabled by a wide range of coordinates and connections from the geological, petrochemical, and atmospheric durations and processes. While these dependencies are in the collection's records under the "Materials, Production" statement, and so forth, they become known, ontologically, as processes of folding and enfolding and as emergence across different temporal durations.

Human populations, and indeed their collections, are part of processes in the present as ongoing acts of creativity and transformation that enmesh and, at the same time, embody elements, including all manner of traces including materials, inventions, political decisions emergent from previous compositions that beget new relations.[23] Collections, therefore, can longer be thought of as bounded human identities and time frames related to significant events locked into objects. That is because all things connect to other things.

While human-centred ecological thought in museological research, and in curatorial practice, invokes thinking in relation to other things it is often one of comparison from a position of hubris. While it is still relational, as a type of ecological thought it does not support the critical requirement of activating communitarian design practices of attunement, and therefore fails to invoke an understanding of relations of deep connection, interconnectivity, intimacy, and radical coexistence, and thus neglects to enact values based on care and respect for others.

Distinctive humanisms emerge in situated contexts, for example, as dynamic bonded biochemical and viral lifeforms but they also spill over and bond with other things. Each is not discrete. These distinctive humanisms are again cut into ontological eco-curating processes. AI lifeforms, as another example, signal new curatorial relations and methods for articulating human agency and embodiment with others. Their potential emerges as a means to articulate intimate and attuned habitability practices that share values and principles of care and attunement, and of respectfully communing, folding, enfolding, and crafting with others.

Museology and curatorial practice, therefore, must abandon visions and knowledge of the world they represent in order to produce new frameworks.

This approach in particular is emerging in art practice where such concepts of interconnectedness are represented through individual artistic creative expressions.[24] Institutions and museology scholars, thus, must question the dominance of euro-centric and modern disciplinary knowledges and strong human agency within their practices. Museums need to institute, as an empirical reality, new forms of curating as eco-curating – that is, a practice and method of composing and crafting with others in respectful communion. Historical and heritage demarcations of fixed time and duration become ecological, compositional, and emerging where demarcations of the past, the present, and future no longer exist. What comes before is a cut, a glimpse into previous emergent ontological processes.

Furthermore, museum practices must embrace non-modern and Indigenous frameworks of understanding and knowledge, and indeed many are starting to do so. Negotiations must be made between different ecological thoughts to develop a fractiversal cosmology that invokes practices of care and attunement. In doing so, museums and museology have the potential to work with and support the establishment of a global collective of attuned ecological thoughts and actions. Of course, such a collective will always be imperfect, partial, and always dependent on context. There will likely be clashes between the concerns and the foci of diverse knowledge practices and this will most likely emerge as distinctive fractiversal politics. These indeed have and will always be present as the legacies of the past subjugation of First Nations knowledges and land. Significantly, curatorial work informed by many different ecological thoughts and acts can promote practices of enmeshment and embodiment, and at the same time remain open to process and ambiguity in our relations with others.

Beyond museological structures and disciplined collections

Museum disciplines and their museologies conceptually demarcate human, more-than-human, and other-than-human agencies and experiences and gather them up into separate, hierarchical knowledge fields. The sciences and its subfields, social history, and art history, for example, are illustrative of these cuts. And such "disciplining" diminishes the ability of museum professionals and scholars to view museum fields of expertise and accomplishment as relational, situated, and contingent. History and anthropology, in particular, are knowledge practices of understanding and of meaning making. Even the natural sciences, typically considered to be sensing disciplines, are directed towards the understanding of specific phenomena, rather than of processes from a standpoint of a broader range of coordinates and planetary communities.

Museum disciplines and their classifications also limit the coordinates that are made visible and obstruct an understanding of the interconnectedness between all these things across categories. Through the use of disciplines museum storytelling has oftentimes been about modern populations, about how they understand the natural world and the things they value. The modern human figure and their disciplines can no longer exist as figures or figurations on their own. Museum narratives of the more-than-human will always be human narratives but will not necessarily be all about humans. Instead, they must open up and explore questions with audiences to respect to how interpretive practices can acknowledge the reality that populations are in communion with others.

Museum scholars and practitioners focus on objects as having linear historical relations, as being original, authentic, auric, as comprising of substances, as being significant locations of cultural expression and memory that fall into disciplinary regimes of knowledge and curatorial practice. Museum disciplines delineate the coordinates that are made visible for research and documentary work and often render invisible cosmological, material complexity, and agency. The notion that identity is locked into objects and forms of classification are humanist cuts into cosmological flows and meshes, although is not acknowledged as so.

For example, while the disciplines of history, natural history, cultural history, social history, biology, geology, and so forth are siloed, collections comprise human and non-human entanglements. They are all more-than-human heritages and disciplines, and thus an opportunity emerges to work with collections as material indices of crises and recuperation that cross the concerns of all disciplines. Significantly, curatorial practice is extended beyond human concerns and also disciplinary and even transdisciplinary predilections.

Objects are cuts into planetary cosmological processes and are a microcosm of planetary circumstances that can be viewed as emerging ecological compositions of infinite extent. Ecological museologies and curatorial practices can ask new questions of the disciplines of science, history, and natural history by reframing disciplinary constellations to do different types of work, such as detailing and understanding the agential dispositions of non-human worlds in communion with human agencies.

Modern keywords and terms

Museums and museology cut and compose according to modern construals and are rendered agential and capable through terms such as culture, nature, society, class, gender, community, nation, cultural diversity, identity, and belonging. Each of these terms and concepts grow into keywords and organising structures over many decades, even centuries, and are oftentimes influenced by political, economic, and policy regimes outside museums. Other terms, such as the Anthropocene, climate mitigation, climate science, and the SDGs rapidly, attain their status as organising structures after the invention of these terms and policy frameworks. Digital cultural heritage, for example, is thought about as heritage in a digital format belonging to an individual, cultural group, nation, or a community (to name a few) and is now a key word within museological practice. In a more-than-human ecological framework of belonging, digital cultural heritage operates beyond itself as a form of human expression, or as something belonging to a person or illustrative of a particular event. It expands out to include a belonging to the Earth, to the terrestrial, earthly processes, to the cosmos, to the flow of life, and to the multiple ecologies of a common world.[25] Eco-curating digital data, in its figuration as an ecological composition also acts as a lever for making visible embodied accountability, and a consideration of how such practices may be promoted to support planetary habitability. Ecological sensibilities, and an attunement to the production and use of digital cultural heritage in a world of data profusion, have become a necessity as well as an opportunity to redress damaging practices and thus formulate survival tactics with others.

New keywords or terms, therefore, must emerge in curatorial practice that can facilitate ecological sensibilities such as multiple belongings in the planetary, embodied accountability, multiple events as ecologies of a common world, and earthly habitability. At the same time, keywords and terms need to expand and engage distinct fractiversal practices with their own knowledge systems, keywords, and methods.

The planetary as curatorial praxis

Conducting museum work from a planetary vision has always been an empirical reality but making that vision visible in a more-than-human framing requires a deep conceptual and pedagogical refiguration of curating and of museology. Media theorist Jennifer Gabrys, in her 2022 thesis, argues for a planetary praxis when she says, "The planetary is the difference, distance, and duration with, within, and against which it might be possible to think differently about being human and becoming collective."[26]

Planetary vision is clearly articulated in my earlier work on digital collections and digital cultural heritage. This planetary optic is mobilised through the articulation of digital cultural heritage as multiple enmeshments and durations of planetary scale illustrated through global computational infrastructures, geomaterialism and its circulation. Therefore museum and more-than-human curation is not just a planetary vision but a distributed planetary curatorial praxis. A planetary curatorial praxis, however, is much more than a technological imaginary or the reconfiguration of planetary life but includes modern human populations as planetary subjects, the planetary as circumstances, collections as ecological compositions of planetary scope and account, and of emerging planetary collectives of increased complexity, extensiveness alongside multiplicity of durations in which curation is enmeshed. Engaged with all of the above curating becomes material and acts of materialising in planetary circumstances of complexity.

Museums are profoundly complicit in the figuration and pedagogical distribution of concepts of planetary life in multifarious ways, an example being the enlightenment praxis of collecting the world through specimens. Collecting and the taxonomic arrangements of all the realms of nature and culture are planetary displays. Henceforth, the planetary in museums becomes multiple collections of planetary processes and life. For example, the geological collection at the National Museum in Prague contains rock specimens that illustrate various internal and external geological phenomena and earthly processes.[27] Plant specimens or herbaria, such as those of Sir Joseph Banks collected during Captain Cook's voyages to the South Pacific in the late eighteenth century, act as a reference point of what was grown in many specific locations in any given era. An examination of their properties enabled the registration and exploitation of these planetary resources to support colonisation and economic growth.[28] Space technologies, such as the Apollo 11 command module in the Smithsonian's National Air and Space Museum enabled a planetary vision from space and represented a desire to conquer, extract and inhabit other planetary realms.[29] Planetary objects, such as the Boulton and Watt "Lap" Engine and the green plastic bucket, are cuts into the emerging, destructive processes of planetary scale and extent, including climate change.

While museums foreground a planetary vision of destruction through the concept of the Anthropocene, curation must be refigured as a planetary praxis of partial recuperation in which modern populations become planetary subjects in eco-compositional situations and eco-curating circumstances. The enrolment of modern populations in the planetary, as a novel figuration of humanism, goes beyond the one-dimensional notion of populations as geological subjects of the Anthropocene.

The difference with a Cartesian planetary praxis of museums of the past is the move away from the human-centric focus and conventions and keywords of the enlightenment. In addition to the refiguration of curation as planetary praxis and the addition of humanisms, neoliberal capitalism must be uncoupled from curatorial practice and museology as an organising logic. A Cartesian articulation of the planetary is another conceptual error because while modern populations are all planetary, they are also sympoietic.

Curating has always been a practice of thinking with, acting with, and composing with other things, narratives, concepts, materials, spatial configurations, people, disciplines, and so forth. It has always been human-relational and eco-systemic incorporating the creative capacities of non-humans. By invoking the planetary as curatorial praxis, human populations in the west are inserted into non-human durations and temporalities of planetary scope and scale.

As a praxis curating for planetary habitability rests on the ecological composition, that is it is a method to think with, act with, work and make with. This conceptual framework also requires a deep level of self-reflexivity – that is, thinking about ourselves in a mode of thinking as, and with, the stuff of planetary strata. That is because modern populations have never been hubristic The Covid-19 pandemic indeed all viruses teach us this important lesson.

Curatorial concepts and methods: eco-curating as curatorial practices

In humanist and social constructivist figurations, museums and museology lack adequate terms to deal with the dynamic and ecological realities of life in which institutions – their fabrics, exhibitions, education programs, collections, staff and audiences, and all manner of stakeholders – are enmeshed within. Indeed, in museums and museological research the non-human world is a stakeholder that must be negotiated with. As I turn from a human-centric to, what I call, a more-than-human ecological framing for museum, curatorial practice, and museologies, the new concepts and terms that emerge include: "the ecological composition," "coordinates," "thingness," "eco-curating," "domains of influencing," "cutting and tracing within and across the ecological compositions," "the planetary as archive and as curatorial praxis," and "community as expanded to include more-than-human and non-human others." Thinking in this way museologies and curatorial practice become experimental forms of communitarian design, and curating as collective and sustaining practices and both are directed to the flourishing of the communal. In making this leap, I came to propose a broader field of influencing.

In this book, and through two previous publications, I developed the concept of curating as an ecological framework and extended curating practice to embrace the earthly and the planetary as distributed agencies and as an expression of the inexplicable circumstances and complexities of life itself. As I argued in *The Future of Digital Data, Heritage and Curation in a More-Than-Human World*, curating is an act by more-than and non-human agencies in eco-systemic processes. Accordingly, conventions of curatorial agency are no longer the expression of curatorial or museum sovereignty and, in the case of digital practice, are no longer solely entangled with the machinic or even an autonomous knowing individual. Rather in an eco-curating figuration, all coordinates curate "influence" and "affect" in different ways and forms.

Curating is conducted by an alliance between all manner of enmeshed entities and processes of infinitely different human and non-human kinds. These include non-human energy, data and data centres, as seen with Trump's tweets, fire and combusted bush in the case of the bucket, and carbon-burning practices and the photosynthetic with the Boulton and Watt "Lap" Engine and its contexts. The artefactual nature of digital cultural heritage, for example, framed solely in terms of human consciousness and cognition, becomes multiple and multiplying in eco-curating processes. It includes cognitive, calculating, and sensing processes as well as the conscious-non-conscious practices of machine learning, elements, algorithmic calculations, infrastructures, electrical currents, their circulation, and so forth. All these different cognitive intelligent and sensing processes make unlikely connections and produce new perceptions and affects that are often not discernible, but at the same time are inextricably vital.

Curating processes, enmeshed in digital cultural heritage, for example, include the strings of materials embedded in life itself, from planetary computational systems to atomic-level elemental agents. Many of these are the result of human design but at the same time also embody their own affordances, intensities, temporalities, speeds, and durations in their enmeshed unfolding. The curation of minerals used for conduction and the circulation of electrical currents, operating when Trump tweets, are both in the archive and out in the world simultaneously and thus are redistributing curatorial agency beyond what would ever be humanly possible as a singular agent.

Curating becomes any activity, or an influencing or transformative process, enacted by many agencies – some known and others less visible, and even unknowable – and with effects that manifest in multifarious ways. Museum spaces, buildings, collections, digital processes and, indeed, all activities are profoundly more-than and other-than-human. The non-human entities enmeshed and embedded in museum spaces, including their collections, insects in collection stores, earthly processes represented as geological and mineral samples, cables, data, and electrical currents and the fabrics of buildings, such as metals and concrete, are all curating agencies. Processes of transformation in conservation practices might be seen as acts of destruction by insects or by ultra-violet light, but decay is an eco-curating process of becoming with others. Curating decay within heritage frameworks is therefore refigured as eco-curating processes. While efforts are made to halt decay through strong human agency, decay is not actually decay. It is rather an eco-curating process of sympoietic becoming and includes its own durations and temporalities.

Bringing to the fore new forms of agency as eco-systemic, I positioned modern human populations as situated, embedded, and enmeshed in ecological circumstances in various ways. I illustrated modern populations as a series of traces, operating across and through ecological compositions from writing code, making hardware and infrastructures, mining minerals, and making and remaking all manner of cultured materials. Other coordinates are attributed agency and exhibit affordances and abilities beyond that of human capability and consciousness. This included algorithims, automations, materials, chemicals, and all these agencies' calculative, sensing capacities and properties, all of which have the potential to change the course of events and entities in unexpected ways.

As an example, the convention of viral curating in museum spaces is an understanding of the material circumstances of viral spread, of masks, and social distancing and is a process informed by scientific research. But viewed according to this new curatorial regime Covid-19 viral agencies and viral control become enfolding processes with museum spaces and surfaces, human bodies, viruses' bonding with human cells, the details of respiratory disruption in those infected, the political and economic change and even the reorientation of scientific investigation of viral contagion to consider the behaviour of the virus itself and its influencing capabilities. Viral museologies, thus, become the practices and dispositions of curating by and with Covid-19 sociobiologically in planetary rhythms and processes.

Trump's tweets as born-digital objects and in eco-curating alliances generated their own data and distributed themselves through electrical cables, currents, algorithms, and machine learning. Embodied in digital cultural heritages are forces that exceed human curatorial agency and in regard to Trump's tweets they distributed through a complex web of interrelatedness humanness, his intent, thoughts, and actions. These interrelations include the effects of algorithmic calculations, of hard drives, AI, and micro and macro memory processes within the archive and out in the world. Such processes involve human crafting and technological innovation, such as the invention of computers, but also the formation of the earth and minerals born of geological processes, as well as the emergent reverberations of the universe and its birth. In this way, it can be seen that digital cultural heritage becomes more-than-human ecological compositions, as well as new types of representational subjects, through an expanded range of producers and collaborators.

Because humanness is embedded in these unfolding and enfolding processes, humanist heritage concepts collapse. That is because the conceptualisation of heritage collections conventionally fails to acknowledge that modern populations exist as relational and porous entities and as part of enmeshed collectives. Through eco-curating processes so-called heritage objects arise to become multi-directional, multi-scalar, and rhizomic processes.

The multiple enmeshed curating agencies – the photosynthetic, fire and combustion, and the viral – all of which curate with other agential coordinates, make visible what Braidotti calls a "posthuman convergence."[30] In the context of digital cultural heritage, this is what I call a more-than-human convergence, and each of these agential coordinates is unique, and greatly expands upon the notion of curatorial agency scaled up to the planetary and beyond. This is also evident in respect to how viral curating continues to mutate through new variants, transform museum spaces, and political and economic circumstances. For example, the surge of Covid-19 infections in China in late 2022 meant that thousands of museums had to be closed, either temporarily or permanently, and staff lost their jobs or were forced to take unpaid leave.[31]

Museum practices and museum institutions themselves are made visible as enmeshed and embedded eco-curating entities in planetary processes in which "community," "audience," "collections" in all their forms, and "documentation" is rethought as more-than and other-than-human, and as material and materialising. Museological and curatorial work becomes experimentations in respect to collections as ecological compositions in which cutting into the multiple reveals many processes and agencies.

And because collection items are composed of different more-than-and non-human coordinates and eco-curatorial processes that make them inherently non-identical and thus inappropriate for assigning humanist-centred categories to them. They become less amendable to classification.

Eco-systemic curating practices of documentation and interpretation activate new methods and terms. The folding out of an object's agential coordinates, and the investigation of their dynamic interactions by tracing the routes of their interrelatedness in their becoming, goes to the heart of collection documentation and interpretative work.

Therefore, curatorial practice operates within and beyond museum spaces. For example, human populations culture the non-human world for themselves in order to produce commodities, to perform everyday practices, and to sustain life. Many of the things we make and do involve the culturing and crafting of the non-human world, some of which eventually become museum collections. All acts of curating, even from a humanist perspective, are metaphorical and material cuts into flows and thus are disruptions of planetary processes in emergence. Interpretive cuts, for example, include those made into data processes, electrical circuits, and data centres, as well as non-human and human agencies such as photosynthetic, viral, and fire ontologies. The photosynthetic and biochemical transformations embodied in collections as more-than and non-human curatorial actions that ^{13}C analysis registers as a method become curatorial cuts into rising emissions and this is illustrated through the interpretation of Ellsworth's graphs outlined in Chapter 3.

New curatorial terms such as the following can support such practices – including eco-curating, influencing, selecting, organising, exhibiting, compositional practice, cutting (documentation and interpretation), and composing. The curator or curatorial agent becomes part of these compositional dynamics. Overall, eco-curating is a cartographic mapping exercise in getting to know who and what a curatorial agent is curating with including their agential affordances and materials, the social collectives and ecological circumstances in which they are enmeshed and accordingly how interventions might be made. It is a process that is responsive to respectful and attuned crafting and composing practices made within others within these expanded and distributed processes.

Communitarian design practices for planetary habitability

Museum and planetary futures are routinely humanist futures characterised by, what theoretical historians Zoltán Boldizsár-Simon and Marek Tamm call, utopian societal design, visions of human progress, and projects of human emancipation.[32] The exhibition "Spark, Australian Innovations Tackling Climate

Change" for example discussed in Chapter 4 promoted teleological futures exploring technological progressive futures based on visions of human progress. Technological futures however must be carefully crafted through acts of both research and communing.

The folding out of collections explicates unintended futures, the complexities and entanglements of these unfolding futures. Unintended futures and crises, for example, are written into the fabrics of collections and made explicit, or at least partially so, through new forms of conceptual and material analyses, for example, studies of the types and forms of humanisms embodied within the fabrics and histories of collections, the historical sources associated with them and their documentation. In refiguring collections as climatic ecological compositions, as illustrated with the Boulton and Watt "Lap" Engine, visions of the engine's progressive history are inverted and become multiple. For example, the Engine's histories have their origins in diversity. In deep cosmological durations in the formation of Earth and its geological materials. In enmeshed social, and economic inequalities, dispossession and the destruction of culture borne out of the Industrial Revolution. And in material indices of extraction and capitalism, fossil fuel intimacies and emerging material circumstances.

The Boulton and Watt "Lap" Engine illustrates not only its multiple histories but also its destructive entanglements through the coexistence of many entities. These include extracted minerals, slave labour, the entanglement of the photosynthetic and fossil-fuelled biochemical lifeforms evident through ^{13}C analyses, masculine power, capital accumulation, and exploitative practices as extended domains of influencing.

The entanglement of these entities expressed in the form of collections enables the illumination of an expanded cosmos agentially and therefore materially as a more diverse and complex expression of how current and troubling planetary conditions arise. The study of these entanglements is a method by which modern populations might eco-curate in a way that is recuperative and sustaining of living practices and methods.

The multiple extended humanisms in these objects, and the perspectives they explicate, are at once a narrative of destructive practices and at the same time expressive in their material embeddedness in these troubling circumstances. Thus the explication of these relations is mobilised in multifarious ways. That is through an investigation of both historically and materially the way they were made and used, how they were conceptualised, how their materials were extracted and moulded, the intent and thinking behind them, how they were rendered obsolete, and the unintended consequences of their operation. They comprise a multitude of processes of varying spatiality, temporality and duration, and form, including computational, neo-capitalist, geological, and quantum atomic processes. Clearly, material embeddedness produces new knowledge.

Eco-curating processes engage collectives in multifarious forms and previously rendered invisible. As a result of this, they not only produce new perceptions of social and material bonding but also expand humans' social relationships and contingencies. Our fleshly, social bodies, and personalities, for example, become entangled in data, storage entities, algorithms, automated systems, and global computational infrastructures to form new types of communions and communities.[33] These participatory modes of understanding become the "we," not mine or ours, to establish a wider community of producers, intelligent forms, and agencies comprising cognitive, calculative, sensing, discursive and materialising processes.

The ecological compositions discussed, including the bucket, the Boulton and Watt "Lap" Engine, and Trump's tweets, are made up of multiple, vital, entangled, and layered ecological processes that impinge on the biosphere, ecosphere, hydrosphere, and lithosphere. Each in its distribution and array of coordinates is part of the planetary archive of energy, elements, the sun, fire, solar energy, and of fossilised plants used to generate electricity. The planetary archive becomes global computational infrastructures but at the same time is of the earthly archive, with its chemicals and minerals extracted from the earth's crust and the immanent cosmological histories of their making. Each example becomes technospheric heritage and an exemplar of humanist damage when the folding out of its elements proves revelatory.

Significantly, each case example is used to illustrate how destructive practices are rendered capable through the convergence of multiple agencies. All the collections' case examples, and their reinterpretation as ecological compositions, are entangled with human-post-natural forces and contaminants, greenhouse gases, mines, toxic soils, water, digital and e-waste decomposition, and even cancer.

If we are to successfully curate for planetary habitability, we must interrogate the ecological circumstances and implications of digital practices, the new materialisms of toxic waste made through e-waste, digital rubbish, data excess, and the changes in atmospheric biochemical compositions that emerge. At the same time, such a refiguration of collections as ecological compositions can promote an understanding of what crafting processes are able to render others capable of.

New recuperative curatorial methods have the potential to emerge by shifting from humanist historical formations to more-than-human collectives, and by explicating collections as ecological compositions in order to make visible processes and durations not normally evident. Thus, eco-curating as a curatorial practice has the potential for framing and co-making habitable worlds as lessons in compositional design. The folding out of the coordinates of collection items is a new method for making visible humanness and its embodiment in a more-than and other-than-human world.[34] Likewise, an opening up of accounts of heritage in all its complexity can make evident our common worldly predicaments and the entities we must attune and negotiate with. Such methods pave the way for new thoughts, humanisms, and, eventually, new acts.

As I argued in *The Future of Digital Data, Heritage and Curation in a More-Than Human World*, digital cultural heritage builds social, ecological, and affective responsibility and when promoted, by making connections between the production of data and its deep entanglements with ecological crises, has the potential to empower ethical practice.[35]

Furthermore as discussed in Chapter 4, the conventional view of modern populations in climate change processes and as a lever for activating mitigation processes is based on the idea of human populations as radiative forcing agents whose carbon-burning practices arise as emissions and these accumulating greenhouse gases in the atmosphere cause heating by absorbing terrestrial infrared radiation.[36] Mitigation therefore is framed around a greenhouse gas-centric logic – the chemical constitution of the atmosphere and how it might be changed through strong human agency by changing our energy sources, our emissions and therefore able to solve climate change. Nature-based solutions especially those that deploy plants and soils seek to exploit their agencies to absorb emissions.

Through ^{13}C analysis human agency is transformed from a consideration of modern populations as separate from the non-human world except for the intermingling of emissions and their chemical signals to one of deep enmeshment through the combining of plants and human populations and their emissions through the photosynthetic becoming dynamic biochemical life lifeforms where fossil-fuelled lifestyles are bonded into wood fabrics. Similarly with the rise of renewable energy solutions, new humanisms emerge in which mineral rare earth elements, cobalt, and lithium tellurium become bonded as emerging mineral lifeforms in which neo-liberal capitalist lifestyles become new forms of energy intimacies with the potential to also be destructive through mining and waste. Through the consideration of modern populations as lifeforms in multifarious forms, the transversal subject emerges.

Furthermore, eco-curating refigures new forms and types of humanisms embedded in collections. Such humanisms enable regenerative curatorial work by illuminating how to think with and act with others in extended communities in a common world. For example, the Boulton and Watt "Lap" Engine, the figuration of fossil-fuel-driven populations as dynamic fossil-fuelled life forms, gestures towards the cultural shifts required to bring about change. This is because it becomes clear that modern populations, in their thoughts and acts, are situated, and at times bonded with others in multiple ways and within diverse social collectives. Rising CO_2 levels, for example, indicate that fossil-fuelled capitalism and the populations inducted into carbon-burning lifestyles are actually deeply bonded with the non-human world, namely, through the photosynthetic process. This suggests that the Boulton and Watt "Lap" Engine can act as a new type of cultural carbon mitigation strategy directed to reducing greenhouse gas emissions by explicating modern populations and their deep bonds with the non-human world not just as a forcing agent through the release of greenhouse gases but as wooding biochemical lifeforms as fossil-fuelled intimacies enmeshed in non-human DNA. This analysis using museum collections acts as a pivot and lever shifting an understanding of ourselves from one of dis-embeddedness to one of deep enmeshment within climate processes from which new mitigation strategies might arise that foreground how to live in the world differently with others including the non-human world. Trees are no longer service providers but stakeholders in new strategies.

Heritage collections, as vital and therefore eco-curatorial, can not only be put to work to show the effects of destructive humanisms but also relations of community and the communal in a common world. From this position, an ethics of communal care and responsibility to both other human populations and the non-human world can be activated through museums as attuned practices in respect to who is involved and with whom we must negotiate might be designed. A focus on greenhouse gas-centric thinking has the potential to promulgate more destruction but of a different kind that museums unconsciously may promote.

Like digital cultural heritage, such as Trump's tweets, the green plastic bucket, and the Boulton and Watt "Lap" Engine are of the earth, of the atmosphere and of life itself. Modern populations share bodily chemicals and minerals with these more-than-human collections. In the case of Trump's tweets, metals act as conductors of electricity within data circulations and these elements reside within the fleshly human, within our own electrical systems. As an ecological entity, digital heritage becomes a more-than-human concept bound to both humans and the planetary.

By rethinking and materially expressing human agencies and their situated positions in complex ecologies, partial recuperative measures can be enacted if an investigation of these relations are inverted. That is not as relations that are hubristic but as ecological contingencies that detail with whom we must negotiate and attune as sustaining practices.

When this occurs, collections assume a new role as an experimental method in which humanisms are highlighted as being embedded in all manner of situations, as shown with the bucket and the Boulton and Watt "Lap" Engine. Viral contagion in museum spaces, for example, leads to respectful curatorial interventions to allow interactions with others that did not invite a hubristic response. This leads to curating and conserving in dynamic systems and with different entities, such as wood, plastic, fabric, stone, and data. Within an ecological compositional framework, strong human agency is put back into its place and eco-curating as a curatorial practice becomes a worldly condition of composing with and in attunement.

An exploration of duration is central to habitability practice and must take account of the coordinates involved and their various speeds and intensities, such as geological, chemical, computational, electrical, plant and atmospheric biochemical processes, and human temporalities, and those emerging through their enmeshment. Understanding modern linear time is no longer the central focus, but rather gaining an understanding of multifarious temporalities that are non-linear and multi-scalar. The problem of fixing time therefore becomes temporal capture, that is, the cut. Curatorial methods need to become empirically grounded and take account of the variable durations of plants and people bonded materially. Such complexities are revealed through ^{13}C analyses, a process that produces new knowledge for real-world applications. A focus on new curating practices for climate empowerment and carbon mitigation strategies brings to light the positionality of modern populations in relation to non-human agencies and durations and demonstrates the need for curating respectfully with cultural carbon.

Here I extend Haraway's use of the sympoietic biological metaphor beyond multispecies to include the folding and enfolding of humanness through agencies of crafting. As dynamic fossil-fuelled life forms these situated transversal figurations of human populations become bounded through the durations and agencies of the eco-curating agencies of combustion, photosynthesis, and atmospheric variabilities.

Thinking in terms of eco-curating processes brings new worlds and economies into being. This is in stark contrast to routine curatorial practices which are directed to analysing, describing, and representing existing objects and collections and narratives as human-centred concerns. The physical sciences, for example, can help understand agential capacities of the non-human world previously left out or rendering invisible in curating practice with the ability to propose a different vision of matter as multifarious agential forms and affects.

And while feminist theorist Donna Haraway suggests that cultural objects can be used to tell new stories of the sympoietic – how things beget other things – I take this matter further by using my concept of ecological composition as a story-making mechanism to show how all planetary life and processes can survive and flourish through an awareness of modern populations and their situatedness in agential formations within newly visible and expanded social collectives. Importantly, the approach is not just about telling stories but also explicating the material histories of human–non-human relations and durations embedded in collections.

An understanding of these entanglements and circumstances – ecological thoughts, material embeddedness, and actions – enables new types of relating which can be used to change the story. Stories can be based on aspects of things and, as is the case of the melted plastic bucket, not to focus the narrative from a purely human perspective. Fire, as an eco-curating agency, for example, in respect to the bucket, provides an entry point, or cut, into the distributed and composing processes through which its multiple stories and material processes are told. Here it is not necessarily about Bill or the bucket as a cultural object but rather its convergence with all manner of material and materialising circumstances in which many different agencies are implicated, from wind and heat, drying organic matter, and electrical wiring to changing atmospheric conditions.

Habitability practices must be fostered bit by bit and that is precisely what eco-curating processes – studies of ecological compositions, material and conceptual analyses, and the retelling of stories –seek to do. Furthermore, building a common liveable world can't be achieved by imposing and implementing one type of ecological thought across all sectors and situations but rather through many attuned and situated ecological thoughts that work together or, alternatively, do similar work. In this scenario, each component embodies the same principles of relational interconnectedness and care is directed towards enacting sustaining living practices. If just one type of ecological thought is privileged and overrules others it becomes just another form of imposition and negation of other worldviews and practices. The compositional practices discussed in this book, alongside the reworking of curatorial work, the institutional forms of museums, relational storytelling and empirical practices of experimentation, and analyses of materials, histories, and concepts based on ecological thoughts can help practitioners to think and act with and in radical coexistence with shared planetary inhabitants.

In terms of Agenda 2030, better futures are humanist ones that support modern projects. But futures have always been more-than-human. Composing better futures, thus, is no longer framed as a neoliberal form of sustainable development but as intent based on principles of value, care and relational attuning. Museum representations and collections become attuned visions of multiple and emerging futures and more-than and other-than-human innovation.

Therefore, anticipatory future-making is reframed through eco-curating frameworks. Curating futures becomes an act of composing worlds we would like to see in recuperation and at least directed to partially mitigating emerging crises or those already upon us. Compositional designing for habitability takes the local politics of the situated human into consideration. When curatorial projects are designed for attunement, such approaches have the potential to foster ethical practice. The activation of new cultural carbon mitigation strategies through collections analysis allows for insights into the circumstances that arose, or might arise, out of the composing and crafting practices that might and are ultimately destructive. And most importantly such practices might render visible what histories of emergence might be crafted and enacted differently to produce more sustaining outcomes. This is seen with the Boulton and Watt "Lap" Engine which was rendered capable as a new type of cultural carbon mitigation strategy by the folding out its multiple dimensions, materials, and histories and by the framing of new forms of bonded biochemical lifeforms as novel humanisms through a consideration of what goes with what, who is entangled with whom and their respective durations. We can see these relations, at least partially, as inverted lessons in attunement.

In order to compose futures for habitability, therefore, the first curatorial task must be to focus on objects' becomings – what went with what and what emerged through these ontological processes. These curating practices are also attuned methods for collecting, interpreting, and documenting. Through the observation and tracing of any ecological compositions' thingness and by cutting into it, it becomes clear what is happening and who is involved.

New futures therefore become a process of attending to who modern populations are, who they are becoming with, and what could happen differently if composed and attuned otherwise . The green plastic bucket, Trump's tweets, and the Boulton and Watt "Lap" Engine, therefore, all become indices for planning better futures.

Through a fractiversal framing of ecological thoughts, ecological compositions and their attuning processes illuminate and activate other practices that have been abandoned or rendered obsolete. The case examples explored in this book show how different types of relatedness and vitalities, when read through diverse knowledge practices and the lens of eco-curating, can become a different type of practice within a fractiversal constellation. Ecological thoughts can be folded with other types of thoughts to form hybrid knowledge practices.

By working with values that promote responsibility and of care, and by framing methods with others that share the same underlying principles, a multiplicity of sustaining futures can emerge – ones that promote habitability and flourishing. These futures may be outside our ability to command and control and therefore an adaptive capacity to respond and act with changing circumstances must always be exercised.

Generally, museologies must investigate different forms of collaboration and knowledge production through eco-curatorial processes and their interrelations. The futures that museums materially enact are empirically more-than-human futures based on relational entanglement. The human futures of progress, strong human agency and power, technology and the advancement of human capacities to enable human flourishing, and the extraction of the non-human to serve the human must be processes of the past. Therefore, while humanist museology has always been more-than-human, explicit accounts and the framing of more-than-human museologies need to be directed towards planetary flourishing and nurturing practices that promote all these things, from the way interpretative documentation is written and rendered in exhibitions to the implementation of pedagogies that are not ones of mastery but of enmeshment, flourishing, and attunement.

Reworking temporalities and durations

Museology and curatorial practice follow what cultural historian Helge Jordheim calls the "temporal structure and organisation of modernity."[37] These temporalities are socio-cultural. Cultural collections, as exemplars of modern heritage time, must be connected to an event or person. Modern heritage times must remain the same and be stabilised through research, interpretative documentation frameworks, and conservation practice.

Conventions of time in humanist museologies and museum practice include the fictions of object times which align with modern heritage times – an example being the time of origin or originating context as one source and point in time. Museology and museum time encapsulate ideas about death, loss, and destruction and the need to pause the ecological composition in its transforming processes through strong human agency. The idea of obsolescence and endangerment becomes collections' documentation time, conservation time, as well as the times of endangerment, colonial time, the time when culture was lost, and the time when salvaged, being just some examples.

Modern time is what Jordheim calls "practices of synchronization."[38] The temporal and historical organisation of modernity operates in a specific direction and according to a defined speed and rhythm.[39] Museums present time in a modern temporal framework and in a way that is homogenous and stable and resonates with what Jordheim calls a presentist regime of temporality. Global crises, such as climate change and its empirical realities, conflate into a linear trajectory. For example, curatorial practice and museology adhere to the years 2030, and 2050 as key greenhouse gas reduction dates in the Paris Agreement policy times. These modern future trajectories mean that curatorial practice and museology, and their concerns in respect to the climate crisis, do not take account of other times and their complexities. That is biochemical and atmospheric times and durations except when mediated through climate science. Indeed, other durations are rendered invisible. Furthermore, the durations of climate change, policy, and museum times clash because the empirical realities of climate change do not neatly fit these schedules.

Collections encompass multiple historical times – times of geology, natural history, species, and so forth. Digital cultural heritage, for example, embodies planetary temporalities, capitalist times, the times of technological obsolescence, and policy times, to name a few. Digital data switches between different temporalities and durations and becomes subject to museum conservation temporalities, data durations and digital heritage, and machine times and durations. Geological specimens reference deep time as a time before the reference point of hominids, a point that demarcates the break between geological and human times. In archaeological collections, time is organised stratigraphically, that is as chronological, successive, progressive, and developmental and is often registered through carbon dating. Often the multiple durations of archaeological objects are rendered invisible. Each is marked or delineated by a substance, for example, stone, bronze, or iron, that indicates a specific type of culture and their culture complexes. The science disciplines establish their own times or durations and try to understand the agencies and durations of the non-human worlds.

The durations and temporal structures of museum curatorial practice change and are multifarious such as the processes of acquisition that follow one temporal framework and that of conservation that follows the temporal durations of many materials and organisms. The temporal frameworks of climate policy, and of Agenda 2030 and its SDGs, for example, are future orientated and linked to teleological time. In teleological times, the past is viewed as deficient or backward or as lacking in sufficient knowledge, and in the case of climate change is only mitigated more effectively through better climate science.

Museum disciplines all adhere to modern progressive times, times in which disciplinary knowledge becomes more precise and more predictive. Writing exhibition text becomes seamless and homogenous, often presenting a linear trajectory of development while remaining closed to other socio-cultural times. Keywords and tenses are used to locate and fix modern times. Historical time is represented as the past, in the past tense. The present is represented by the present tense. The future is represented by the future tense. Museology and curatorial practice and the semantic practices, terms, and keywords used in respect to time might be thought about as a product of what Jordheim calls, in modern time, a complex set of synchronic conceptual, and technological practices.[40]

The types of museological temporalities that guide practice have contributed to producing modern futures through speculative practices, visions of technological futures, biodiversity schemas of extinction, climate mitigation measures, and so forth. More-than-human museologies can frame different types of futures. Museum temporalities are not ones of fixed time and presentism. No one temporal regime exists on its own, but rather there are multiple times in sympoiesis.

Museological temporal figurations must be rethought. Modern historical thinking, heritage, museology, and museum and curatorial times fold all things, and their multiple temporalities durations and emergent eco-curating processes, into one linear temporal frame. This is erroneous because objects, curatorial and conservation practice, exhibition schedules and the exhibitions themselves (including museum buildings) comprise more-than and other-than-human temporalities and durations. For example, the curatorial register includes the photosynthetic, viral, fire, and the rhizomic durations of all these entities.

The time of modernity, and therefore humanist museology, is on the verge of collapse because this temporal framework is being challenged by other times. The times of climate periodisation, or "the Anthropocene," for example are challenging past social relations and political decisions.[41] Indeed the times of the pandemic, or the Viroscene, challenge the very foundations of neoliberal capitalism and humanist-centred museology. In more-than-human times, the dominance of the present is a different method. That is, being present and presentism, are the cut, or cutting processes, that act as interpretative interventions in eco-curating processes. They involve observing what is emerging through the tracing of coordinates, humanisms, interactions, durations, and multiple temporalities, rather than as inert representations of the past.

Refiguring economy

Museological and curatorial representations of economy are not always evident until the principles and practices in which economy is enacted are closely examined. Equally, museums tend to foreground capitalocentric representations that include business and private accumulation, waged labour, commodity production, private enterprise, exchange markets, private property, the enclosure of land, and institutional finance. The promotion of neoliberal economies and practices is evident across all the case examples. This includes, for example, the Boulton and Watt "Lap" Engine and its links to the Industrial Revolution, waged labour, the factory system, colonial dispossession, and the design and the harnessing of the non-human world for economic growth. The melted bucket has links to economies and the notion of private property. Collections and narratives at the Smithsonian National Air and Space Museum, and Deutsches Museum are exemplars of commodity production, technological innovation, and business, finance, and private accumulation. These capitalocentric imperatives are seen most powerfully with climate engagement and the implementation of the Paris Agreement and the SDGs. For example, through SDG implementation museums articulate that a paid job is the best way to survive and thrive. Agenda 2030 suggests that capitalist growth benefits all, or that capital exchange and the movement of traded goods is for the greater good and has the potential to pull people out of poverty. While the Museum of Capitalism critiques the formation of capitalism and its consequences, little work has been done in critiquing museum capitalisms. Museums are necessarily born out of capital and economy, and especially those that have colonial legacies, such as economic, technical, and science museums. Museums are essentially transactional and finance is at their centre and anything else, for example, the representation of Indigenous First Nations economies is presented as pre-modern.

The reframing of the economy through curatorial practice inspired by diverse economies and frameworks promotes a shift in the economic formations that frame museology and curatorial practice and can act as a lever to curate for planetary habitability. This shift in perspective necessarily involves museums and museology to promote critical thinking strategies that challenge the framing of capitalism as the ultimate real.

Museology and museums, therefore, have important work to do in reformulating economies in order to work with exchanges between more-than and non-human earth others. Economy takes place in ecologies. According to Bruno Latour, the incredible power that these place-based and Earth systems contribute to economic activities in order to nurture the planet in the western world is only now being understood.[42] Such processes are often threatened by the actions of modern humans, including the destructive economic practices museums promulgate.

Post-capitalist economies visionaries, Gibson Graham, state that by engaging with community economies, economy becomes an ethical practice and vision for changing ourselves, changing our thinking, and changing our world[43] and, as I argue in this book, eco-curating formations. Forms of enterprise become eco-curating processes concerned with more-than-human care, and with flourishing enacted through attunement, through crafting, and composing, with others in respectful ways.

As part of a wider call for alternative economies, made by Gibson Graham, museums have the potential to support emancipatory, ethical, solidaristic, and diverse economic practices.[44] In this way, economy becomes entanglements in the broader social collectives of Earth others. This involves rethinking the situated place of human agencies in transversal collaborations towards respectful crafting, composing, and influencing processes. Reinterpreting collections representative of economies deemed pre-modern leads to great insight and the telling of new stories according to Indigenous epistemologies. By disrupting assumptions embedded in Agenda 2030 and engaging with the principles and practices of diverse economies this shift would transform assumptions about developing economies as under-developed into understandings about the need for promotion of capital to habitability and to living practices as more-than and other-than-human livelihoods. As represented in museums, non-modern knowledge economies can work with modern ones in new ways to form economy ecologies in which knowledge practices of attunement include studies of emergence, polities, entanglements, and of agencies.

Conclusion

As curatorial practices become more attuned to the empirical realities of planetary enmeshment and move from objects to ontological understandings, museology must change from its nineteenth-century underpinnings. That is the promotion of thoughts of static objects and descriptions as distinct standard categories of origin, history, and significance. Time must no longer be categorised as singular and linear. Analyses of substances need to be about more-than human power and the utility of substances for human agency. Communities and audiences should no longer be considered as merely curatorial agents and distinct human-centric forms within neo-capitalist economies. All these categorisations need to be reconceptualised because they do not operate in radical coexistence and they assume the world is divided into discrete bounded things.

Traditional curation –according to curatorial divisions, interpretations, documentation and types and forms of expertise – hinders ecological thoughts and acts. Ontologically all these things change. Historicity changes, aura changes, temporality changes, the notion of authenticity, transformation and significance changes, and the practices and values that inform the idea of heritage, museum curatorial practice, and economy radically transforms. Furthermore, who is involved in curatorial work changes and really becomes a question about who is missing or has been sidelined. Museums are places that tell stories, collect humans, and other-than-humans but they, alongside museology, must become ecologising entities and practices that extend beyond a physical edifice directed to create a more-than-human imagination, that embodies thinking acting, feeling, and caring. That is, more-than-human, planetary and cohabitation practices, and differentiated politics.

Museums and museologies, both new and old, are built on the concept of modern humanism and human agency. That is human agency as the curatorial agent directed to reforming populations and the definition of the individual in terms of social and biological categorisations. More-than-human museologies is a thinking, acting, and composing with the practice of becoming, and in using eco-compositional design to curate for planetary habitability where the humans doing the curating, and composing are not the only ones involved. Importantly, under this model, heritage institutions and curation are no longer floating above society as a hubristic knowledge complex. This is a fundamental error of concept. Museum institutions, curatorial practice, and museology are coordinates in metabolic processes, and examples of more-than-human crafting and compositing in social, political, and biological contexts within fractiversal cosmologies.

More-than-human futures require radical more-than-human museologies that extend beyond humanist thinking and engage with the empirical realities of life. That is, through ontologies that emphasise entanglements and other knowledge practices, including those of the non-human, interdependence, embodiment, embeddedness, distributed responsibilities and accountabilities, planetary histories, attunement with others, and relations that nurture. Central to this is the promotion of new forms of humanism, knowledge, and experience.

Radicalising museologies as more-than-human, and as enmeshed and embodied, promotes the empirical reality of ecological belonging as operating in an extended frame that is dispersed and embedded. Questions become by whom, what, and how are different affects and knowledge produced and what are their materialising potentials.

Even museums and their narratives of organisation must be rethought as liquid, enmeshed, and embedded in planetary predicaments. Liquid museums, as enmeshed and embodied institutional forms, offer a planetary optic on museum and curatorial practices in which all activities, including buildings, are made up of cultured materials and practices including non-humans, the earth, minerals, and concrete. Such a vision of institutions deeply embeds curating practices in planetary figurations of significance. Museums are ecological compositions of various forms held together by all manner of dynamic agencies. Together the more-than and non-human are becoming museological and curatorial practices.

For the contemporary world, museologies are necessarily post-anthropocentric. That is, they are beyond practices that promote environmental destruction and hubristic mindsets but in a way that actually extends beyond practices of the Anthropocene because this concept is too Eurocentric and is located within a scientific geological framework. More-than-human museologies are speculative, experimental, and directed to designing more equitable futures. Fundamentally, more-than-human museologies are and must, in practice, be regenerative.

In this regard, museologies become sensing, biochemical and calculative and involve many different forms of cognition and affects beyond human discursive considerations. Storytelling through documentation, interpretative and pedagogical practices, and practical experimentations are required to illustrate the embeddedness of humans in radical ways. Notions of time and space must engage with the interrelatedness of any given object or thing, as well as its coordinates and agencies. Eco-curating, thus, becomes thingness, historical becoming, and interpretive cuts.

More-than-human museologies, collections, and museum buildings are all curating entities. Other-than-human museologies involve non-human curating agencies. For example, viral museologies have never been humanist but the pandemic is humanist in how it is conceptualised socially and, to a degree, medically. Museological investigation must critique museums' exercise of power, their representation of non-human and more-than-human histories, their complicity in the pandemic, and, through an investigation of speciesism and ecocide, curatorial humanisms' idea of progress. In terms of the pandemic, museums need to curate "with" the virus in a form of sympioetic curating and representation because the virus itself curates outside of human control. The documentation of pandemic collections really must be reworked as more-than-human inhabitations.

The same humanist predilections govern the climate crisis when it is conceptualised as human and social and is directed towards expanding human power and agency. Rather in the face of climate change humans need to be humbled and commune. Human populations and their roles in the climate crisis need to be thought of in terms of museological formations that investigate humanisms as bonded with the non-human, for example, photosynthetic or wooding museologies, and thereby radically transform modern views of complicity and attuned responsibility.

Both the climate crisis and pandemic may be presented as bonded humanisms by promoting museological practices of representation, through, for example, exhibitions, public programming, and education, that represent an "in-habiting" of climate and viral worlds.

All museum practices involve non-human actors. Even policy implementation is more-than-human even though policies may appear as human and discursive. Through more-than-human museologies there are no distinctions between human life, organic entities, earthly processes, viruses, and microbes – all life and "so-called non-life" is rightly understood as having always been ecological and more-than-human and part of situated compositions that are each unique. Human agencies are seen to become humanness, as crafting, composing, gathering, non-conscious, and as traces of more-than-human eco-curating processes. Nothing is ontologically discrete and museum categorisations collapse. Ecological compositions, as seen in each of the chapters of this book, are understood as cuts into worldly life, as worlding processes, and as examples of distributed humanness.

Enmeshed museologies are relational and include concepts such as ubiquitous museums and buildings, human populations as transversal subjects, heritage as ecological compositions embedded in the very fabric of society and life, and humanisms that are not markers of identity but as plural, situated, and enmeshed. Modern populations are dynamic and made up of multiple coordinates at once both planetary and distributed. No distinct individual exists. Modern populations, their thoughts and actions, are always distributed and compositional and involve many coordinates in sympoietic fusion or melding. Indigenous people could be described as sym people, conceptually and practically enmeshed in radical coexistence and non-sym are those that are separate or dis-embedded. It is the role of museums, museologies, and curatorial practices to propagate this change and for such thoughts and acts to go viral.

Eco-curating as a practice of worlding is extended in all manner of directions and is vast, vertical, horizontal, three dimensional, and inside and outside in ways that are not bounded and undefinable. This figuration of curating promotes thinking with the planet as a whole, a means to simultaneously think locally and practically. Museum curatorial work emerges as experimental, collective, communal, and as cutting into ontological processes.

Museum work is not just restricted to storytelling. Informed by more-than-human figurations museologies and curatorial practice can locate their investigations in a broader horizon of living, acting, and dying through an investigation of the material aspects of climate change, and a consideration of human subjects as material subjects. New vocabularies and methods can show, interpret, and represent the world in a way by which material evidence becomes even more compelling. Fundamentally, curatorial practice can shift from saving for the future, or representing the past and the power of human agency, to composing respectful futures through a consideration of what goes with what, the role of collections in representing human, more-than-human and non-human predicaments and analyses who we must negotiate with.

Beyond anthropocentric museology, different museologies, such as Indigenous, post-capitalist, and decolonial ones emerge alongside the more-than-human museologies detailed in the book. Fractiversal museologies will arise that represent different cuts into worldly life under the stewardship of different knowledge and habitability practices.

I hope that these diverse museologies, and museum and curatorial practices, will trigger different futures as an antidote to the way that new museology, or its predecessor the so-called old museology, re-enforced, through a governing assemblage of institutions, thoughts, and acts, what has turned out to be very destructive. The future legacies that museums promote need to go beyond the concerns of saving and passing on their collections to future generations and strive to promote practices of care for planetary habitability.

Notes

1 Peter Van Mensch, "Towards a Methodology of Museology," Ph.D Thesis, University of Zagreb, Croatia, 1992, 1.
2 Van Mensch, "Towards a Methodology of Museology," 9.
3 Kenneth Hudson, *A Social History of Museums: What the Visitors Thought* (London: Macmillan, 1975): 15.
4 Duncan F. Cameron, "The Museum, a Temple or the Forum," *Curator* 4, no. 1 (1972): 11–24.
5 Peter Vergo, *The New Museology* (London: Reaktion, 1989).
6 Ólöf Gerður Sigfúsdóttir, "Blind Spots: Museology on Museum Research," *Museum Management and Curatorship* 35, no. 2 (2020): 196–209.
7 Ólöf Gerður Sigfúsdóttir, "Blind Spots."
8 Anwar Tlili, Sharon Gewirtz and Alan Gibb, "New Labour's Socially Responsible Museum, Roles, Functions and Greater Expectations," *Policy Studies* 28, no. 3 (2007): 269–89.
9 Fiona R. Cameron, *The Future of Digital Data, Heritage and Curation in a More-Than-Human World* (London: Routledge, 2021): 276.
10 Cheng Chang, "Community Involvement & Ecomuseums: Towards a Mutual Approach to Ecomuseology and Landscape Studies," Ph.D Thesis, Swedish University of Agricultural Sciences, Uppsala, Sweden, 2015, 9.
11 Melanie Arenson and John D. Coley, "Anthropocentric by Default? Attribution of Familiar and Novel Properties to Living Things," *Cognitive Science* 42, no. 1 (2018): 253–85, 254.
12 Cameron, *The Future of Digital Data*, 368.
13 Fiona R. Cameron, "Ecologizing Experimentations: A Method and Manifesto for Composing a Post-Humanist Museum," in *Climate Change and Museum Futures*, eds. Fiona R. Cameron and Brett Neilson, Routledge Research in Museum Studies (New York: Routledge, 2014): 16–33.
14 Cameron, *The Future of Digital Data*, 52.
15 Cameron, *The Future of Digital Data*, 368.
16 Fiona R. Cameron, "Object-Orientated Democracies: ConceptualisingMuseum Collections in Networks," *Museum Management and Curatorship* 23, no. 3 (2008): 229–43.
17 Zoltán Boldizsár Simon, Marek Tamm and Ewa Domańska, "Anthropocenic Historical Knowledge: Promises and Pitfalls," *Rethinking History* 25, no. 4 (2021): 406–39.
18 Cameron, *The Future of Digital Data*, 271, 205.

19 *Kiruna in Swedish Lappland*, accessed September 19, 2022, https://kirunalapland.online.citybreak.com/en/to-do/1811991/lkabs-visitor-centre-guided-tours/showdetails.
20 Anne and Bernard Spitzer Hall of Human Origins, American Museum of Natural History, accessed September 19, 2022, https://www.amnh.org/exhibitions/permanent/human-origins.
21 *Fiona R. Cameron, Shaping Histories and Identities: Collecting and Exhibiting Māori Material* Culture at the Auckland and Canterbury Museums, 1850s–1920s. Ph.D thesis, Massey University, New Zealand, 2002.
22 James Williams, "Process Ontology," in *Posthuman Glossary*, eds. Rosi Braidotti and Maria Hlavajova (London and New York: Bloomsbury Academic, 2018): 371–73, 372.
23 Cameron, *The Future of Digital Data*, 273.
24 Fiona R. Cameron, "Theorizing More-Than-Human Collectives for Climate Change Action in Museums." *L'Internationale Online Climate Change Special Issue Ecologizing Museums* 2015, accessed September 19, 2022, http://www.internationaleonline.org/research/politics_of_life_and_death/48_theorising_more_than_human_collectives_for_climate_change_action_in_museums.
25 Cameron, *The Future of Digital Data*, 381.
26 Jennifer Gabrys, "Becoming Planetary," *e-flux Architecture*, accessed September 19, 2022, https://www.e-flux.com/architecture/accumulation/217051/becoming-planetary/.
27 National Museum, Czech Republic, Geological Collection, accessed September 19, 2022, https://www.nm.cz/en/about-us/science-and-research/geological-collection.
28 Kerry Lotzof, "Collections: Joseph Banks – Scientist, Explorer and Botanist," *Natural History Museum*, London, accessed September 19, 2022, https://www.nhm.ac.uk/discover/joseph-banks-scientist-explorer-botanist.html?
29 "Apollo 11," *Smithsonian* (National Air and Space Museum), accessed September 19, 2022, https://www.si.edu/spotlight/apollo-11#:~:text=The%20National%20Air%20and%20Space,lunar%20landing%20mission%2C%20Apollo%2011?
30 Rosi Braidotti, *Posthuman Knowledge* (Cambridge: Polity Press, 2019): 14.
31 "Beijing Shuts Parks and Museums as China's Covid Cases Rise," *The Guardian, Australian Edition*, November 22, 2022, https://www.theguardian.com/world/2022/nov/22/beijing-shuts-parks-and-museums-as-chinas-covid-cases-rise.
32 Zoltán Boldizsár Simon and Marek Tamm, "Historical Futures," *History and Theory* 60, no. 1 (2021): 3–22, 8.
33 Cameron, *The Future of Digital Data*, 274.
34 Fiona R. "Posthuman Museum Practices," in *Posthuman Glossary*, eds. Rosi Braidotti and Maria Hlavajova (London: Bloomsbury Academic, 2018): 349–51, 351.
35 Cameron, *The Future of Digital Data*, 254.
36 *Climate Change Science: An Analysis of Key Questions*, Committee on the Science of Climate Change, National Research Council, National Academy Press, Washington DC, 2001, 9.
37 Helge Jordheim, "Introduction: Multiple Times and the Work of Synchronization," *History & Theory* 53, no. 4 (2014): 498–518, 499, 505.
38 Jordheim, "Introduction,"505.
39 Jordheim, "Introduction," 500.
40 Jordheim, "Introduction," 505.
41 Jordheim, "Introduction," 500.
42 Bruno Latour, "Agency at the Time of the Anthropocene," *New Literary History* 45, no. 1 (2014): 1–18, 10.

43 Julie Katherine Gibson-Graham, "Diverse Economies: Performative Practices for 'Other Worlds'," *Progress in Human Geography* 32, no. 5 (2008): 613–32, 618.
44 Julie Katherine Gibson-Graham, *The End of Capitalism (As We Knew It): A Feminist Critique of Political Economy* (Minneapolis: University of Minnesota Press, 1996).

References

"Apollo 11." *Smithsonian* (National Air and Space Museum), accessed September 19, 2022, https://www.si.edu/spotlight/apollo-11#:~:text=The%20National%20Air%20and%20Space, lunar%20landing%20mission%2C%20Apollo%2011.

Arenson, Melanie. "Anthropocentric by Default? Attribution of Familiar and Novel Properties to Living Things." *Cognitive Science* 42, no. 1 (2018): 253–85.

"Beijing Shuts Parks and Museums as China's Covid Cases Rise." *The Guardian, Australian Edition*, November 22, 2022, accessed February 26, 2023, https://www.theguardian.com/world/2022/nov/22/beijing-shuts-parks-and-museums-as-chinas-covid-cases-rise.

Braidotti, Rosi. *Posthuman Knowledge*. Cambridge: Polity Press, 2019.

Cameron, Duncan F. "The Museum, a Temple or the Forum." *Curator* 4, no. 1 (1972): 11–24.

Cameron, Fiona R. *Shaping Histories and Identities: Collecting and Exhibiting Māori Material Culture at the Auckland and Canterbury Museums, 1850s–1920s*. Ph.D thesis, Massey University, New Zealand, 2002.

Cameron, Fiona R. "Object-Orientated Democracies: Conceptualising Museum Collections in Networks." *Museum Management and Curatorship* 23, no. 3 (2008): 229–43.

Cameron, Fiona R. "Ecologizing Experimentations: A Method and Manifesto for Composing a Post-Humanist Museum." In *Climate Change and Museum Futures*, edited by Fiona R. Cameron and Brett Neilson, 16–33. Routledge Research in Museum Studies. New York: Routledge, 2014.

Cameron, Fiona, R. "Theorizing More-Than-Human Collectives for Climate Change Action in Museums." *L'Internationale Online Climate Change Special Issue Ecologizing Museums* 2015, accessed September 19, 2022, http://www.internationaleonline.org/research/politics_of_life_and_death/48_theorising_more_than_human_collectives_for_climate_change_action_in_museums.

Cameron, Fiona R. "Posthuman Museum Practices." In *Posthuman Glossary*, edited by Rosi Braidotti and Maria Hlavajova, 349–51. London: Bloomsbury Academic, 2018.

Cameron, Fiona R. *The Future of Digital Data, Heritage and Curation in a More-Than-Human World*. London: Routledge, 2021.

Chang, Cheng. "Community Involvement & Ecomuseums: Towards a Mutual Approach to Ecomuseology and Landscape Studies." PhD Thesis, Swedish University of Agricultural Sciences, Uppsala, Sweden, 2015.

Climate Change Science: An Analysis of Key Questions, Committee on the Science of Climate Change, National Research Council, National Academy Press, Washington DC, 2001, 9.

Gabrys, Jennifer Gabrys. "Becoming Planetary." *e-flux Architecture*, accessed September 19, 2022. https://www.e-flux.com/architecture/accumulation/217051/becoming-planetary/.

"Geological Collection." *National Museum*, Czech Republic, Geological Collection, accessed September 19, 2022, https://www.nm.cz/en/about-us/science-and-research/geological-collection.

Gibson-Graham, Julie Katherine. *The End of Capitalism (As We Knew It): A Feminist Critique of Political Economy*. Minneapolis: University of Minnesota Press, 1996.

Gibson-Graham, Julie Katherine. "Diverse Economies: Performative Practices for 'Other Worlds'." *Progress in Human Geography* 32, no. 5 (2008): 1–25.

Hudson, Kenneth. *A Social History of Museums: What the Visitors Thought*. London: Macmillan, 1975.

Jordheim, Helge. "Introduction: Multiple Times and the Work of Synchronization." *History & Theory* 53, no. 4 (2014): 498–518.

Latour, Bruno. "Agency at the Time of the Anthropocene." *New Literary History* 45, no. 1 (2014): 1–18.

"LKAB:s Visitor Centre – Guided Tours." *Kiruna in Swedish Lappland*, accessed February 18, 2023, https://kirunalapland.online.citybreak.com/en/to-do/1811991/lkabs-visitor-centre-guided-tours/showdetails.

Lotzof, Kerry. "Collections: Joseph Banks – Scientist, Explorer and Botanist." *Natural History Museum*. London, accessed September 19, 2022, https://www.nhm.ac.uk/discover/joseph-banks-scientist-explorer-botanist.html.

Sigfúsdóttir, Ólöf Gerður. "Blind Spots: Museology on Museum Research." *Museum Management and Curatorship* 35, no. 2 (2020): 196–209.

Simon, Zoltán Boldizsár and Marek Tamm. "Historical Futures." *History and Theory* 60, no. 1 (2021): 3–22.

Simon, Zoltán Boldizsár, Marek Tamm and Ewa Domańska. "Anthropocenic Historical Knowledge: Promises and Pitfalls." *Rethinking History* 25, no. 4 (2021): 406–39.

Tlili, Anwar, Sharon Gewirtz and Alan Gibb. "New Labour's Socially Responsible Museum, Roles, Functions and Greater Expectations." *Policy Studies* 28, no. 3 (2007): 269–89.

Van Mensch, Peter. "Towards a Methodology of Museology." PhD Thesis, University of Zagreb, Croatia, 1992.

Vergo, Peter. *The New Museology*. London: Reaktion, 1989.

Williams, James. "Process Ontology." In *Posthuman Glossary*, edited by Rosi Braidotti and Maria Hlavajova, 371–73. London and New York: Bloomsbury Academic, 2018.

INDEX

Note: *Italic* page numbers refer to figures.